Introductory Plant Nematology

P. PARVATHA REDDY
Former Director,
Indian Institute of Horticultural Research, Bengaluru, India

SCIENTIFIC® PUBLISHERS
New Delhi • Jodhpur

SCIENTIFIC®
PUBLISHERS

Published by:

Jodhpur
5 A, New Pali Road, P.O. Box 91
Jodhpur - 342 001 Rajasthan (India)
E-mail: info@scientificpub.com
Website: www.scientificpubonline.com

New Delhi
4806/24, Ansari Road, Daryaganj,
New Delhi - 110 002 (India)

Print : 2024

ISBN: 978-93-87893-91-7 (H/B)
eISBN: 978-93-87991-68-2

Visit the Scientific Publishers (India)
website at http://www.scientificpub.com
Printed in India

Dedicated to all of my teachers, who took time to instill the knowledge of Nematology in me.

Dr. V.G. Perry, Univ. of Florida, Gainesville, USA

Dr. D.J. Raski, Univ. of California, Davis, USA

Dr. M. Oostenbrink, Intn. Agri. Centre, Wageningen, Netherlands

Dr. Abrar M. Khan, Aligarh Muslim Univ., Aligarh, India

Dr. A.R. Seshadri, Indian Agri. Res. Inst., Delhi, India

Dr. R. Mankau, Univ. of California, Davis, USA

Dr. G.C. Smart, Univ. of Florida, Gainesville, USA

Dr. M.S. Jairajpuri, Aligarh Muslim Univ., Aligarh, India

Dr. S.K. Saxena, Aligarh Muslim Univ., Aligarh, India

Dr. S.K. Prasad, Indian Agri. Res. Inst., Delhi, India

Mr. Van Berkum, Intn. Agri. Centre, Wageningen, Netherlands

PREFACE

Several species of nematodes are now recognized in India as parasites of agricultural crops which cause considerable damage. Some of these have been placed in quarantine (for example, potato cyst nematodes (*Globodera rostochiensis, G. pallida*) to prevent the free exchange of plant material from other countries. The farmers spend large sums of money on control measures against such pests as the root-knot (*Meloidogyne* sp.), cyst (*Heterodera* and *Globodera* sp.), burrowing (*Radopholus similis*), reniform (*Rotylenchulus reniformis*), lesion (*Pratylenchus* sp.), citrus (*Tylenchulus semipenetrans*), and other nematodes. Besides causing direct damage, the nematodes are involved in causing complex plant diseases in association with fungi, bacteria and viruses.

There has been a long felt need in India by the students and teachers for a comprehensive book on plant nematology. This book has been designed as a text and guide for students, teachers, researchers and extension workers in plant nematology and related fields.

This book, entitled "**Introductory Plant Nematology**" gives a comprehensive account of all aspects of plant nematology and should be of profound help to the students, teachers, researchers and extension workers alike. The syllabus of ARS Net – Nematology has also been fully covered in this book. Hence, persons appearing for ARS Net – Nematology can also refer to this book.

The book is divided into eight sections. The first section describes the importance of nematodes in agriculture, presents a historical review, records the contributions of great authorities, nematode as biological models, entomopathogenic nematodes, training of Nematologists, and lists the professional societies and their publications.

Information on the nematological techniques like microscopy, sampling nematode populations, extraction of eggs and nematodes from soil, plant material, culturing of nematodes, handling nematodes, staining, and microtome sectioning are outlined in section two.

The morphology of nematodes is described and presented in clear schematic drawings in section three. The taxonomic classification along with keys for identification of nematodes up to generic level is also provided. In addition, monographic references have been included for the identification of species belonging to different genera.

In section four, the biology, physiology and ecology of nematodes are described and the influence of various factors on population dynamics are explored.

The mechanism of action of parasitic nematodes on plants is analyzed in section five of the book from the view point of the mutual physiological reaction of both organisms, i.e. host and parasite. The symptoms of aerial and underground infestation by different nematodes are described and depicted in many photographs.

In section six, the interrelationships between nematodes and fungi, bacteria and viruses are discussed.

Management of nematode diseases by host resistance and by suppression of nematode population through regulatory, physical, cultural, chemical, biological, and integrated methods have been presented in section seven.

The last section of the book discusses the most important nematode induced diseases of horticultural, plantation and spices, commercial and field crops and their management.

Quotable quotes, list of common names of nematodes together with the glossary of nematological terms is presented in Annexures. A comprehensive bibliography provides convenient entry to both current and older literature.

I earnestly hope that this book will provide students, researchers and extension workers with an overview of the entire field of plant nematology. I thank Mr. Tanay Sharma of Scientific Publishers, for extending his valuable help in processing the manuscript and getting the book published.

P. Parvatha Reddy

Bengaluru
January 25, 2018

CONTENTS

Chapter 1

NEMATODES – AN OVERVIEW

1.1. INTRODUCTION

Nematology, like entomology and plant pathology, is an important branch of biological science, which deals with the study of a group of animals known as nematodes. Nematodes are also called 'eelworms' in Europe, 'nemas' in the United States and 'round worms' by most Zoologists. By tradition, the study of nematodes that are parasites to man and his domesticated animals is called `Helminthology', and the study of plant-parasitic and free living nematodes is called `Nematology'. The name nematodes (*Nema* - thread, *oides* - resembling) indicates the external appearance of the organisms of this group.

Nematodes are found in almost every kind of environment, from the arctics to the tropics and from the ocean depths to the tops of high mountains. Rarely is any crop free from their attacks. They are so numerous that Cobb (1914 a) aptly remarked " --- *if all the matter in the universe except the nematodes were swept away, our world would still be dimly recognizable --- we would find its mountains, hills, valleys, rivers, lakes, and ocean represented by a film of nematodes.*"

1.2. IMPORTANCE OF NEMATODES IN AGRICULTURE

According to Thorne (1961), ``*Each year these minute organisms exact an ever increasing toll from almost every cultivated acre in the world: a bag of rice in Burma, a pound of tea in Ceylon, a ton of sugar beets in Germany, a bag of potatoes in England, a bale of cotton in Georgia, a bushel of corn in Iowa, a box of apples in New York, a sack of wheat in Kansas, or a crate of oranges in California*". Estimated overall average annual yield loss of the world's major horticultural crops due to damage by plant parasitic nematodes is estimated to be 13.54%. For the 7 horticultural crops (banana, cassava, coconut, field bean, potato, sugar beet and sweet potato) that stand between man and starvation (life sustaining horticultural crops), an estimated annual yield loss of 12.77% is reported (Sasser and Freckman, 1987).

The cereal cyst nematode, *Heterodera avenae* causes the `molya' disease of wheat and barley in Rajasthan, Punjab, Haryana, Himachal Pradesh and Jammu and Kashmir. The loss due to this nematode is about 32 million rupees in wheat and 25 million rupees in barley in Rajasthan State alone.

The seed gall nematode, *Anguina tritici* causes the ear cockle disease of wheat in Northern India. This nematode along with a bacterium, *Rathayibacter tritici* causes `tundu' or `yellow slime' disease. The overall damage is about one per cent, but sometimes the damage is as high as 80 per cent. Annual loss due to this nematode in North India is about 10,000 tons of wheat costing 70 million rupees.

Potato cyst nematodes, *Globodera rostochiensis* and *G. pallida* are a serious problem in Nilgiris and Kodai Hills. About 3,000 hectares are infested by this nematode. Total failure of the crop has been recorded under severe infestation conditions. This nematode has recently been reported from Himachal Pradesh also.

The root lesion nematode, *Pratylenchus coffeae* is a serious pest of coffee in South India. About 1,000 hectares are infested by this nematode. Annual loss is about 20 million rupees.

The burrowing nematode, *Radopholus similis* causes `spreading decline' of citrus in Florida, `pepper yellows' in Indonesia and `banana rhizome rot' in various parts of the world. This nematode is becoming a serious pest in India on plantation crops like pepper, coconut, areca nut, ginger and banana. It is responsible for `slow wilt' disease of pepper in Karnataka and Kerala (Venkitesan, 1976).

The citrus nematode, *Tylenchulus semipenetrans* is responsible for `slow decline' disease of citrus. It is suggested that the citrus nematode is also one of the factors responsible for dieback disease of citrus trees in India. The total annual reduction in the citrus crop due to nematodes is estimated to be 15 per cent (Anon, 1971).

The root-knot nematodes (*Meloidogyne* species) produces galls on the roots of many vegetable crops, pulses, some fruit crops, tobacco ornamental crops thus causing severe losses. This nematode is known to attack more than 3,000 separate host plants. *M. incognita* causes 33 per cent loss in vegetables in Bihar and 60 per cent loss in Delhi. Reddy (1981) reported that *M. incognita* caused 39.7 per cent loss in tomato yield at a pre-plant nematode population density of 20 larvae per gram of soil. The avoidable yield losses due to *M. incognita* were estimated to be 28.08, 33.68, 43.48, 28.60 and 20.00 per cent in okra, brinjal, French bean, cowpea and peas, respectively (Parvatha Reddy and Singh, 1981). Krishnappa *et al.* (1981) found that *M. incognita* was responsible for 44.87 per cent yield loss in brinjal. The per cent losses in yields of Okra, tomato and brinjal due to *M. incognita* were 90.9, 46.2 and 27.3, respectively (Bhatti and Jain, 1977)

The reniform nematode, *Rotylenchulus reniformis* has been reported to cause 14.9, 6.0, 4.8, 13.2 and 8.7 per cent loss in yields of cotton, maize, finger millet, cowpea and black gram, respectively (Palanisamy and Balasubramanian, 1981).

The above examples represents only a small portion of the nematode problems confronting India. Besides acting alone the nematodes are also known to be associated with bacteria, fungi and viruses in causing complex plant diseases.

Besides affecting plants, nematodes are also known to infect human beings, domestic animals and insects. Ascarids (*Ascaris lumbricoides*) lives in the small intestines of humans and competes for nourishment from the food eaten by the host. In many places, people still suffer from nematodes which cause elephantiasis (*Wuchereria bancrofti)*. The adult hook worms (*Ancylostoma duodenale*) suck blood from the small intestine of humans and cause anemia. Pin worm (*Enterobius vermicularis*), guinea worm (*Dracunculus medinensis*) and *Trichinella spiralis* also infect humans. *Strongyoides papillosus* and *Haemonchus contortus* infect sheep; gape worm (*Syngamus trachea)*, *Heterakis gallinae* infect poultry; lung worm (*Metastrongylus apri*), *Gongyloneam pulchrum* infect pig; *Probstmayria vivipara* infects horse; kidney worm (*Dioctophyme renale*) infects dogs. Mermithids are mainly parasites of the body cavity of insects and certain of these entomopathogenic nematodes are useful for biological control, an important consideration in view of the limitations of some chemical pesticides.

1.3. HISTORICAL

The wheat seed gall nematode, *Anguina tritici*, was discovered by Needham (1743) and appears to have been the first nematode parasite of plants to come to the attention of the early investigators. It was not until 1855 that Berkeley determined a root-knot nematode, *Meloidogyne* sp. to be the cause of galls on the roots of greenhouse grown cucumber plants in England. Kuhn (1857) found the stem and bulb nematode, *Ditylenchus dipsaci,* infesting the heads of teasel. Schacht (1859) reported the sugar beet cyst nematode, *Heterodera schachtii,* from Germany. These are some of the early records of plant parasitic nematodes. Some of the later findings are listed below in chronological order (Table 1.1).

Table 1.1. The significant early findings on plant parasitic nematodes listed in chronological order

Year	The early records of plant parasitic nematodes
1871	Kuhn-First soil fumigation experiments using CS_2 for the control of the sugar beet nematode, *Heterodera schachtii.*
1873	Butschii-First detailed descriptions of the morphology of free-living nematodes.
1884	DeMan-Taxonomic monograph of soil and fresh water nematodes of the Netherlands.
1888	Strubell-Detailed morphology of *H. schachtii.*
1889	Atkinson and Neal - Independently published on root-knot nematodes in the United States.
1892	Atkinson - Root knot nematode and *Fusarium* complex in vascular wilt of cotton.

Year	The early records of plant parasitic nematodes
1901	Barber- Root-knot nematode on tea in South India.
1907	Cobb (Fig. 1.1) - Joined the USDA, considered to be the father of American Nematology.
	Titus - Reported *H. schachtii* in the United States.
1913, 1919	Butler – Reported 'Ufra' disease on rise in Bengal caused by *Ditylenchus angustus*.
1914	Cobb – Book on "*Contributions to a Science of Nematology*".
1918	Cobb - Developed methods and apparatus used in Nematology.
1926, 1933, 1934	Ayyar - Root knot of vegetables and other crops in India.
1933	T. Goodey - Book on "*Plant Parasitic Nematodes and the Diseases they Cause*".
1934	Filipjev - Book on "*Nematodes that are of Importance for Agriculture*", translated from Russian to English in 1941 by S. Stekhoven under the title "*A Manual of Agricultural Helminthology*".
1936	Dastur - White tip of rice caused by *Aphelenchoides besseyi* in Central India.
1943	Carter - Nematicidal value of D-D which initiated the era of soil fumigation.
1945	Christie (Fig. 1.1) - Nematicidal value of EDB.
1948	Allen - Taught the World's first formal university course in Nematology at the University of California, Berkeley.
1950	Oostenbrink (Fig. 1.1) - Book on "*The Potato Nematode, A Dangerous Parasite to Potato Monoculture*".
1951	Christie and Perry - Role of ectoparasites as plant pathogens.
	T. Goodey (Fig. 1.1) - Book on "*Soil and Fresh Water Nematodes*".
	FAO - First International Nematology Course and Symposium held at Rothamsted Experiment Station, England.
1954	Holdeman and Graham - *Fusarium* wilt of cotton augmented by *Belonolaimus longicaudatus*.
1955	European Society of Nematologists founded.
1956	*Nematologica* - First journal devoted entirely to Nematology papers.
	Thirumala Rao Reported root-knot nematodes on citrus for the first time from India.
1958	Hewitt, Raski (Fig. 1.1) and Goheen - Transmission of a soil-borne plant virus (grapevine fan leaf) by a nematode (*Xiphinema index*).
1959	Prasad, Mathur and Sehgal – Reported the cereal cyst nematode for the first time from India.
1961	G. Thorne (Fig. 1.1) – Book on *Principles of Nematology*

Year	The early records of plant parasitic nematodes
	M.R. Siddiqi (Fig. 1.1) - Reported the citrus nematode for the first time from India.
	Society of Nematologists founded in the United States.
	Jones - Reported the potato cyst nematode for the first time from India.
1966	Nair, Das and Menon- Reported the burrowing nematode on banana for the first time from India.
1967	Organization of Tropical American Nematologists founded.
1968	First South East Asia Post- graduate Nematology course held in India.
1969	*Journal of Nematology* - first published by the Society of Nematologists, USA.
	The Nematological Society of India founded. First all India Nematology Symposium held in New Delhi.
1971	*Indian Journal of Nematology* - first published.
1972	First All India Nematology Workshop held at New Delhi.
1973	*Nematologia Mediterannea* published from Italy.
1976	Summer Institute in Phytonematology held at Allahabad (India).
1978	*Revue de Nematologie,* published from France.
1979	Second All India Nematology Workshop and Symposium held at Bhubaneshwar.
1981	Third All India Nematology Workshop and Symposium held at Coimbatore.
1983	P. Parvatha Reddy published book "*Plant Nematology*" comprehensively covering the subject for the first time from India.
1983	Fourth All India Nematology Workshop and Symposium held at Solan, Himachal Pradesh.
	Summer Institute in Phytonematology held at Hissar, Haryana.
1985	Fifth All India Nematology Workshop and Symposium held at Udaipur, Rajasthan.
1986	National Conference on Plant Parasitic Nematodes - Problems and progress held at New Delhi.
1987	P. Parvatha Reddy published book entitled *A Treatise on Phytonematology*
	All India Nematology Workshop held at Pune.
	Group Discussion on Nematological problems of Plantation crops held at Sugarcane Breeding Institute, Coimbatore.
1992	Summer Institute on "Management of Plant Parasitic Nematodes in Different Crops" organized by ICAR at Haryana Agricultural University, Hisar.
1992 –1995	All India Nematology Workshop and National Symposium on Nematode problems of India held at IARI, New Delhi.

Year	The early records of plant parasitic nematodes
1997	Summer School on "Problems and Progress in Nematology during the past one decade" was organized by ICAR at Indian Agricultural Research Institute, New Delhi.
1998	Afro - Asian Nematology Conference held during April 1998 at Coimbatore.
1999	National seminar on "Nematological Research in India: Challenges and preparedness for the new millennium" held at C.S. Azad University of Agriculture and Technology, Kanpur.
2000	National Nematology Symposium on "Integrated Nematode Management" held at Orissa University of Agriculture & Technology, Bhubaneshwar, Orissa.
2001	National Congress on "Centenary of Nematology in India: Appraisal and Future plans" at IARI, New Delhi.

Fig. 1.1. Eminent Nematologists: *Top row* - N.A. Cobb, T. Goodey, G. Thorne, J.R. Christie. *Middle row* - A.L. Taylor, D.J. Raski, M. Oostenbrink, S. van Gundy, J.N. Sasser. *Bottom row* – A.C. Tarjan, F.G.W. Jones, D.W. Dickson, A.M. Khan, M.R. Siddiqi, M.S. Jairajpuri

1.4. NEMATODES AS BIOLOGICAL MODELS - *CAENORHABDITIS ELEGANS*

Caenorhabditis elegans is a nematode of microscopic size which, due to its biological characteristics, has been used since the 1970s as a model for research in molecular biology, medicine, pharmacology, and toxicology. It was the first animal whose genome was completely sequenced and has played a key role in the understanding of apoptosis and RNA interference. The transparency of its body, short lifespan, ability to self-fertilize and ease of culture are advantages that make it ideal as a model in toxicology. Due to the fact that some of its biochemical pathways are similar to those of humans, it has been employed in research in several fields. *C. elegans*' use as a biological model in environmental toxicological assessments allows the determination of multiple endpoints. Some of these utilizes the effects on the biological functions of the nematode and others use molecular markers. Endpoints such as lethality, growth, reproduction, and locomotion are the most studied, and usually employ the wild type Bristol N2 strain. Other endpoints use reporter genes, such as green fluorescence protein, driven by regulatory sequences from other genes related to different mechanisms of toxicity, such as heat shock, oxidative stress, CYP system, and metallothioneins among others, allowing the study of gene expression in a manner both rapid and easy. These transgenic strains of *C. elegans* represents a powerful tool to assess toxicity pathways for mixtures and environmental samples, and their numbers are growing in diversity and selectivity. However, other molecular biology techniques, including DNA microarrays and MicroRNAs have been explored to assess the effects of different toxicants and samples. *C. elegans* has allowed the assessment of neurotoxic effects for heavy metals and pesticides, among those more frequently studied, as the nematode has a very well defined nervous system. More recently, nanoparticles are emergent pollutants whose toxicity can be explored using this nematode. Overall, almost every type of known toxicant has been tested with this animal model. In the near future, the available knowledge on the life cycle of *C. elegans* should allow more studies on reproduction and transgenerational toxicity for newly developed chemicals and materials, facilitating their introduction in the market. The great diversity of endpoints and possibilities of this animal makes it an easy first-choice for rapid toxicity screening or to detail signalling pathways involved in mechanisms of toxicity.

1.5. ENTOMOPATHOGENIC NEMATODES (EPN)

The entomopathogenic nematodes particularly of the genera *Steinernema* and *Heterorhabditis* characterized by their symbiotic association with bacteria of the genera *Xenorhabdus* and *Photorhabdus* (harbor certain bacteria which act as toxins to insect systems) have been found promising particularly against soil insects such as cut worms (*Agrotis* sp.) and white grubs (*Holotrichia* sp.) on potato, ginger, etc. They are particularly more virulent against lepidopterous and coleopterous pests (Fig. 1.2). Their potential as biocontrol agents is already proved and are being used to control insect pests in high value crops for example, in turf, plant nurseries, tree fruit orchards and groves, cranberries, and mushrooms. They can be used on a large scale in integrated pest management, organic farming and sustainable horticulture systems to control insect pests. Although mass production can be done on insect larvae, commercial production is done using either bedding method or liquid fermentation (Gaugler, 2002).

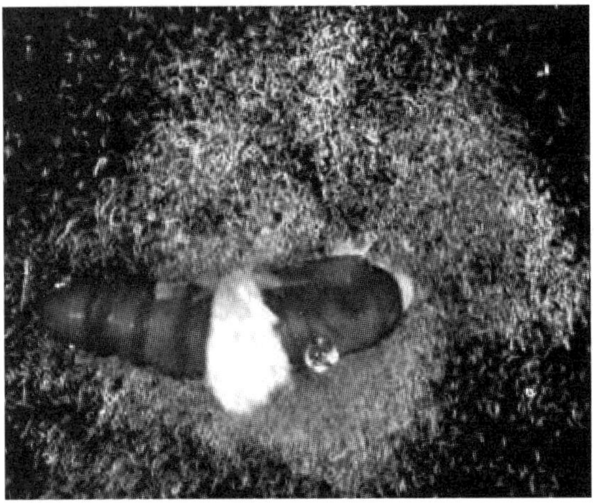

Fig. 1.2. Infective juveniles of insect-parasitic nematodes (*Steinernema*) emerging from a beet armyworm pupa.

The major EPN species used for biological control of insect pests include:

1.5.1. *Steinernema carpocapsae*

1.5.1.1. Turf grass pests: Parasites of cutworms, armyworms, chinch bugs, crane flies.

***1.5.1.2. Orchard, ornamental and vegetable pests*:** Parasites of banana moths, codling moths, peach tree borers.

1.5.2. *Steinernema feltiae*

Parasites of immature flies including mushroom flies, western flower thrips, leaf miners.

1.5.3. *Steinernema glasseri*

Parasites of white grubs (scarabs, especially Japanese beetle, *Popillia* sp.), banana root borers.

1.5.4. *Steinernema riobrave*

Parasites of citrus root weevils (*Diaprepes* sp.).

1.5.5. *Heterorhabditis bacteriophora*

Highly parasitic on lepidopterous and coleopterous insect larvae - white grubs (scarabs), cutworms, flea beetles, corn root worms, citrus root weevils.

1.5.6. *Heterorhabditis megidis*

Parasites of weevils and various other soil insects.

1.5.7. *Heterorhabditis indica*

Parasites of root mealy bugs, grubs.

1.5.8. *Heterorhabditis marelatus*

Parasites of white grubs (scarabs), cutworms.

1.6. REASONS FOR THE SLOW PROGRESS OF THE SCIENCE OF NEMATOLOGY

Nematology has been, and is still a much neglected science, in spite of the fact that nematodes are one of the largest taxonomic groups of the animal kingdom. The progress in this field has been very slow because of the following reasons:

- Nematodes are microscopic, soil-borne organisms which are not visible to the naked eye.
- The expression of symptoms due to nematodes is not dramatic unlike fungi and bacteria.
- Till recently, the science of Nematology was not included in the curriculum of our universities.
- Nematicides used for controlling nematodes are imported and costly.
- Initially there were not many trained personnel to work with these tiny, soil-borne organisms.

1.7. FACTORS RESPONSIBLE FOR THE DEVELOPMENT OF NEMATOLOGY AS A SEPARATE DISCIPLINE

The following are the several significant discoveries during 1941 - 1958 that accelerated the development of Nematology as a separate discipline:

- The introduction of practical nematicides like D-D (Carter, 1943) and EDB (Christie, 1945). These nematicides demonstrated yield losses due to nematode infestation of crop plants.
- The discovery of the potato cyst nematode, *Globodera rostochiensis,* in the major potato-producing region of Long Island, New York, USA (Cannon, 1941). Plant Quarantine Act was enforced in 1944 in New York State against this nematode.

- The demonstration that the burrowing nematode, *Radopholus similis*, was the cause of spreading decline disease of citrus in Florida State, USA (Suit and DuCharme, 1953).
- Recognition of the ectoparasitic nematodes as plant pathogens (Christie and Perry, 1951).
- Recognition of many interactions between nematodes and other soil inhabiting fungi and bacteria in complex plant diseases, including breakdown of disease resistance.
- Discovery of the transmission of soil borne plant virus (grapevine fan leaf) by a nematode vector (*Xipninema index*) (Hewitt *et al.,* 1958).

Because of these and other developments, research in nematology received increased attention and financial support. The introduction of undergraduate and postgraduate courses for the training of nematologists at several universities, the availability of text books emphasizing plant and soil nematodes, new research equipment and techniques further accelerated the progress of plant nematology.

1.8. TRAINING OF NEMATOLOGISTS

Teaching programs in nematology were developed or expanded, primarily within the Department of Plant Pathology, or Entomology in Agricultural Universities and in Botany, Zoology or Biology Departments in traditional Universities. Several State Agricultural Universities/Indian Agricultural Research Institute, New Delhi (ICAR) in India have set up separate Departments of Nematology and award Masters and Doctoral degrees. In some universities, undergraduate and postgraduate students receive a few formal courses in Nematology. Most postgraduate students receive training on an informal basis, working with a nematologist at a laboratory.

In India, the first training course in plant nematology was conducted by F.G.W. Jones in the botany department of the Aligarh Muslim University. This was followed by a second course in 1965 at the Indian Agricultural Research Institute, New Delhi, sponsored by the Rockefeller foundation and run by F.G.W. Jones, J.B. Goodey and D.J. Raski. From 1967-68, "Southeast Asia Post Graduate Nematology Course" was organized for a few years at New Delhi and Aligarh, through the co-operation of International Agricultural Centre, Wageningen; Indian Agricultural Research Institute, New Delhi and Aligarh Muslim University, Aligarh, to train nematologists in India and other south-east asian countries. An "All India Short-term course in Recent Advances in Nematological Techniques and Control" was held at the University of Agricultural Sciences, Bangalore in 1973. The Indian Council of Agricultural Research organized a "Summer Institute in Phytonematology" at Allahabad Agricultural Institute, Allahabad, during 1976 and at Haryana Agricultural University, Hissar during 1983.

1.9. PROFESSIONAL SOCIETIES AND PUBLICATIONS

The Helminthological Society of Washington was founded in 1910. The Society publishes its journal *Proceedings of the Helminthological Society of Washington*, in which many important research papers, in all areas of nematology appear. A volume is issued in two parts each year.

Helminthological Abstracts published by Commonwealth Agricultural Bureau, England, does the abstracting service of the world literature on nematology. Review articles are also included in some issues. One volume is issued in four parts each year.

The ' Society of European Nematologists' was founded in 1953. The Society conducts a symposium every two years and issues a newsletter *Nematology News* twice a year.

Nematologica, an International Journal of nematological research which includes research papers on nematodes and general papers on morphology, taxonomy ecology and physiology. One volume is issued in four parts every year.

In 1962, the ' Society of Nematologists' was formed in the United States. The Society publishes *Nematology News Letter,* and *Journal of Nematology,* and holds annual meetings at which papers representing many aspects of nematology are presented. One volume per year, issued in four parts is published of both journals.

The ' Organization of Tropical American Nematologists' was founded in 1967. This society publishes a newsletter, *Nematropica* (one volume per year issued in two parts) and holds annual meetings.

The ' Nematological Society of India' was formed in 1969. The Society publishes the *Indian Journal of Nematology* (one volume per year issued in two parts starting from 1971).

Nematologia Mediterranea is published from Italy (one volume per year issued in two parts beginning in 1973).

Revue de Nematologie is published twice yearly from France, beginning 1978.

The *Pakistan Journal of Nematology* is published twice yearly from Pakistan, beginning 1984.

The *Bibliography on Plant, Soil and Freshwater Nematodes* is a card title service published by E.J. Brill from the Netherlands. This covers records of literature held at Ottawa, Wageningen and Beltsville.

The ' Commonwealth Agricultural Bureau' publishes the *CIH Descriptions of Plant Parasitic Nematodes* (series starting 1972). Each year one folder containing a set of 15 standardized descriptions of plant-parasitic nematodes is published.

The *Japanese Journal of Nematology* is published from Japan.

Fungicide and Nematocide Tests is published from the American Phytopathological Society, one volume per year.

1.10. EARLY BOOKS ON PLANT NEMATOLOGY

Anon, 1958. *Control of Plant-Parasitic Nematodes*. Vol 4. Principles of Plant and Animal Pest Control Series 1696, Nat Acad Sci, Washington DC, 172 pp.

Chitwood, B.G. and Chitwood, M.B. 1974. *Introduction to Nematology*. Univ Park Press, Baltimore, Maryland, USA, 334 pp.

Christie, J.R. 1959. *Plant Nematodes. Their Bionomics and Control*. Univ. Fla Press, Gainesville, USA, 256 pp.

Crofton, H.D. 1966. *Nematodes*. Hutchinson Univ Library, London, 160 pp.

Decker, H. 1981. *Plant Nematodes and their Control* (*Phytonematology*). Amerind Publishing Co Pvt Ltd. New Delhi, 540 pp.

Dropkin, V.H. 1980. *Introduction to Plant Nematology*. Wiley-Intern Sci, New York, 293 pp.

Goodey, T. 1963. *Soil and Freshwater Nematodes* (2nd edn.) (revised and rewritten by J.B. Goodey, Methuen, London); John Wiley and Sons Inc., New York, 544 pp.

Jenkins, W.R. and Taylor, D.P. 1967. *Plant Nematology*. Reinhold Publishing Corp., New York, Amsterdam, London, 270 pp.

Lamberti, F. and Taylor, C.E. (eds.), 1979. *Root-knot Nematodes* (Meloidogyne *species*) *Systemics, Biology and Control*. Acad Press, New York, 477 pp.

Mai, W.F. and Lyon, H.H. 1975. *Pictorial Key to Genera of Plant Parasitic Nematodes*. Cornell Univ Press, Ithaca, London, 211 pp.

Nickle, W.R. (ed.) 1984. *Plant and Insect Nematodes*. Marcel Dekker Inc, New York and Basel, 925 pp.

Parvatha Reddy, P. 1983. *Plant Nematology*. Agricole Publishing Academy, New Delhi, 287 pp.

Parvatha Reddy, P. 1987. *A Treatise on Phytonematology*. Agricole Publishing Academy, New Delhi, 381 pp.

Peachey, J.E. (ed.) 1969. *Nematodes of Tropical Crops*. Tech Commun No 40, Common Agric. Bur., Farnham Royal, England, 355 pp.

Sasser, J.N. and Jenkins, W.R. (eds.) 1960. *Nematology Fundamentals and Recent Advances with Emphasis on Plant-Parasitic and Soil Forms*. Univ N Carolina Press, Chapel Hill, USA, 480 pp.

Siddiqi, M.R. 1986. *Tylenchida - Parasites of Plants and Insects*. CAB International, Farnham Royal, U.K., 645 pp.

Smart, G.C. and Perry, V.G. (ed.), 1968. Tropical Nematology. Univ. Fla Press, Gainesville, USA.

Southey, J.F. (ed.) 1970. *Laboratory Methods for Work with Plant and Soil Nematodes*. Tech Bull No. 2, Minist Agric Fish Fd., London, 148 pp.

Southey, J.F. 1965. *Plant Nematology*. Tech Bull No. 7, Minist Agric., London.

Thorne, G. 1961. *Principles of Nematology*. McGraw Hill, New York, Toronto, London, 553 pp.

Wallace, H.R. 1973. *Nematode Ecology and plant Disease*. Arnold, London, 228 pp.

Webster, J.M. (ed.) 1972. *Economic Nematology*, Acad Press, London, 563 pp.

Zuckerman, B.M. and Rohde, R.A. (eds.), 1982. *Plant Parasitic Nematodes. Vol III*, Acad. Press, New York.

Zuckerman, B.M. Mai, W.F. and Rohde, R.A. (eds.) 1971. *Plant Parasitic Nematodes. Vol. II*, Acad Press, New York, 347 pp.

Zuckerman, B.M., Mai, W.F. and Rohde, R.A. (eds.) 1971. *Plant Parasitic Nematodes. Vol I*. Acad Press, New York, 345 pp.

Section II
NEMATOLOGICAL TECHNIQUES

Chapter 2

MICROSCOPY

2.1. INTRODUCTION

Most plant nematodes are very small in size and a microscope is necessary to study them. Proper use of the microscope plays an important role in identification of nematodes and other studies in plant nematology. Nematological work requires the most exacting microscopic techniques for the resolution of certain minute diagnostic morphological characters. Accurate identifications cannot be made by superficial, casual observation under low powered magnification except for a few very common and well known species. Generally, two types of microscopes are used in the Nematology Laboratory as aids in the diagnosis of nematode diseases: the compound microscope and the Stereoscopic microscope.

2.2. COMPOUND MICROSCOPE

This instrument employs a series of lenses arranged in such a way so as to give a sharp, clear and highly magnified image of very small objects. The amount of magnification depends on the lenses used and is roughly the product of the magnifications of the eye piece and the objective. Thus, a 10x eye piece and a 40x objective would yield a magnification of approximately 400 diameters.

2.2.1. Components

The body of the microscope carries in the upper part the eye-piece tubes, in the middle part the revolving nose piece; and the object stage and in the lower part the condenser and a lamp attachment (Fig. 2.1).

For nematological work, a binocular tube is very useful; one eye-piece tube, mostly the left, carries a ring for focusing the eye-piece for differential visual acuity. Between the eye-piece tubes is a lever for setting the interpupillary distance.

The revolving nose piece serves to carry objectives. For nematology work, the achromatic objectives with the magnifications 3.5, 10, 40 and oil 100 can be used. The magnification is changed by rotating nose-piece. Each objective is arranged to click home, when it is arranged in its proper position. Each objective is labelled or marked (Fig. 2.2). The oil immersion can be distinguished by a black ring. When this objective is used, a drop of oil is added on to the coverslip. When the oil immersion objective is in its proper position the drop of oil is between coverslip and front lens. The oil serves to prevent bending of light rays out of the field and has the same refractive index (n = 1.515) as that of the coverslip and the front lens of the objective (Fig. 2.3).

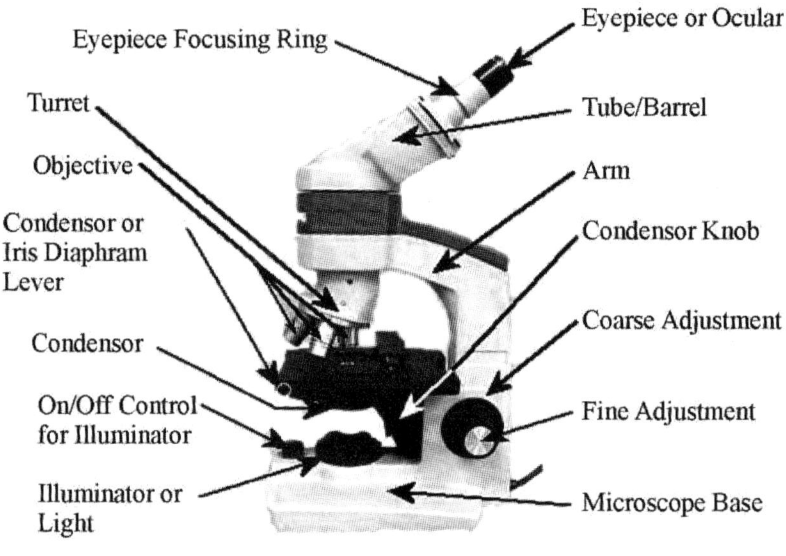

Fig. 2.1 Parts of a compound microscope

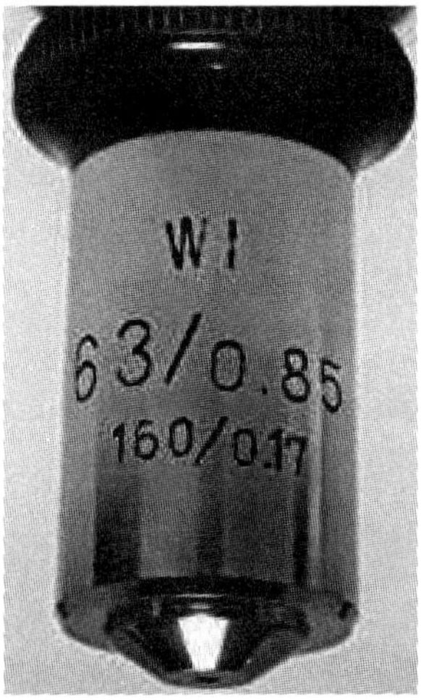

Fig. 2.2. Objective Markings. WI of course indicates the important fact that the objective is designed to work with its front lens immersed in water. The next line of figures is the magnification and the N.A. separated by a slash, and the bottom figures are the tube length (160 mm) and cover glass. The figure 0.17 is the thickness in millimetres of the cover glass for which the objective is corrected.

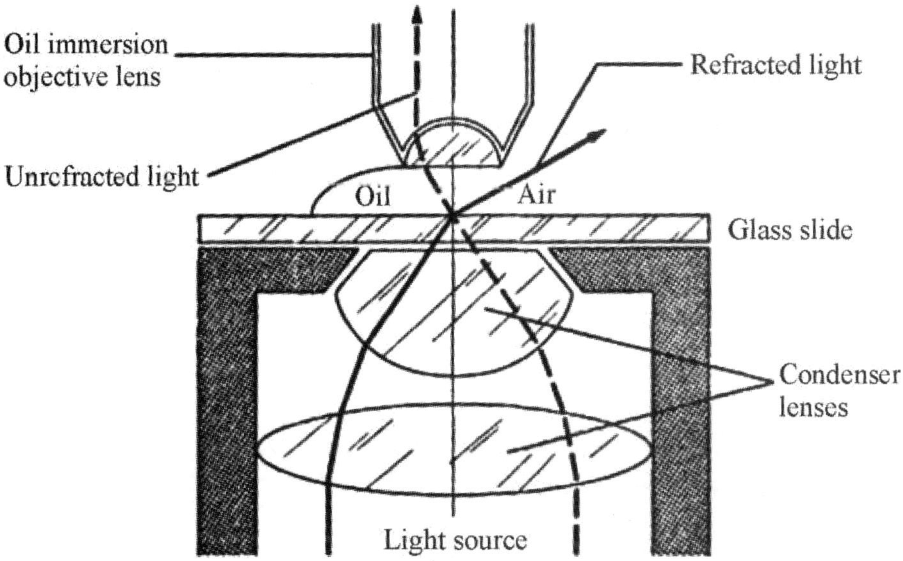

Fig. 2.3. How immersion oil gathers more light for use in the microscope.

The objective stage carrying an object holder can be adjusted by means of a coaxial knob. The object holder is for object slides of 100 mm maximum length.

The condenser has a condenser lens, an aperture diaphragm and a filter holder. The condenser can be adjusted with two centering screws. The body of the microscope contains the focusing mechanisms, which consists usually of two knobs, one for coarse ad another for fine adjustment.

2.2.2. Use

As a matter of practice, the following procedure should be used in bringing specimens up to the greatest magnification possible with the microscope.

- Place a specimen slide on the stage, using the slide holder to secure it in place.
- Focus the low power objective on the specimen with the coarse adjustment knob. Then use the fine adjustment knob for exact focus.
- Now place the high power objective into position and again focus the specimen with the fine adjustment knob.
- Place a small droplet of immersion oil on the specimen slide and bring the oil immersion objective into position. It is a good procedure to watch carefully, to make sure that the objective will not touch the slide. Focusing must always be done with the fine adjustment knob when using the oil immersion objective.
- Regardless of the objective used, adjust the iris diaphragm to control contrast and definition of the specimen.

2.2.3. Alignment and care

For maximum effectiveness, the mechanical and optical components of the microscope must be clean and aligned. Follow this list of procedures each time a microscope is used:

- With the coarse adjustment knob, crank the body tube all the way up.
- Clean each objective lens with lens paper and a solvent (xylene) to remove any oil or grease. Never use alcohol for cleaning the microscope.
- Clean the eye piece lens.
- Clean the slide stage.
- Remove the light source.
- Lower the condenser lens all the way down and clean it.
- Make sure the lamp is not plugged in and then open the lamp box and remove the diffusion filter. Clean the filter.
- Close the iris diaphragm. Tip the microscope on its side. Look through the condenser lens from the bottom and move the lens up and down. Note that the iris diaphragm goes in and out of focus.
- Set the microscope back on the desk and the lamp assembly back in place without the diffusion filter. Turn on lamp.
- Place a piece of filter paper tightly over a slide and observe the image of the lamp filament on the paper. Do not look at the filament through the eye piece. See if the filament is centered. Move the condenser lens up and down and observe the filament image on the filter paper. Leave the lens at a point where a peripheral ring of light begins to appear. At this point the image should be in focus on the filter paper. Now the light source will be focused on any object mounted for microscopic examination. This point is also the point at which the iris diaphragm was in focus when observing through the bottom of the condenser lens.
- Do not change the position of the condenser lens after it has been set.
- Replace the diffusion filter.
- Put a slide with a specimen on the stage and secure with the slide holder.
- With the low power objective lens in place, focus the specimen with the coarse and fine focusing knobs.
- Open the iris diaphragm fully while looking through the eye piece, slowly close the diaphragm until the light begins to fade. Light from the condenser now just fills the back lens of the objective.
- Keep the microscope free from dust by covering it with a plastic cover.

2.3. DISSECTING MICROSCOPE

The dissecting microscope should be of the low power wide-field binocular type. Proper magnification and working space beneath the objectives are two of its most important qualities. Select an objective lens system which gives one to four magnification and paired eye pieces of 15 x or 20 x (Fig. 2.4). This combination will give sufficient magnification for the nematologist to recognize common genera of nematodes and determine if specimens are young or mature.

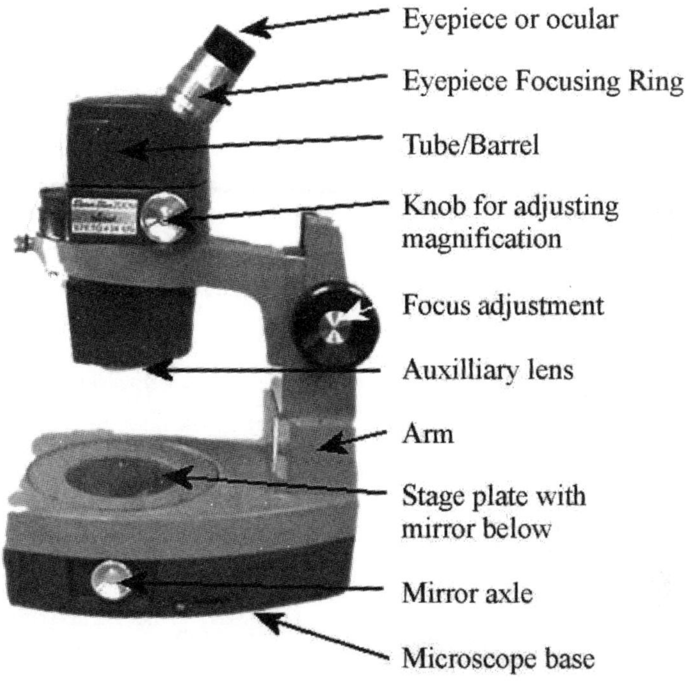

Fig. 2.4. Parts of a dissecting microscope microscope

For most work in nematology, diffuse lighting is satisfactory. Therefore, use day light or a strip light with two electric tubes as close as possible in holders, with a shade above and a stand to give height adjustment for transmitted or incident illumination.

2.3.1. Use

The following procedures should be used when observing a specimen:

1. Clean the stage plate, eye pieces, mirror and objective lenses with lens paper.
2. Place a specimen or object in the center of the stage.
3. Illuminate the object (top light for opaque objects and transmitted light for transparent or translucent objects).

4. Raise the body above the focus point with focusing knob. Then lower the body to bring the object into sharp focus.
5. Start with the magnification knob in the lowest position and adjust upwards as necessary.

2.4. SCANNING ELECTRON MICROSCOPE (SEM)

This instrument is rapidly becoming a common adjunct to the light microscope. A beam of electron is focused on the surface of a specially prepared specimen. Secondary electrons are emitted in such a way that a sharp image of the surface of the object is created (Fig. 2.5). Photographs of nematodes produced by scanning electron microscopy (SEM) are becoming important in several ways in phytonematology: in taxonomy, in studies of host-parasite relations, and to a limited extent in studies of the internal structures of plants and nematodes. Moreover, with additional instrumentation, some chemical analysis of surface materials can be made. The techniques, although very different from those of light microscopy, are accessible to students, and the instrumentation is now available in many laboratories. Wergin (1981) published a thorough account of SEM.

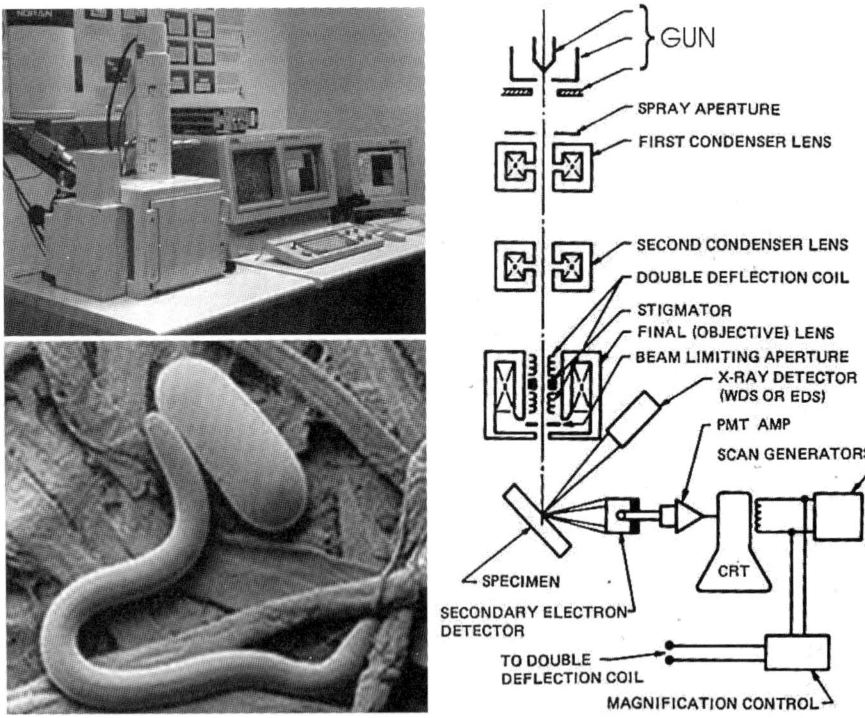

Fig. 2.5. *Clock-wise* - A typical SEM instrument, showing the electron column, sample chamber, EDS detector, electronics console, and visual display monitors; Electron and X-ray optics of a SEM – EPMA; SEM image of soybean cyst nematode and egg.

2.4.1. Advantages

SEM offers five advantages over conventional light microscopy:

- Images are readily comprehended.
- The range of magnification extends from 20-50,000x and the images are sharp throughout.
- The depth of field is 300-500 times greater than that of the light microscope.
- The scanning electron microscope can be used for relatively larger specimens.
- Because the stage is easily moved, the object can be viewed from several angles, and three-dimensional images are obtainable.

2.4.2. Use

Mastery of the details of the technique, requires special instruction. In essence, specimens must be fixed, dried, mounted on a stub, and coated with a fine film of metal before insertion into the microscope for observation. One system of fixation in common use is to subject the specimens to a primary fixative made of 3% glutaraldehyde in 0.05M phosphate buffer, pH 6.8. Fixation proceeds at room temperature for 1½ hr. Special systems have been developed for handling individual phytonematodes to avoid losing the specimens.

Fixed nematodes should be washed in about six complete changes of the above buffer solution, each change lasting 10-15 minutes. Post-fixation at room temperature requires a 2 hour immersion in 2% osmium tetroxide in 0.05M phosphate buffer.

For dehydration, the organisms are transferred successively through 20, 40, 60, 80, 95 and 100% ethanol during 1-2 hours. Following this, three changes of absolute ethanol are required. Following the last ethanol, specimens still in ethanol are transferred to the chamber of a critical-point dryer, which is then sealed. Liquid carbon dioxide replaces the ethanol in a gradual procedure, and the chamber is slowly heated until the critical-temperature and pressure are reached. Then the chamber is kept at the critical-temperature, pressure is released slowly, and the dehydrated nematodes are ready for mounting on aluminum stubs with a special tape.

Finally, the stub is transferred to another instrument, the sputter coater, in which a 200-300 angstrom layer of gold or gold-palladium alloy is deposited over the specimen. The preparation is now ready for examination by SEM.

2.5 DRAWING AND MEASURING NEMATODES

2.5.1. Direct measurement

The measurement of minute objects by means of the microscope is an interesting and valuable feature of microscopic study. The simplest form of measuring device is an eye piece micrometer, a disc of glass having an engraved scale, a series of accurately spaced lines. The spaces do not have a standard value and each disc must be calibrated for each given eye-piece and set of objectives. Place the disc upon the metal diaphragm in the eye piece. If the diaphragm is in the correct position, the lines on the disc will be in sharp focus.

The stage micrometer with which the calibration is made is a slide bearing an engraved scale with known values (1 mm divided into 100 parts or 2 mm divided into 200 parts), usually in tenths and hundredths of a millimeter. When the stage micrometer is brought into focus, the scale of the eye piece will be seen superimposed on the scale of the stage micrometer. Shift the stage micrometer and revolve the eye-piece until the two scales are in such position that the values may be compared (Fig. 2.6). A specific case using a 43x objective and a 10x eye piece is shown in Fig. 2.6. It will be seen that the 10 small divisions of the eye piece scale, are equivalent to 21 small divisions of the stage micrometer scale. The computation is:

21 stage micrometer divisions = 10 ocular micrometer divisions

1 stage micrometer divisions division = 21/ 10 = 0.0021 mm = 210 μ

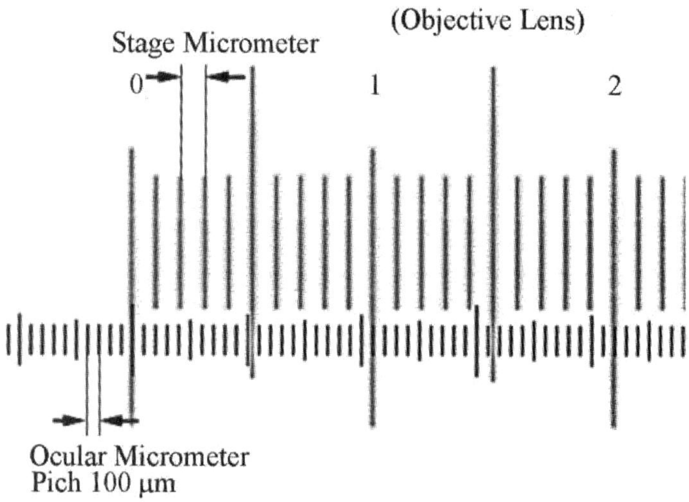

Fig. 2.6. Calibration of an eye piece micrometer disc. *Top* - stage micrometer scale, *Bottom* – ocular micrometer scale.

Short, straight measurements such as stylet length or body width can be made using an eye piece micrometer. Body length, vulval position on straight nematodes can also be estimated by the use of an eye piece scale.

2.5.2. Drawing equipment

To draw and measure nematodes accurately, it is essential to use camera lucida that gives a large, sharp image of the specimen superimposed on the drawing area. A camera lucida essentially consists of an adjustable ring which can be fitted to the tube of any microscope. This is done by first removing the eye piece, fitting the ring and then reinserting the eye piece into the tube of the microscope. The prism head which consists of a prism fitted in such a manner that it comes to lie exactly over the eye piece and the structure on the slide can be viewed through this prism and the eye piece. The body on which the prism is fitted can be rotated and the height is also adjustable. The prism is so fitted that it also reflects the image of the objective on to a mirror. The reflecting mirror is attached to a long lever that can be fitted and screwed to the body holding the prism. The image which is reflected by the prism falls on this mirror, it is reflected down on the table, to the side of the microscope (Fig. 2.7).

Fig. 2.7. Camera lucida apparatus attached to microscope for drawing nematodes

2.5.3. Procedure

Fit the camera lucida on to the microscope. Place the slide bearing the object and focus the structure, viewing through the prism of camera lucida. Adjust the artificial source of light properly (neither bright nor too dim), place a white paper on the right hand side of the microscope on the table and hold a lead pencil in your right hand. The object as well as the white paper and the pencil (also the fingers holding it) should be visible to your eye, even though you are looking into the eye piece of the microscope and not sideways towards the paper.

Once the object as well as the paper and pencil point is clearly focused, the outline and the details of the object can be sketched. Looking through the microscope, it appears that the pencil point is actually moving on the object while in reality the image of the object is being reflected, down on the paper.

Such a sketch is later made permanent using India ink. The various parts should then be labelled and the legend giving details of the object stenciled on the bottom of the sketch. When final figures are being made, these should better be drawn directly on to white card sheets about 28 cm x 22 cm. The magnification should invariably be mentioned after the legend.

2.5.4. Drawing nematodes for measurements

Freshly processed specimens that have been suitably mounted are drawn using camera lucida. Choose a magnification such that the nematode can be drawn in one piece, or with larger specimens, in two or three pieces. An outline of the nematode can be drawn but frequently it suffices to draw a median line along the body length. Mark on this line the positions of organs and orifices. As the body width, stylet length and sometimes tail length are very short relative to body length, it is advisable to measure them direct at higher magnification.

Straight sections of the drawings are measured in mm with a ruler. Curved lines are measured by fitting lead fuse wire to the curvature straightening it and measuring against a ruler. The actual length of the specimen is then calculated by comparison with the magnification. Most camera lucida drawings magnify the object more than does the microscope alone. It is therefore, essential to determine the magnification by drawing the scale of a stage micrometer for each combination of eye-piece, objective and drawing apparatus used.

■■■■

Chapter 3

SAMPLING NEMATODE POPULATIONS

3.1. SAMPLING PROCEDURES

The primary objectives of nematode sampling include: general detection, surveys, diagnosis of nematode diseases, providing advice in integrated nematode management programs and fulfilling research needs.

3.1.1. Detection and survey

The quarantine or phytosanitary regulations of many countries require that planting materials be produced on land certified free from nematodes such as the golden nematode or potato. Soil sampling for certification of widely distributed planting materials requires extreme precision for detection of quarantined pests. A combination of sanitation and sampling greatly reduces chances of introducing nematodes into new areas.

Sampling still serves as the basis for determining the occurrence and distribution of many nematodes.

3.1.2. Diagnosis or detection

Sampling for nematodes is an increasingly important component of plant disease diagnosis, especially for perennials and various ornamentals. Without sampling, poor plant growth due to nematodes may be misinterpreted as nutrient deficiencies or other maladies. The art of collecting representative root and/or foliage as well as soil samples is the primary component of sampling for diagnosis. For example, sampling from adjacent healthy plants may be as important as collecting from the severely affected hosts.

3.1.3. Advisory

The fact that initial numbers of nematodes can be related to the yield of annual crops has enabled nematologists to develop functional advisory programs, even though relationships between nematode numbers and crop damage may be modified by environment. Because of the importance of reliable detection, much sampling for advisory purposes is conducted when population densities are near maximum, often after harvest. Sampling at the time of planting, however, theoretically gives a better estimate of the nematode problem where a reliable inventory can be made. Repeated sampling is often necessary with perennials, since non-detectable populations may increase to damaging levels.

3.1.4. Research

Much effort must be invested in collecting soil or plant material for experimentation. Relative numbers and stages of nematodes may be greatly affected by sample handling or extraction. The design and management for sampling may need to be modified for specific types of experimentation.

3.2. COLLECTION OF SAMPLES

Before samples are collected, a detailed sampling plan, including the collection pattern, number and size of cores per sample and the number of samples must be developed, based on theoretical and practical considerations (Fig. 3.1).

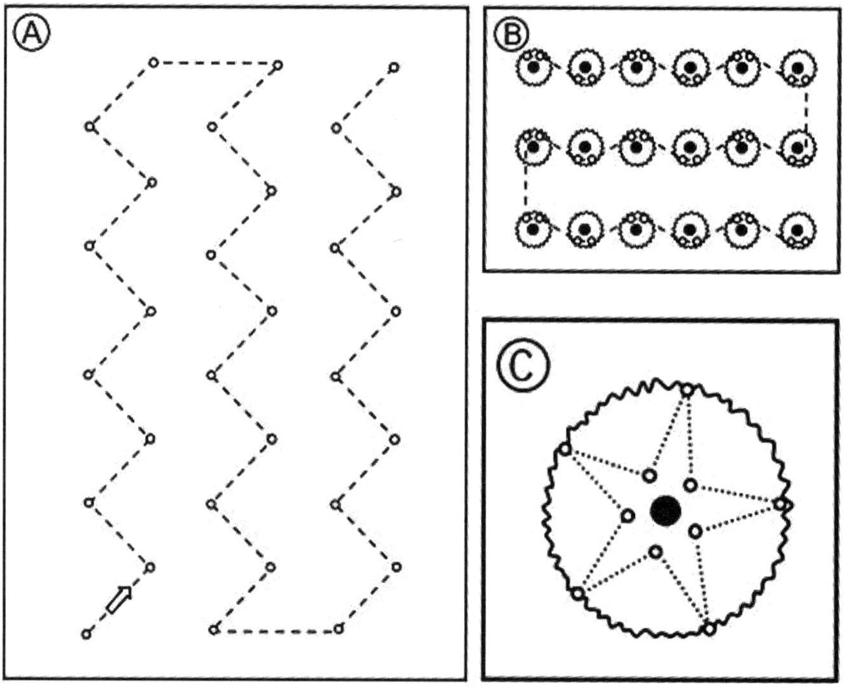

Fig. 3.1. Different patterns of sampling. (A) On fallow land (B) in an orchard (C) for a single plant/tree (Barker *et al.*, 1978)

3.2.1. Sampling equipments

Equipment for collecting soil samples for nematode assays includes shovels, soil augers or tubes (Fig. 3.2). The typical cylindrical tube type sampler or auger serves as standard equipment for many nematologists. The smaller tubes 20-25 mm in diameter are used more frequently; 50 cores from these tubes give about 2 kg of soil.

Fig. 3.2. Soil sampling equipment

3.2.2. Timing of sample

For population data to be of value, samples must be collected when the target organisms are at detectable levels. Repeated or sequential sampling may be necessary to give a reliable census of polyspecific communities in which different species change differently.

3.2.3. Sample size

Sample size may be prescribed on the basis of population densities. Ten sampling units may provide an adequate estimate of a large population, whereas 100 units may be necessary for low densities.

3.2.4. Sampling patterns

Possible sampling procedures for nematodes includes simple random, stratified random, systematic, haphazard, cluster and two stage sampling. The specific sampling procedure is determined by the given distribution pattern. Generally, systematic sampling gives more reliable results than random sampling. Guidelines for sampling have been offered for evaluating nematicide efficacy on various types of crops.

3.3. CARE AND CONDITIONING OF SOIL AND PLANT TISSUE SAMPLES

Plant parasitic nematodes are aquatic micro-organisms. The cuticle and body walls, while complex, are quite thin and provide very limited protection. Individuals of most species are short lived, even under ideal conditions. Exposure to the sun or high temperatures usually kills them.

3.3.1. Handling of samples during collection and transit

Soil samples must be handled properly to maintain the nematodes in the state at which they were collected. Extraction methods that depend on nematode motility rely on maintenance of physiologically active specimens. The nematodes must be in sufficiently good condition so that they can be identified.

3.3.2. Storage

Laboratory storage of samples may allow considerable hatching and increase of yields as compared to extraction immediately after collection. Improper storage often results in death and deterioration of nematodes.

3.3.2.1. Temperature and moisture: Temperature and soil moisture are the two most critical parameters that affect nematode survival in storage. Numerous investigators have found that storage temperature of 10°-15°C usually provide maximum recovery rates that often deviate little from yields of non-stored samples. Warm temperature storage results in a rapid depletion of lipids, the source of energy for plant parasitic nematodes. Excessive moisture can result in rapid decline of nematode populations in storage.

3.3.2.2. Time: The composition of nematode populations at the time of collection may influence apparent survival rates in storage. In a soil sample with large numbers of eggs, the quantity of extractable nematodes would increase with time, especially at moderate temperatures.

■■■■

Chapter 4

EXTRACTION OF NEMATODES FROM SOIL

4.1. PRINCIPLES

Most of the extraction methods are indirect, making use of a number of properties to separate nematodes from the surrounding medium:

4.1.1. Weight and rate of settling

In water, nematodes are separated from particles that settle faster and can subsequently be poured off (decanted). This principle is the basis of a number of applications, such as the use of an undercurrent that keeps nematodes afloat while other particles settle (elutriation) and use of a liquid with higher specific gravity than nematodes, which keeps them afloat while other particles (with a higher specific gravity than the liquid) sink to the bottom. This is applied in centrifuge floatation techniques. Dried cysts contain air bubbles, making them to float on the water surface, which separates them from sinking particles.

4.1.2. Size and shape

Because of their size and elongated shape, nematodes can be separated from other (soil) particles by using a set of sieves with different mesh size.

4.1.3. Mobility

Because of their mobility, living nematodes can be separated from other particles. When a sample is placed on a sieve with a moist filter paper, positioned in a shallow water-filled tray, nematodes will crawl from the sample into the water where they can be collected as a clear suspension.

4.2. FREE-LIVING NEMATODES

4.2.1. Cotton wool filter technique

This method is usually used for soil samples of 50 ml (20 to 100 ml). The procedure followed is Mix - 50 ml soil thoroughly in about 600 ml of water in container A, allow to stand for 10 seconds, then decant into second container B.

Rinse remaining soil in container A twice more and add suspension to container B. Quickly pour the whole of suspension in container B (about 1.8 liter) on to a double cotton wool filter, about 18 cm in diameter supported on a coarse-mesh sieve (Fig. 4.1), followed by one liter of clean water. The suspension and the additional water should be poured on to a watch glass on the filter, which is just under water in a shallow dish, to prevent damage to the filter and to get more even distribution. Then, transfer the filter to a shallow extraction dish which contains just enough clean water to keep the filter wet. Carefully remove the filter after 12 to 24 hour when most of the active nematodes will have collected in the extraction dish.

Fig. 4.1. Cotton wool filter technique

4.2.1.1. Advantages

The most economic and most efficient extraction method for soil samples up to 50 ml.

4.2.2. Decanting and sieving technique

This method takes advantages of the differences in size and specific gravity between nematodes and other soil components. The step-by-step procedure of this method is as follows (Fig. 4.2).

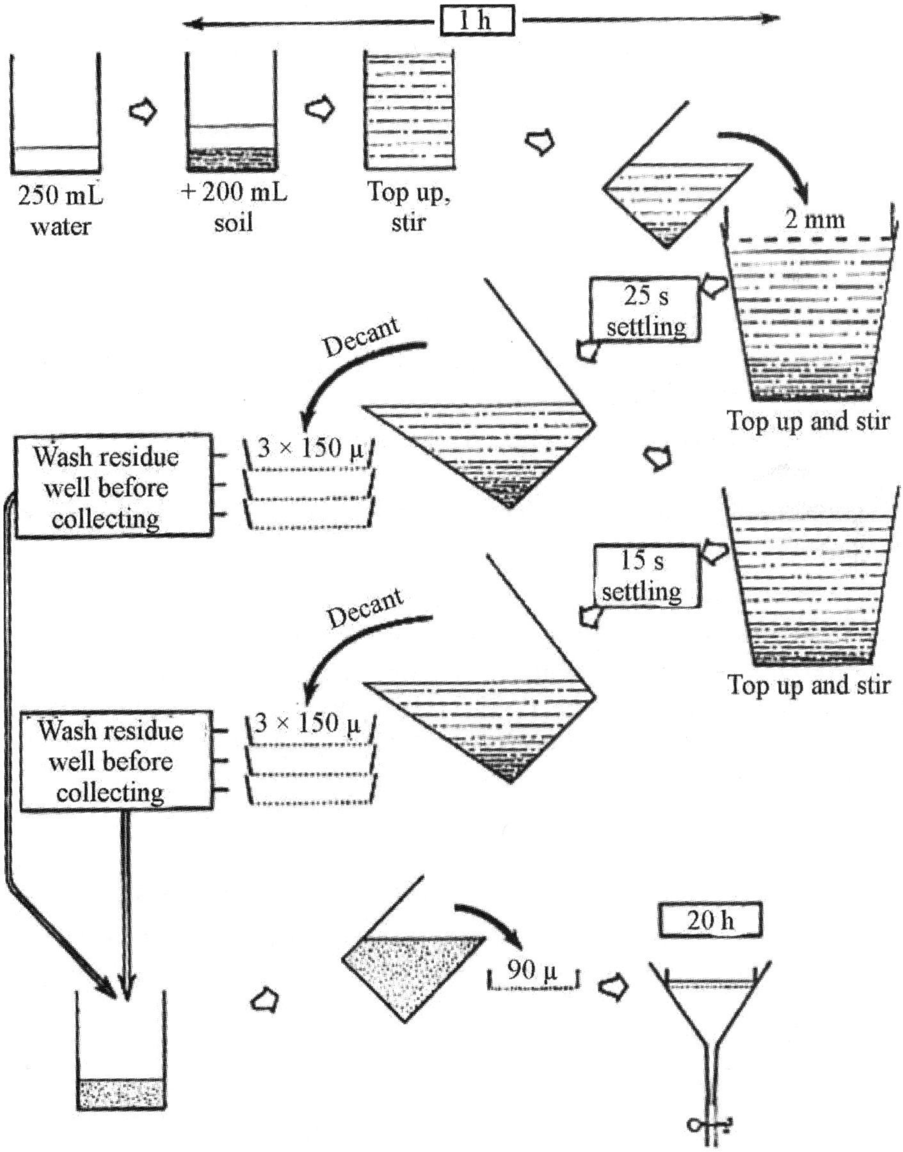

Fig. 4.2. Decanting and sieving technique for extraction of nematodes from soil.

4.2.2.1. Procedure

- Fill a 1 L beaker with about 250 mL of water.
- Add the prepared soil sample (approx. 200 mL soil) and soak for about 30 min (loamy soil) to 60 min (clay soil), stir 2–3 times during soaking period.
- Place the 2 mm aperture sieve on the 5 L plastic bucket.

- Wash the soil suspension through the 2 mm sieve into the 5 L bucket.
- Remove the sieve and top up the bucket with water, then agitate solution by stirring.
- After 25 seconds sedimentation time, the supernatant suspension is decanted through a bank of three 150 μm aperture sieves; ensure that the sediment stays in the bucket.
- Gently wash the residue on the sieves with a wash bottle to a clean 1 L beaker.
- Top up the residue in the bucket again with water and swirl thoroughly.
- After 15 seconds sedimentation, decant the supernatant through the same bank of three 150 μm aperture sieves.
- After washing, the residue on the sieves is added to that collected previously.
- Pour the complete content of the liter beaker onto a 45 μm aperture sieve.
- For cleaning the nematode suspension from soil particles, use the Baermann funnel or Oostenbrink dish method.
- Examine nematodes within a counting slide at 40× magnification using a dissecting microscope or transfer nematodes with a handling needle to a microscope slide for inspection at higher magnification in a compound microscope.

4.2.2.2. Advantages

- Produces clean, healthy specimens for inoculation purposes.
- Some genera such as *Trichodorus* and *Helicotylenchus* are readily yielded in larger numbers.

4.2.2.3. Disadvantages

- Does not always yield specimens of *Criconemoides, Criconema* and *Hemicriconemoides* even when these are abundant and living in the soil sample.
- The tissue paper acts as a barrier to swollen females and some slow-moving forms.
- If the residues are dense and abundant, such as from clay and silt, nematode activity is inhibited due to oxygen suppression.
- Very fine soil particles sometimes pass through the tissue papers making the sample difficult to read.
- Length of time involved is considerably more.

4.2.3. Simple extraction method for large nematodes like *Xiphinema* and *Longidorus*

- 100 ml of soil is covered with water in basin I. After stirring thoroughly, the supernatant is decanted into basin II.
- After stirring, the suspension in basin II is poured over the 100 mesh sieve.
- The catch on 100 mesh sieve is transferred into a beaker.
- The nematode suspension in the beaker is poured onto a wire gauze sieve containing two layers of tissue paper kept in a funnel holding sufficient water to remain in contact with the bottom of the wire gauze.
- The next day the nematodes can be collected in clear water at the bottom of the funnel stem.

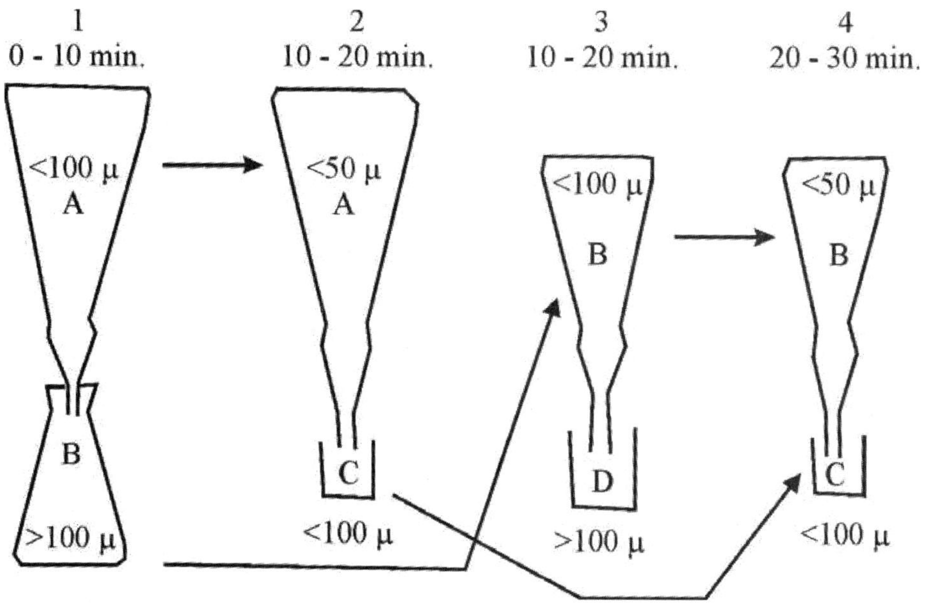

Fig. 4.3. Seinhorst's two flask extraction technique

4.2.4. Two-flask technique

This method is based on the principle of counter flow (Fig. 4.3). Place 500 ml of the soil sample in a vessel with 700 ml of water. The resultant suspension is passed through a coarse sieve (to remove pieces of roots and stones) and poured into a 2-litre Erlenmeyer flask (A), which is filled with water and capped. After a vigorous shaking, flask (A) is overturned and placed on a second Erlenmeyer flask (B) which has already been filled with water. The heavy particles of soil will settle in the lower flask (B) and the displaced water pass into the upper flask (A). The counter flow in the first few minutes prevents the lighter soil particles and nematodes from settling. After approximately 10 minutes, flask (A) is removed from flask (B), agitated and placed for another 10 minutes in a Belcher glass (C) filled with water. Flask (B) is likewise capped, overturned and placed in another Belcher glass (D). Within 10 minutes (20 minutes from the commencement of the procedure) flask (A) is removed from the Belcher glass (C) and flask (B) placed in it (C) for another 10 minutes. Both the flasks now contain small nematodes and soil particles less than 50 μ in size, while the Belcher glass (C) contains larger nematodes and soil particles less than 100 μ in size. The contents of both flasks are poured through a sieve with a 50 μ aperture and the contents of the Belcher glass (C) poured through a sieve with 100 μ aperture. The sediment from both sieves is put aside while the contents of the Belcher glass (D) are agitated, left to rest for a minute or two, then poured through a sieve with a 250 μ aperture. This sediment is then combined with that from Belcher glass (C) and both flasks are placed on a sieve with a cotton wool filter. The nematodes pass through the cotton wool filter and collect in the water below.

4.2.5. Elutriation Techniques

Two elutriation systems are in use, each based on the principle that a measured flow of water in an upward direction will support nematodes in a given range of specific gravities but will allow heavier soil debris to pass downward so the nematodes can be collected in a relatively clean state. The two elutriation systems are similar in principle but differ markedly in that each requires specialized apparatus and processing methods peculiar to the system.

The chief use of the elutriation system at present is to secure for inoculation purposes those nematode genera, which are not readily yielded by sieving and decantation technique, such as *Criconemoides*, *Criconema* and *Hemicriconemoides*.

4.2.5.1. Oosternbrink's elutriation system

The step by step procedure of this system is as follows (Fig. 4.4):

- Before starting, fill the apparatus with clean water until the outlet of the funnel by a constant water stream of 1000 ml/min through a perforated pipe from the bottom of the can (Fig. 4.4).
- Mix the soil thoroughly. A sample of 100 to 500 ml moist soil is placed in the 1 mm pore size top sieve.
- Wash the sample into the can via the funnel by means of a nozzle delivering about 700 ml/min until two-thirds of the column is filled up.
- Turn off the top nozzle. Reduce the constant water stream from the bottom to 600 ml/min until the water reaches the base of the funnel stem.
- The suspension is poured onto a set of 4 sieves of 325 or 400 mesh, 30 cm diameter, placed on top of one another.

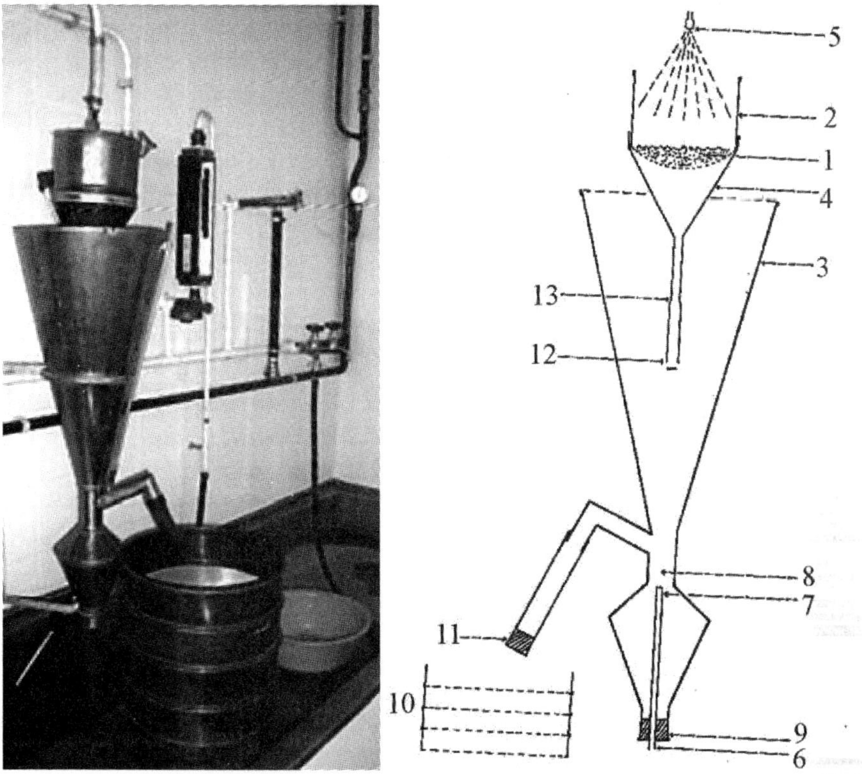

Fig. 4.4. Oostenbrink elutriation system for extraction of nematodes from soil. 1 - Sample of moist soil, 2 - 1 mm top sieve, 3 - Flotation apparatus, 4 - Funnel-pipe with baffle plate, 5 – Nozzle of about 700 ml of water/min, 6 – Upward water current coming from reservoir tank near the ceiling of the room and adjusted through a Turbo meter, 7 – Insertion pipe with perforated tip, 8 – Neck of the apparatus where the nematodes are stemmed, 9 – Rubber plug which can be removed for flushing the apparatus, 10 – Nest of sieves, 11 – Rubber plug of the outlet, 12 – Water level at start, 13 –Water level when the nozzle 5 is closed and the upward current is adjusted to 400 ml/min.

- The catch is immediately washed into a 250 ml beaker.
- Pour the nematode suspension on to a wire gauge sieve containing 2 layers of tissue paper. Place the wire gauge with tissue paper in a funnel holding sufficient water to remain in contact with the bottom of the wire gauge.
- Leave the funnel overnight. The final suspension containing the nematodes is ready for analyzing at the bottom of the funnel stem.

4.2.5.2. Seinhorst's elutriation system

500 ml soil sample is stirred in 750 ml of water and strained through a coarse sieve into a 2 liter Erlenmeyer flask (A) provided with a funnel cap (C) and a

rubber plug (D) (Fig. 4.5). The flask is closed with the cap and so placed that only the outlet protrudes into the funnel C_1 + C_2, which is connected with a second funnel of similar shape but smaller in size (D_1 + D_2). C_1 has an overflow pipe at the upper end (L) which discharges into a small funnel. C_2 and D_2 have outlet pipes provided with rubber sleeves and stopcocks (K, J). The end of funnel D_2 discharges into funnel E_1. The outlet pipe (I) is fitted with a clamp. A feed pipe (N) is connected to the upper end of funnel E_1. E_1 discharges into E_2 which is a straight tube. The lower end of E_2 is closed with a plug provided with a device (M). A cable-connects E_1 and E_2 and a free plug is used to close the passage between them if E_2 has to be emptied.

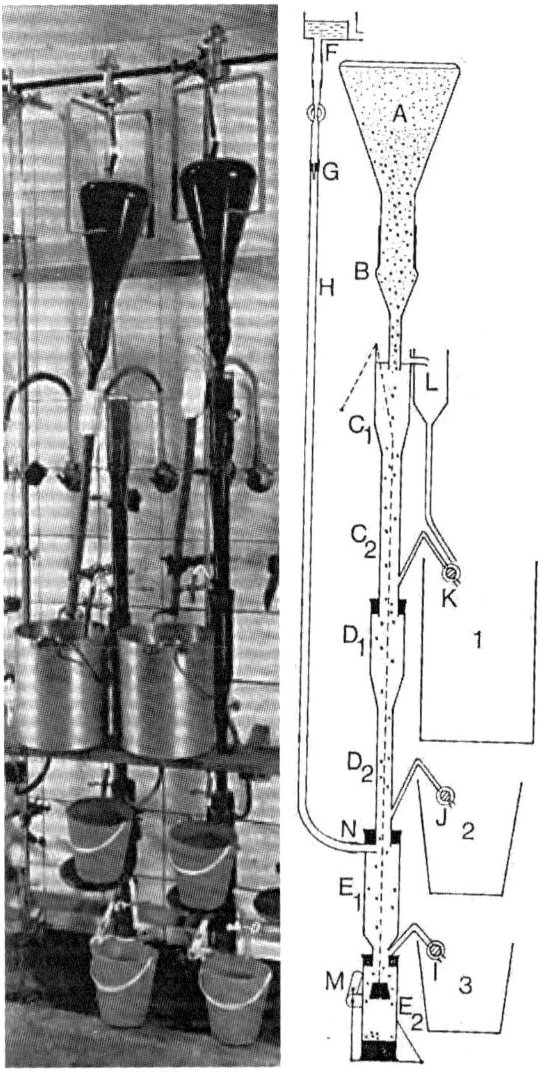

Fig. 4.5. Seinhorst's elutriator.

The flotation apparatus is filled with tap water up to the overflow pipe. The Erlenmeyer flask is opened and the sample poured through a funnel using a length of wire. By regulating the counter flow of water the nematodes and small soil particles are maintained in a suspended state or settle slowly, while the heavier particles of soil settle relatively more rapidly in container E_2.

The flow rate of water must be so regulated that the rate of rise in D_2 is 975 cm per hour or 380 cm per hour in C_3. The reserve tank ensures the constant flow of water through an outlet pipe located above the overflow pipe (F) and connected by a hose with syringe (G).

The small nematodes with a settling rate below 380 cm per hour remain in the funnel $C_1 + C_2$, whereas particles over 50 μ in size settle to the bottom. In the funnel $D_1 + D_2$ the nematodes with a settling rate of 380 to 975 cm per hour are separated from particles over 100 μ in size. The large nematodes (more than 2 mm) settle within 7 to 9 minutes in E_1 and the largest particles of soil are collected in E_2.

Within 20 minutes (sandy soil) or 30 minutes (loam) all the particles which are more than 50 μ in size will have settled from the Erlenmeyer flask and it may be removed. After an additional 10 or 15 minutes, the clamp may be opened and the contents of $C_1 + C_2$ poured into vessel 1. Finally, $D_1 + D_2$ are emptied into vessel 2 and E_1 into vessel 3. E_2 is then emptied and the whole apparatus washed out.

Sieve the contents of vessel 2 through a bank of seven 100 μ aperture sieves of 10 cm diameter, washing the residues on the sieves into vessel 1. Sieve the contents of vessel 1 through a bank of seven 50 μ aperture sieves of 20 cm diameter, collecting the residues in as small a bulk of water as possible. Pour the contents of vessel 3 through a bank of 250 μ aperture sieves. Wash off all the residues and concentrate the suspension of them by pouring through a 50 μ aperture sieve and washing the residues off this into a beaker with a small quantity of water.

(a) Advantages

- It can deliver the nematodes in different categories according to size.

(b) Disadvantages

- Cumbersome and time consuming (30-40 min per sample).
- Expensive due to breakage.
- Difficult to repair the essential glass units.

4.2.6. Centrifugal flotation technique (Fig. 4.6)

Flotation of nematodes by the use of a sugar solution of greater specific gravity (1.18) than that of the nematode (1.05) can be used when an estimate of mere numbers is required. The step by step procedure of this technique is as follows:

- Obtain nematode suspension in the beaker after sieving as described in sieving and decanting technique.
- Pour the contents of a beaker (retaining the homogenous state by swirling in the beaker) into the centrifuge tubes distributing the contents equally between two or four tubes.

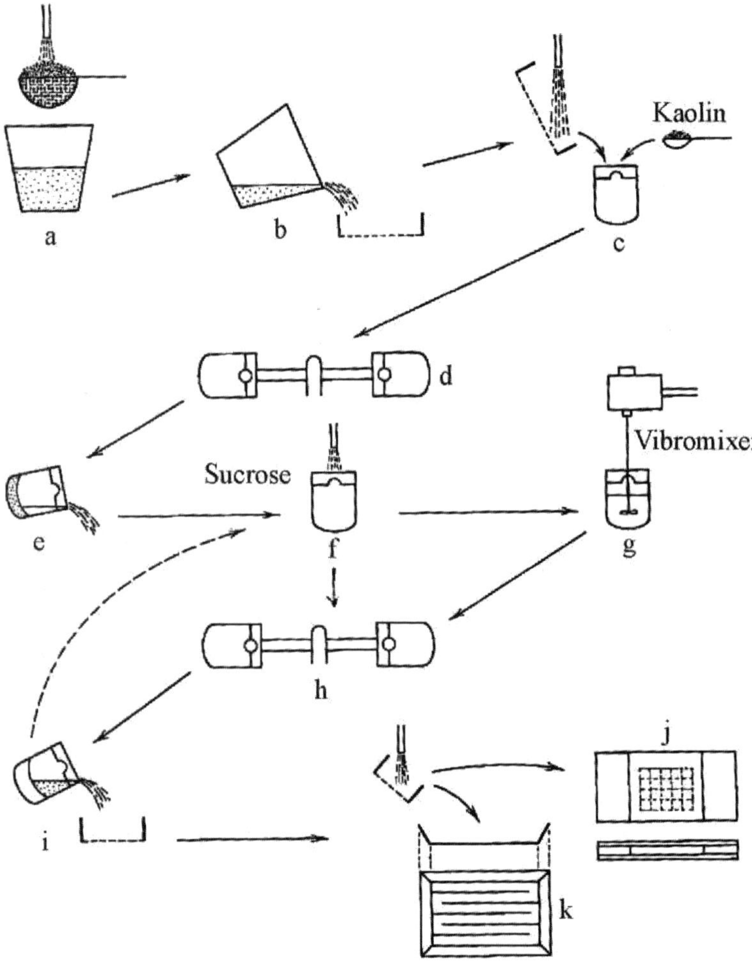

Fig. 4.6: Centrifugal flotation method for extraction of nematodes from soil. **(a)** soil washed through 1000 μm sieve, **(b)** filtrate from **step a** poured off slowly (2-3 min) through a 40 μm sieve, **(c)** sieve rinsed into a 500 ml centrifuge cup with addition of kaolin to retain the nematodes in **step e**, **(d)** first centrifugation at 3000 rpm for 4 min, **(e)** elimination of supernatant water: nematodes are held in the residue under the layer of kaolin, **(f)** addition of sucrose solution with a density of 1.22 (500 g sucrose + 1 liter water), **(g)** mechanical agitation puts residue into suspension in sugar solution, **(h)** second centrifugation at 3000 rpm for 2 min, **(i)** filtration through 350 mesh sieve, **(j)** and **(k)** collection from sieve and enumeration on a counting slide or in a counting dish.

- Add water if needed to bring the level to within 0.5 cm of the tube top.
- Centrifuge at 3000 rpm for four minutes.
- Nematodes and debris of nearly equal or greater specific gravity than that of the nematodes will be compacted at the bottom of the centrifuge tube. Carefully discard the supernatant solution by slowly pouring off with a rotating motion of the hand so as to pour off most of the debris around the lip of the tube. If a ring of this debris should still be present around the lip of the tube after pouring off the supernatant solution, remove the debris with a clean fingertip.
- Fill each tube half full with the sugar solution (dissolve 500 g of sugar in one liter of water. Add 10 ml of 10% lactic acid to the sugar solution to inhibit bacterial and mold growth). Then close the tube with a rubber stopper and shake until the compact mass at the bottom goes into suspension.
- Remove rubber stopper and fill tubes with sugar solution to within 0.5 cm of the top. Wash equally around sides to put all adherent debris into suspension.
- Centrifuge at 3000 rpm for two minutes.
- Nematodes and debris of nearly equal specific gravity including insect larvae, tardigrades, mites, oligochaetes and various fungal spore containing structures will be in the supernatant sugar solution. The compact mass at the bottom of the tube will be heavier debris and larger invertebrates. The nematodes are removed by decanting the supernatant suspension from each tube through the 350 mesh sieve held at a sharp angle. Wash residues on the sieve very gently to remove sugar solution.
- Gently flush residues on the fine sieve into a 250 ml beaker.

Caution: Once sugar solution has been added all processing must be as fast as possible. Never leave nematodes in sugar solution for long periods of time.

4.2.6.1. Advantages

- Fastest method known for extracting nematodes from soil. Entire processing operation should not exceed 12 minutes.
- All types of soil are processed easily.
- Maximum number of individual nematodes would be obtained from a given soil sample.
- Greatest diversity of genera will be obtained.
- Swollen and sessile females are revealed.
- Certain migratory coarse annulated forms such as *Criconemoides, Criconema* and *Hemicycliophora* are yielded in some instances where other methods fail to expose them.
- This is a standardized, fixed time and method based technique.

4.2.6.2. Disadvantages

- The sugar solution has a detrimental effect on the nematodes and the post processing mortality after 12 hours is very high. However, the nematodes are in excellent shape for taxonomic work or counting up to 6 hours after being removed from the sugar solution. The nematodes obtained by this technique cannot be used for inoculation purpose.
- A few genera such as *Trichodorus* and *Helicotylenchus* are yielded in lesser numbers.
- Occasionally, excessive minute debris other than nematodes makes the sample difficult to read.

4.2.7. Estimating nematodes in aqueous suspensions

The suspensions containing the nematodes are poured into 100 ml beaker and water is added till they are filled. Mixing of this suspension is done with compressed air (aquarium pump). Two 10 ml aliquots are taken from each beaker with a standard pipette and counted in Perspex counting dishes. A wide enough aperture to the sampling pipette is essential to avoid blocking or differential filtration. Counting and analyzing of nematode genera in samples is done by using a stereoscopic microscope at 25-50 x magnification (Fig. 4.7). Illumination is conveniently provided by a suitable strip light. Different genera are counted and noted with the help of a multi-unit tally counter with 6 to 9 keys.

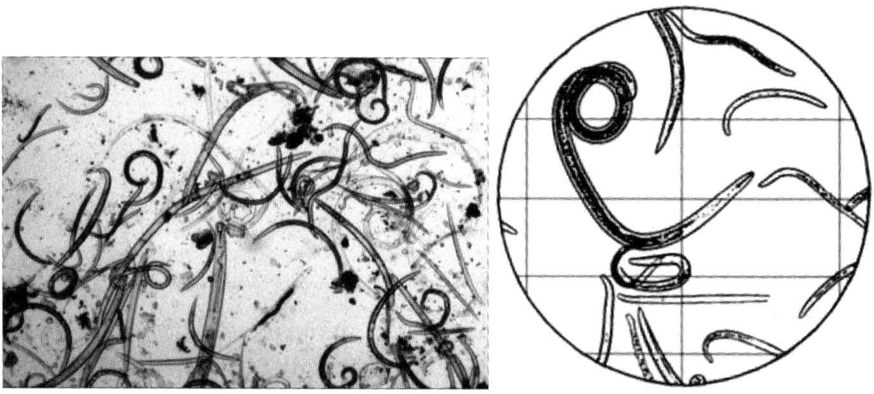

Fig. 4.7. Nematode suspension under a dissecting microscope

4.3. EXTRACTION OF CYST NEMATODES

Size, shape and floating capacity of the cysts makes it possible to collect them from air dried soil. Drying of soil sample is essential since most methods of extraction are based on the separation of cysts from the soil particles by flotation. Dry cysts float in water whereas wet ones sink to the bottom and cannot be retrieved.

4.3.1. A simple extraction method (Fig. 4.8)

- 50 ml well mixed dried soil is placed in a 60 mesh sieve. Wash the soil carefully with water.
- The catch of the sieve is poured into a white bowl with a sloping edge. The cysts float along the edge of the bowl and can be separated with a camel hair brush No. 1.

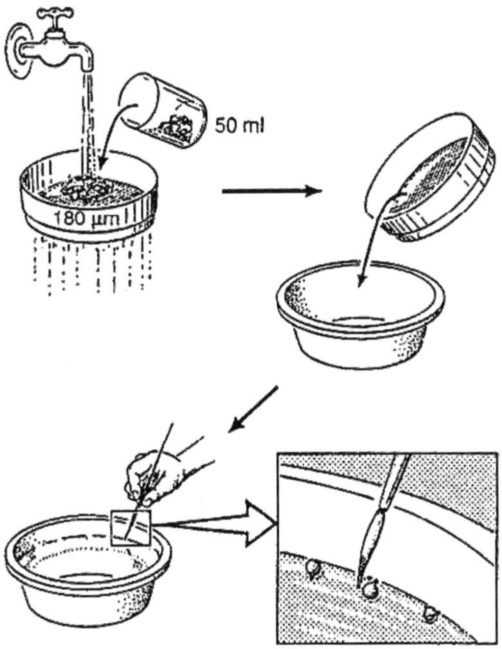

Fig. 4.8. "White bowl" method for extraction of cysts from soil

4.3.2. Acetone or acetone: CCl_4 method (Fig. 4.9)

- 50 ml well mixed dried soil is placed in a 60 mesh sieve.
- Wash the soil carefully with water. The debris on the sieve is dried at room temperature and transferred into a 250 ml volumetric flask.
- Half fill the flask with acetone or a mixture of 3 parts acetone and one part carbon tetrachloride and shake this flask vigorously. Then fill the flask till just below its lip and allow it to stand.
- After one minute, the floating cysts and debris are decanted through a filter paper in a funnel into a second flask, while rotating the original flask. The filter is placed in a petri dish and viewed through a dissecting microscope with overhead light; the cysts are picked up with a camel hair brush No. 0 or 1.

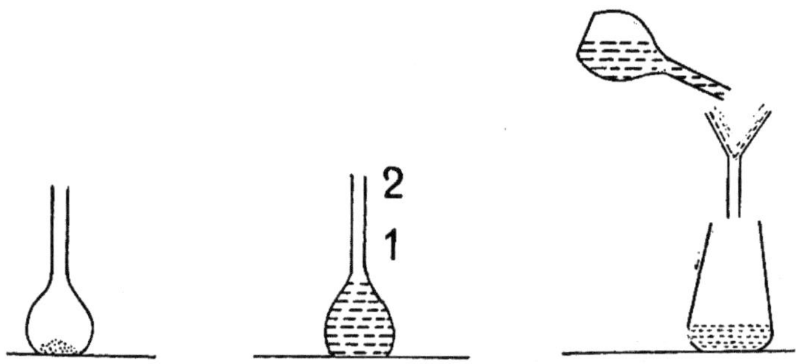

Fig. 4.9. Acetone or Acetone: CCl_4 Method for extraction of cysts

4.3.3. Paper strip method (Fig. 4.10)

A 500 ml long beaker is lined with a strip of filter paper, down to 1 to 2 cm from the bottom, i.e. the area which will be filled with soil is left uncovered. The ends of the filter strips should overlap a few centimeters. The beaker is then filled two-thirds full with water and the soil sample (100 cm^3) carefully poured in, preferably through a funnel. The soil will begin to settle after repeated stirring. After 20 to 30 minutes, raise and lower the paper strip along the wall 2 or 3 times. Then remove the paper strip from the beaker and spread out to dry on a glass plate. The cysts together with the floating contents of the soil will mainly occur on the surface of the water in the beaker, and are collected on the paper strip and counted.

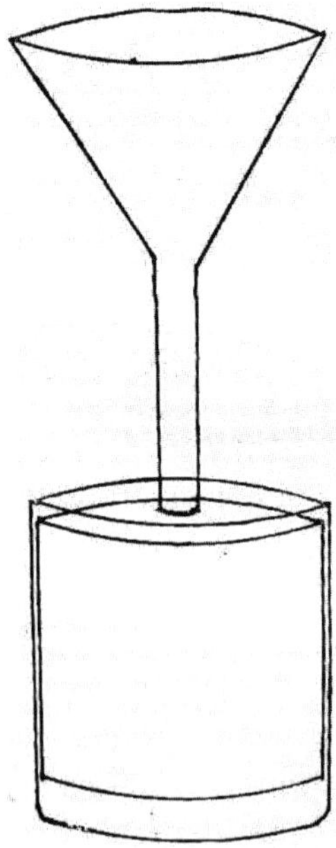

Fig. 4.10. Paper strip method of extraction of cysts.

4.3.4. Kirchner's funnel method

The equipment consists of a glass funnel with an etched surface. The tube of this funnel is bent at an angle of about 80° (Fig. 4.11).

The capacity of the funnel is approximately 400 ml and its gradient 60°. The funnel is supported on a stand and the flow end stoppered with a conical rubber plug to which a hook is attached for easy removal. Approximately 50 ml of dry, sieved soil is placed in the closed funnel and washed with a jet of water, which causes the soil to swirl and forces the cysts to adhere to the walls of the funnel. When the water is shut off, the rubber plug is removed from the tube. The water and soil particles (mud) in it flow out slowly because of the 80° angle, and thus the cysts adhering to the walls on the funnel are not disturbed and can subsequently be counted.

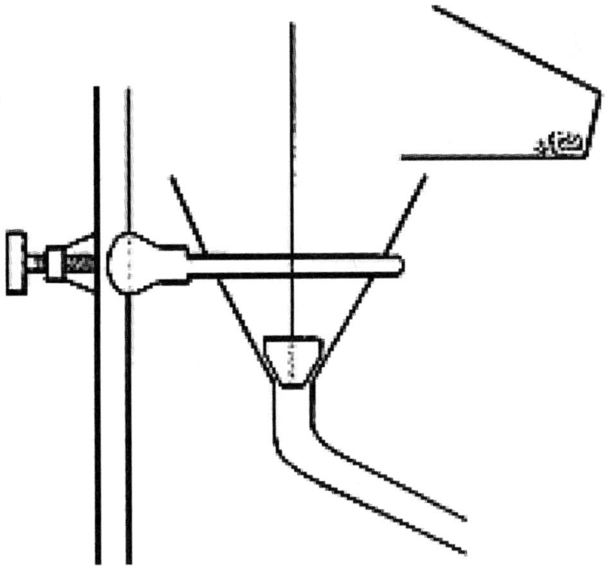

Fig. 4.11. Kirchner's funnel method of extraction of cysts

4.3.5. Modified Fenwick can method

Fill the modified Fenwick can with water (Fig. 4.12). A sample of 100 ml well mixed dried soil is placed in the top sieve of 1 mm pore size. Wash the sample into the apparatus via the funnel; the coarse material is retained on the top sieve. Heavy soil particles like sand sink to the bottom of the apparatus. The floating cysts are carried off over the overflow collar. Cysts, root debris and other particles are collected on a 60 mesh sieve. Particles less than 60 mesh and water pass through the sieve. The catch of the sieve is poured into a white bowl with a sloping edge. The cysts float along the edge of the bowl and can be separated with a camel hair brush.

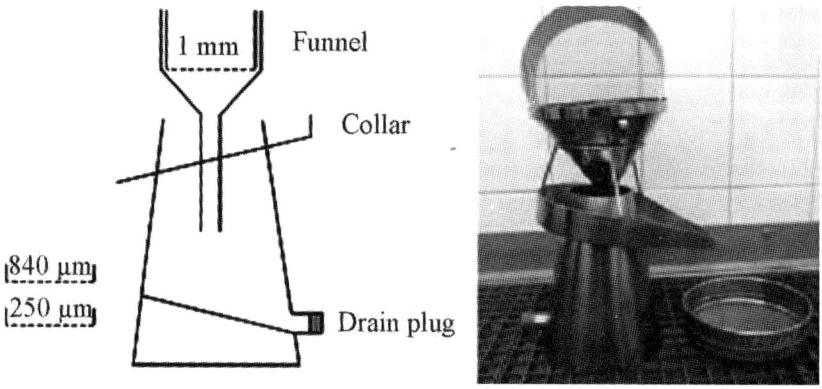

Fig. 4.12. *Left*: Vertical-section diagram of Fenwick can. *Right*: Fenwick can made of stainless steel with 60 mesh sieve.

4.3.6. DPI cyst separation technique

Introduce air dried soil into a 12 L spout bucket. Add water up to 2 cm above soil level. When soil is well soaked, add water, at the same time mixing the suspension with a free hand. Break up large soil clods with fingers and discard large pieces of debris. When water is up to 10 cm of the top, remove your hand. Wash debris off hand and off the sides of the bucket into suspension. Insert a 20 mesh sieve into a 60 mesh sieve and place both in the sink basin directly beneath the bucket spout. Gently flush floating debris from water surface and the bucket sides until clear water is leaving the spout. Decant water from bucket onto sieves until cloudy debris reaches bucket lip, the stop. Hard spray thoroughly debris on 20 mesh sieve so small matter is flushed through to the 60 mesh sieve. Discard bucket residues and 20 mesh sieve debris. Flush debris on 60 mesh sieve into a 50 ml beaker, fill beaker with water and let it stand for 10 minutes. Decant floating debris onto 7.5 cm 60 mesh sieve taking care not to pour out debris which has sunk to the bottom. The sample is ready to read using standard cyst reading techniques.

4.3.7. Oostenbrink's flotation apparatus (Fig. 4.13)

This apparatus is suitable for washing out cysts from both dry and wet soil. Place the 100 to 200 ml soil sample on the 1 mm mesh sieve and wash it with a jet (0.7 L/min) sprayed through a shower head, while a stronger flow of water (4 L/min) passes simultaneously through the perforated tube from the lower end of the apparatus. This creates an upward current that washes the smaller soil particles, cysts and organic matter over the edge onto the overflow collar and then into the sieve with a smaller meshes (0.18 mm).

The flotsam is then lightly dried in the air and transferred to a small bottle with a 1: 1 mixture of alcohol and CCl_4 or acetone. The cysts and debris remain floating and are collected on a filter paper.

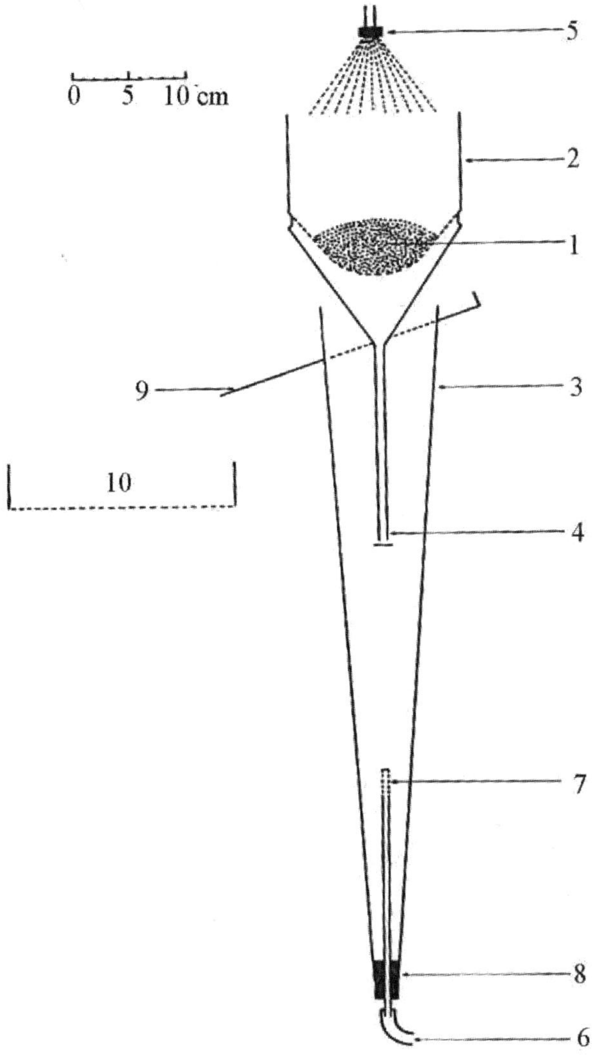

Fig. 4.13. Oostenbrink's flotation apparatus for extraction of cysts

4.3.8. Reinmuth's centrifuging method

10 ml dried and sieved soil is placed in a graduated centrifuging test tube and an equal quantity of water is added. The uniform movement of the centrifuge can be ensured by equilibrating the centrifuge tubes with a few drops of water. After vigorously shaking the sample, stoppered by the thumb or a rubber plug, release the mouth and place the tube in a centrifuging stand. Close the apparatus and set in motion. The centrifugal force (4,500 rpm) separates the cysts from the soil. While the soil settles, the cysts collect on the surface of the clear water and can be removed to a white dish.

4.3.9. Separation of cysts from admixtures

In almost every method of soil analysis some soil particles, mainly of organic origin, are trapped with the cysts. To separate cysts from these admixtures, many people have tried to develop a satisfactory method of cleaning.

4.3.9.1. Separation of round cysts: The simplest method is to spread the buoyant material on a sheet of filter paper. In lifting the paper the round cysts of the potato nematode will roll off and the filliform organic particles remain.

Cooper (1955) constructed a vibrating apparatus wherein the round cysts roll downward on a slightly tilted plate, while the other particles remain at the top.

4.3.9.2. Separation of elongated, lemon-shaped cysts: Andersen (1956) designed a device for separation of lemon shaped cysts from admixtures, the main feature of which is a wide, rotating belt drive 45 cm in length. The belt moves on 2 end rollers at an angle of 30 to 35° and is powered by a small electric motor (40v). The rate of movement is 3.4 m per min. The buoyant material is placed on the moving belt and the cysts roll downward against its movement into a container, while the filliform particles are carried upward and propelled into a second container.

4.3.10. Estimation of the egg content of cysts

The number of eggs or larvae within the cysts is of great importance in scientific investigations, especially in studies on population dynamics. Hence, the number of eggs and larvae per unit of soil is more important than the number of cysts.

A Perspex rod with a round end is fixed to the metal shaft of a small electric stirrer. The rod is of such a diameter that it fits closely into a polyethylene test tube. The cysts are placed in the tube with a little water and the tube is pushed over the revolving rod. The water acts as a cushion to prevent damage to the eggs and larvae, but the cysts are crushed. Transfer the cysts of one species with the camel hair brush into a small drop of water in the glass tube of the homogenizer with a rough inner surface. The piston is placed in the tube and carefully rotated by hand.

The eggs and larvae, released from the cysts, are poured with water into a beaker and filled up to 100 ml. Mix this suspension with compressed air (aquarium) and take two aliquots of 10 ml with a standard pipette and count the viable eggs and larvae in a counting dish under the dissecting microscope (50 x magnification) with front light.

■■■■

Chapter 5

EXTRACTION OF NEMATODES FROM PLANT MATERIAL

5.1. DISSECTING PLANT MATERIAL IN WATER

Some selected plant material e.g. leaves of the chrysanthemum showing angular leaf spots caused by the foliar nematode or wheat seed galls infected with *Anguina tritici,* is placed in water in a dish and is carefully shredded with a dissecting needle. This process is best carried out under the dissecting microscope with front light using a magnification of 50x. The nematodes work their way down into the water and can be collected with the help of a handling needle. Set the dish aside and re-examine after about one hour.

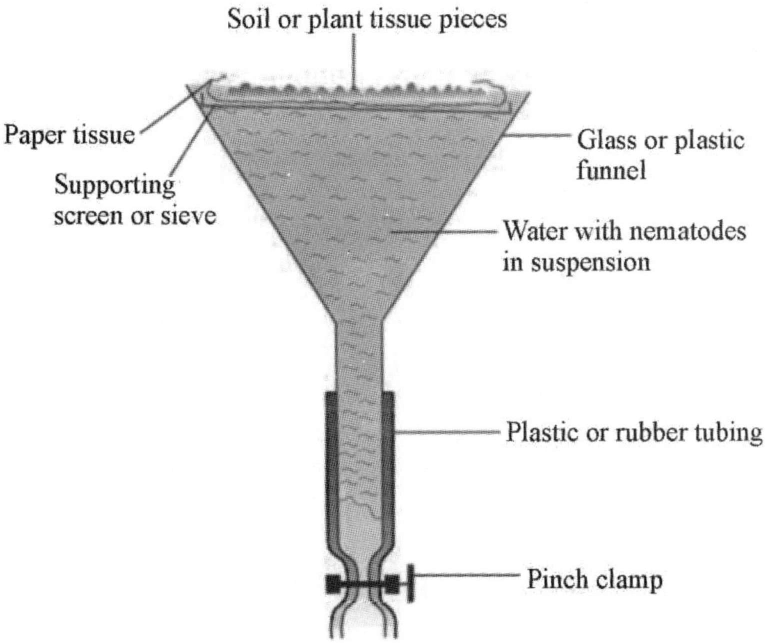

Fig. 5.1. Modified Baermann funnel technique for extraction of nematodes from plant material

5.2. MODIFIED BAERMANN FUNNEL TECHNIQUE

A Baermann funnel is often used to separate nematodes from roots or soil. It consists of a conical glass funnel (10 cm in diameter) with a short piece of rubber tubing attached to the stem and closed by a clamp (Fig. 5.1). A wire gauge sieve is placed inside the top of the funnel, resting about 2 cm below the top rim. A wet strength facial tissue, to retain roots or soil, is placed on the wire gauge sieve. The funnel is filled with tap water to a level above the wire gauge sieve.

Wash 10 g of the fine rootlets or any other plant material and cut into small pieces. Place the plant material onto a tissue paper. Leave the funnel for about 2 to 5 days. By then the nematodes will have migrated from the roots or other plant material and passed through the tissue paper to collect in the bottom of the funnel stem.

5.3. COTTON WOOL FILTER TECHNIQUE

The other modification is to use a shallow tray, dish or bowl to further improve aeration and reduce the number of nematodes remaining on the wall of the funnel. Similar to the above, a cotton wool filter or facial tissue is placed on a support and the chopped plant material or soil is placed on it. The support and the material for nematode extraction is placed in a tray (or dish, bowl, etc.) filled with tap water. Small feet (for example cut pieces of polyethylene) can be attached to the support ring to give a space of about 2 mm between the base of the support and the collecting tray. The material on the sieve should be moist (but not flooded) and it may be kept moist by placing another facial tissue on top or using a relatively large tissue and folding it over the material (Fig. 5.2).

Fig. 5.2. Cotton wool filter technique for extraction of nematodes from plant material.

Wash 10 g of the fine rootlets or any other plant material and cut into small pieces. Place the plant material onto a cotton wool filter or tissue paper. Leave the tray for about 2 to 5 days. By then the nematodes will have migrated from the roots or other plant material and passed through the tissue paper to collect in the bottom of the collection bowl.

5.4. MECHANICAL MACERATION TECHNIQUE

Five to 10 g of clean roots or shoots are placed in homogenizer, together with 50 to 100 ml of water (Fig. 5.3). The motor is run at full speed for 20 seconds. Most plant material seems to be sufficiently macerated in this time. The suspension is then poured onto a tissue paper supported by a wire gauze sieve kept in a funnel holding sufficient water to remain in contact with the bottom of the wire gauze. Leave the funnel for about 1 to 2 days. By then the nematodes will have passed through the tissue paper to collect at the bottom of the funnel stem.

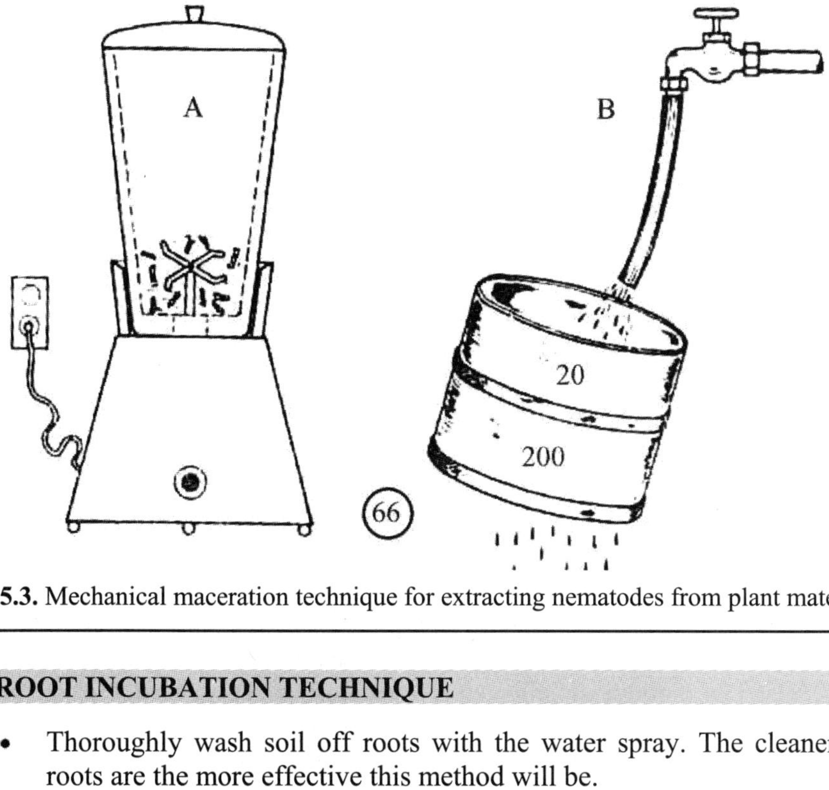

Fig. 5.3. Mechanical maceration technique for extracting nematodes from plant material.

5.5. ROOT INCUBATION TECHNIQUE

- Thoroughly wash soil off roots with the water spray. The cleaner the roots are the more effective this method will be.
- Place washed roots in polyethylene bags or glass jars where the atmosphere remains humid. Roots should be wet but not dripping. Seal jar by screwing lid just a few loose turns - do not tighten snugly (Fig. 5.4). If polyethylene bag is used, secure with a rubber band. The container should be at most only one-half full of roots. Avoid over filling.
- After 72 hours some of the endoparasitic nematodes will have migrated out of the roots onto the root surface. These may be removed by flushing the roots in the container with water 3 times, each time pouring the wash water onto a 325 mesh sieve held at an angle. Flush nematodes off sieve into a 250 ml beaker.

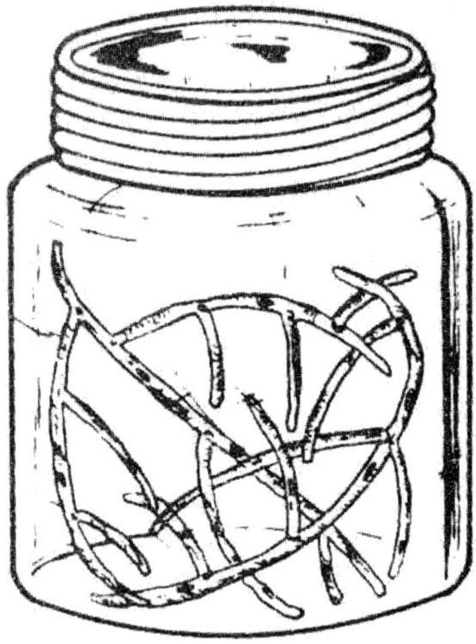

Fig. 5.4. Root incubation method for extracting nematodes from roots (Courtesy : A.L. Taylor)

5.5.1. Advantages

The root incubation method is excellent for experimental work involving endoparasites. It is simple, standardized and very effective.

5.6. MISTIFIER TECHNIQUE

The apparatus consists of a cabinet or enclosure in which Baermann funnels are set up under an intermittent mist spray (Fig. 5.5). The mist is applied for 1.5 minutes in a 10 minute cycle. Mist nozzles are suspended over the enclosure (50 - 70 cm above the funnels). Funnel stems are inserted deeply into the collection tubes so that the long stems will deliver the nematodes near the bottom of the tubes. Water in the tubes is allowed to over flow but at a rate too slow to carry away the nematodes, which settles in the bottom. At the end of the extraction period, the tubes are carefully removed from the stems and the excess water is slowly drawn off by aspiration. Nematodes in the remaining water (10 - 12 ml) are ready for examination.

5.6.1. Advantages

- A continuous supply of oxygen is provided for the nematodes in the sample, maintaining nematode activity for a longer period of time.
- A downward flow of water assists nematode movement down through the sample.
- A more constant and near optimum temperature is maintained in the sample.

- The system eliminates loss of nematodes which occurs in the jar incubation technique when rinse water is poured through the screens.
- Materials toxic to nematodes, emanating from roots and associated micro-organisms, cannot accumulate in the samples during the incubation period.
- This method can be used both for root samples and for screened soil samples.

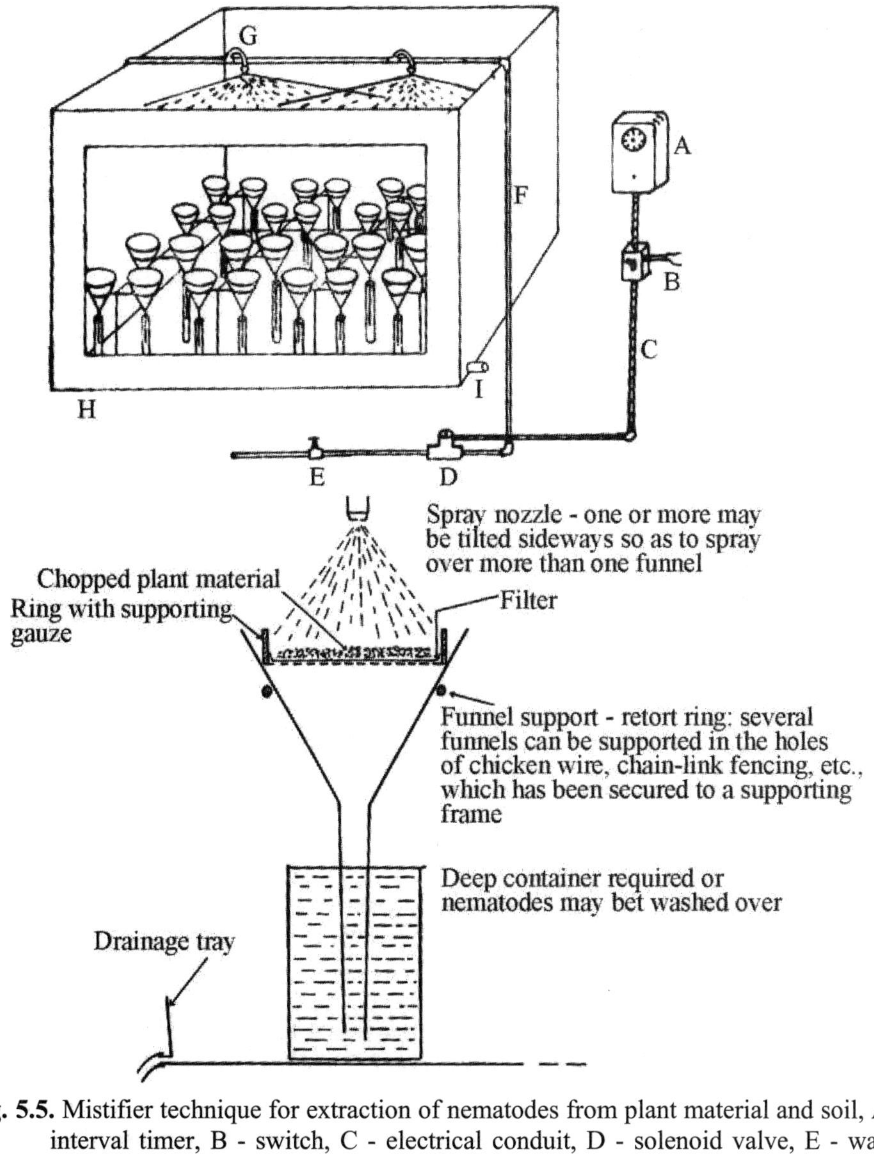

Fig. 5.5. Mistifier technique for extraction of nematodes from plant material and soil, A - interval timer, B - switch, C - electrical conduit, D - solenoid valve, E - water valve, F - water pipe, G - mist nozzle, H - enclosure, I - drain.

5.6.2. Disadvantages

- Fine particles from clay or silt soils are sometimes washed into the collecting tube, making the samples cloudy and difficult to examine.
- The mist nozzles tends to become fouled with lime deposits, causing them to drip or spit water droplets.
- Algae may grow in the constant film of water at the bottom of the enclosure.

Chapter 6

EXTRACTION OF EGGS

6.1. FROM ROOTS

6.1.1. With a few exposed eggs

Infected roots are washed and cut into pieces about 5 mm long. A 5 g portion is placed in a mixer with 500 ml of water and run at a slow speed for 15 seconds. The resulting suspension is poured through a coarse sieve followed by a fine sieve and then washed thoroughly. The residue on the fine sieve is washed into a centrifuge tube to which 1 cm^3 of kaolin powder is added. After thorough mixing, it is centrifuged for 5 minutes and the supernatant is poured off. A sugar solution of 1.15 sp. gr. is added, the mixture again stirred and again centrifuged for 4 minutes. This brings the eggs to the top and they are poured on a fine sieve.

6.1.2. With many exposed eggs

Egg sacs from tomato roots of 50 plants were obtained by kneading the washed roots in water and recovering the egg sacs on a 60- mesh sieve, Eggs sacs were placed in an electric blender in 100 ml tap water, 500 ml of 1% NaOCl (20% commercial bleach) added and the blender operated at a maximum speed for 40 seconds. The resulting suspension of eggs and root debris was poured quickly through nested 100 and 400 mesh sieves and eggs which were retained on the 400 mesh sieve were washed with several volumes of distilled water.

Eggs were eluted from sieves, transferred in 40 ml of water to a 50 ml polycarbonate centrifuge tube and centrifuged at 1000 g for 5 min. The pellet containing eggs and some debris was resuspended in 40 ml of sucrose solution (454 g sucrose in one liter distilled water) and centrifuged at 1000 g for 40 seconds. The supernatant was quickly decanted on to a 400-mesh sieve and the eggs retained on the sieve were washed 3 times with 100 ml portions of distilled water. Eggs were concentrated by centrifugation at 1000 g.

The eggs treated by this method hatched normally and the larvac readily penetrated and developed in roots of tomato seedlings.

6.2. FROM SOIL

6.2.1. Root-knot nematode eggs

A 500 ml soil sample is placed in a 10 L plastic bucket and 6 L of water are added. The resulting slurry is swirled vigorously for about 30 seconds and the supernatant liquid containing suspended material is poured through a 20 cm 40 mesh (420μ) sieve. Water is again added to the soil in the bucket and the swirling and pouring are repeated once.

The material caught on the 40 mesh sieve is washed into a 500 ml beaker by a water spray, the volume made up to 200 ml and 20 ml of 5.25% NaOCl (Clorox) are added to the suspension. The suspension is stirred at 1600 rpm for 10 minutes to facilitate the dissolution of the egg mass matrices and uniformly disperse the suspended eggs.

The stirrers are raised and rinsed into the beakers with a wash bottle jet of about 10 ml water. An aliquot of 2 or 5 ml is taken from the now 230 ml suspension of eggs and organic material to determine the number of eggs present and added to a 150 ml beaker containing 100 ml water (thus giving multiplication factors of 115 and 46, respectively for the number of eggs in the original 500 ml soil sample). This dilutes the NaOCl and prevents it from disintegrating the eggs.

The contents of the 150 ml beaker are swirled and poured through a 7.5 cm 500 mesh (25 μ) sieve to collect the eggs. An additional 50 ml of water is poured over the sieve to wash off any remaining NaOCl which would interfere with egg staining. The eggs are washed into a clean 150 ml beaker by a wash bottle jet and suspended in 25 ml water. To facilitate counting, the eggs are stained by adding 2 drops of a 0.35% acid fuchsine solution in 25% lactic acid to the beaker and boiled for 1 minute.

Examination of the eggs by a dissecting microscopic at 30 x magnification is facilitated by tilting the sub-stage mirror to a nearly horizontal position, directing a diffuse light to the counting dish. This provides excellent contrast between the bright red color of the eggs and non-stained debris.

6.2.2. Free living nematode eggs

Thoroughly shake 100 ml of finely crumbled soil for 2 minutes in 200 ml sucrose solution of sp. gr. 1.18 (484 g of sucrose per liter of water). After standing for 30 seconds to allow larger particles to settle, the supernatant is decanted into 250 ml centrifuge bottles which are balanced by the addition of sucrose solution and spun for 6 minutes at approximately 1700 g. The supernatant from each flask is poured immediately into a 1 liter measuring cylinder of water to prevent osmotic disruption of the eggs. After settling for 1-2 hour, all but about 100 ml of water is removed by a pipette attached to a suction pump and the remainder is examined for eggs, which continue to develop normally.

■■■■

Chapter 7

HANDLING NEMATODES

7.1. PICKING AND TRANSFERRING INDIVIDUAL NEMATODES

The picking of nematodes is best carried out under the dissecting microscope. A bamboo splinter, sharpened to a very fine point fixed in a needle holder or pig's hair mounted at the end of a bamboo splinter, are the most useful instruments for picking up slender nematodes. A dentist needle can also be used for picking nematodes. Lift a nematode from the bottom of the dish with the pick. When you get it to the surface of the water pass the pick under the nematode and lift it through the air water interface. It will be held on the pick by surface tension. Then submerge the tip of the pick in a drop of water on a slide or in a small volume of water in a vial. The nematode will be released from the pick.

7.2. ANAESTHETIZATION

The structural characters are generally more distinct in live nematodes than in killed and fixed specimens. However, it is difficult and usually impossible to determine these characters due to the high mobility of the worm; hence immobilization is required. A good relaxant is a solution of 2 drops of dichloroethyl ether in 50 ml of water. In this fluid the nematodes lose their mobility for a period of several minutes, becoming active again when placed in fresh water. Alternatively, a 0.5 to 1.0 per cent solution of propylene phenoxetol or carboxyl methyl cellulose in tap water can be used.

The nematodes can also be temporarily immobilized by heating the slide briefly to about 50°C.

7.3. KILLING AND FIXING

For the purpose of identification and in-depth studies, it is often necessary to transport nematode collections. The dispatch of live plant parasitic nematodes is an infraction of thc established rules since several nematodes are included in the quarantine lists of many countries. However, there is no objection to the dispatch of dead or fixed nematodes.

The easiest method of killing few nematodes is to heat them in a small quantity of water on a plain or cavity slide which is then held over a small flame for 4 to 6 seconds until the nematodes suddenly assume the form which is characteristic of heat death. Heating should not be prolonged as the structure of the nematode may be severely damaged by overheating. The nematodes are immediately transferred to a fixative or fixed on the slide by adding "double strength" fixative.

A safer method of killing a large number of nematodes is to heat them in a tube of water. The volume of suspension is decreased by centrifuging at about 1500 rpm for 1-2 minutes or by allowing the nematodes to settle under gravity. Then the supernatant water is poured so that they remain in a few ml only. Shake the tube to distribute the nematodes freely in suspension and plunge it into a beaker of hot water maintained at 65°C for 2-4 min, until the nematodes are dead. Then add an equal volume of a ``double strength" fixative.

A larger number of nematodes may be put into 2-3 ml of water in a cavity block which is then put into an oven at 43°C. After about 12 min the cavity block is filled with fixative at 43°C from a bottle that has been kept in the oven. The cavity block is removed from the oven and kept at room temperature for as long as desired. The nematodes can also be killed at 60°C for 5 to 10 minutes.

Nematodes can be killed with chemical reagents. One drop of diluted KI solution (1: 1000) in a drop of water containing nematodes ensures their rapid death. The nematodes should be fixed immediately after killing as decomposition sets in soon and makes subsequent identification difficult. Fixed specimens can be stored for months and years with no marked change in their quality.

One of the best methods is that which kills and fixes nematodes in one process. The nematodes are collected in a very small drop of water in a cavity block and 3-4 ml of Formal acetic Acid fixative 4: 1 heated to 100°C is quickly added to the nematodes. This method fixes glands and gonads well and nuclei are often clearly visible. Better results have been claimed when propionic acid was used in place of acetic in the killing solution.

7.3.1. Fixatives

The fixatives listed below have been tested and approved.

7.3.1.1. Common fixative

Ethanol (50%)	-	100 ml
Formalin (40%)	-	10 ml
Acetic acid	-	10 ml

7.3.1.2. FA fixative

Formalin (40%)	-	10 ml
Acetic acid	-	10 ml
Distilled water	-	80 ml

7.3.1.3. FAA fixative

Ethanol (96%)	-	100 ml
Formalin (40%)	-	30 ml
Acetic acid	-	5 ml
Distilled water	-	200 ml

7.3.1.4. TAF fixative

Triethanolamine	-	2 ml
Formalin (40%)	-	7 ml
Distilled water	-	91 ml

7.3.1.5. Formalin - glycerol fixative

Formalin (40%)	-	8 ml
Glycerol	-	2 ml
Distilled water	-	90 ml

7.3.1.6. Formalin

Formalin (40%)	-	4 ml
Distilled water	-	96 ml

7.3.1.7. Double strength formalin

Formalin (40%)	-	8 ml
Distilled water	-	92 ml

7.4. PREPARATION OF NEMATODE MOUNTS

7.4.1. Temporary mounts

Place a drop of 3% formalin on a clean glass slide. Pick up 8 nematodes from the suspension and arrange them in 2 lines in formalin drop. Arrange 3 glass wool pieces radially around the nematodes. Apply a cover slip over the drop carefully. Drain-off the excess fluid. Seal the edge of the cover slip with a ringing medium like nail polish, `Zut' or `Glyceel'. Label the slide.

7.4.2. Semi-permanent mounts

Collect nematodes, kill and fix them as described in the above method. Remove the nematode into a drop of warm, vaporizing lacto-phenol and allow cooling. Fix the nematodes in a small drop of lacto-phenol on a final glass slide. Place a cover slip on the top and ring with nail polish, `Zut' or `Glyceel'.

7.4.3. Permanent mounts

In fixed nematodes, much of the internal body contents, especially gonad structure, may be obscured by the granular appearance of the intestine. Specimens can be cleared by processing to lacto-phenol or glycerol, which are also suitable mountants.

It is quicker to process and make mounts in lacto-phenol and they will often last several years, but they degenerate with time especially when left in the light. Well prepared specimens in glycerol will keep almost indefinitely and their optical quality is often better than those in lacto-phenol, but processing to glycerol takes longer.

7.4.3.1. Lacto-phenol method: A cavity slide should be almost filled with lacto-phenol, containing 0.01% cotton blue stain if desired and heated on a hot plate at about 65°C. When the lacto-phenol is hot, transfer the nematodes that have been fixed for several hours or overnight. Keep the slide hot and examine under a stereoscopic microscope after 2-3 minutes by which time the specimens should have cleared. The slide can be reheated if more clearing is required. The nematodes should then be transferred to clear lacto-phenol or to lacto-phenol with 0.0025% cotton blue stain. Fixed nematodes should not be put into cold lacto-phenol as they are often irreversibly distorted.

7.4.3.2. Glycerol method: Transfer fixed nematodes to about 2 ml of a solution of 1.5% glycerol in distilled water in a small cavity block. A trace of $CuSO_4$ or thymol should be added to the solution to prevent the growth of molds. Place the cavity block in a small air tight container together with a small tube of desiccant such as $CaCl_2$ or silica gel. Keep the chamber at 25-30°C. The water should have been absorbed after about 4 weeks. It is important not to rush the process or specimens may become distorted.

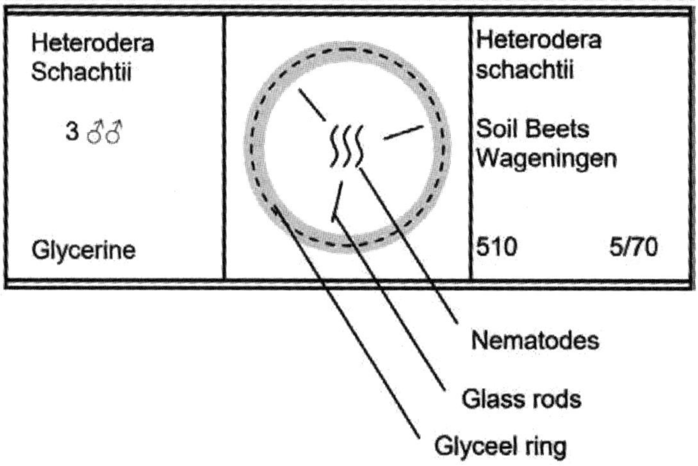

Fig. 7.1. Permanent mount of nematodes.

Place a drop of dehydrated glycerol on a clean slide. Pick up 3 nematodes from the dish and arrange them in a line in glycerol drop. Arrange 3 glass wool pieces radially around the nematodes. Apply a cover slip over the drop carefully. Drain off the excess fluid with a piece of blotting paper. Seal the edge of the cover slip with nail polish, `Zut' or `Glyceel'. Label the slide (Fig. 7.1).

7.5. FIXATION AND PRESERVATION OF PLANT MATERIAL

The most suitable method for fixing and preserving plant tissue (roots or whole plants) containing nematodes is by immersing them in cold 5-10 per cent formaldehyde, FA 4: 10 or TAF. Better nematode specimens will often be obtained if the infested tissue is immersed in hot (70-80°C) fixative. For histological studies fixation in FAA is more suitable. If it is intended eventually to stain it in lacto-phenol, it is often best to use lacto-phenol in the first instance, because tissues are softened and nematodes (especially *Meloidogyne* sp.) are more easily dissected from them.

Plant material fixed in formaldehyde or FAA can be stored and displayed for museum purposes in suitable air tight jars.

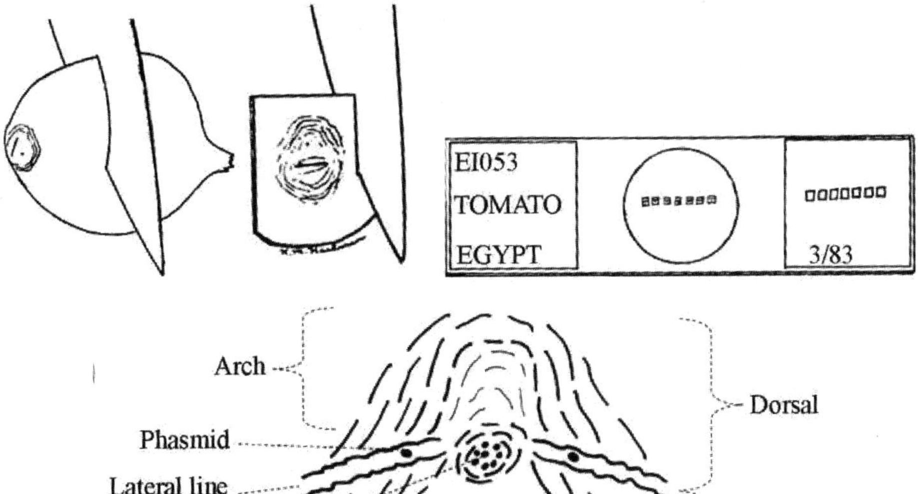

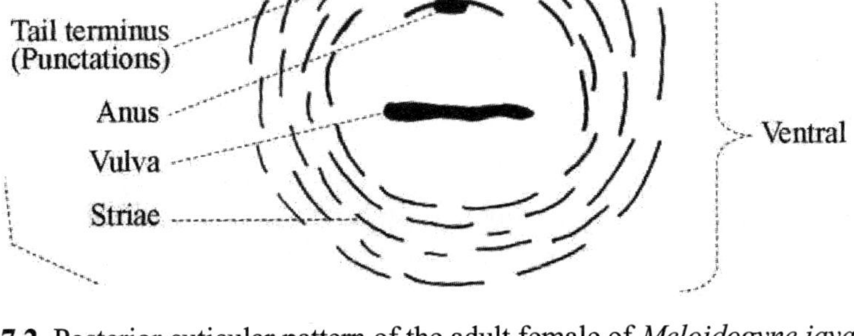

Fig. 7.2. Posterior cuticular pattern of the adult female of *Meloidogyne javanica*

7.6. PREPARATION OF POSTERIOR CUTICULAR PATTERN OF ROOT-KNOT NEMATODES

The identification of root-knot nematodes can be carried out from observation of the posterior cuticular pattern of the adult females (Fig. 7.2). Stain galled roots in lacto-phenol - cotton blue and allow them to differentiate in lacto-phenol. Dissect out about 20 females and transfer them to fresh lacto-phenol on a Perspex slide. Cut out the posterior end with an optical scalpel. Remove the body tissues by lightly brushing the inner surface of the cuticle with a nylon bristle. Carefully trim the cuticle so that it is only slightly larger than the perineal pattern. Transfer the perineal pattern to a drop of lacto-phenol or glycerin on a clean glass slide. Apply cover slip and seal it with a ringing medium.

7.7. PREPARATION OF VULVAL CONE OF CYST NEMATODES

Besides the general shape of cysts, the pattern of the perineum and surrounding area, together with certain associated internal structures, are of use in identifying the different species of cyst nematodes (Fig. 7.3).

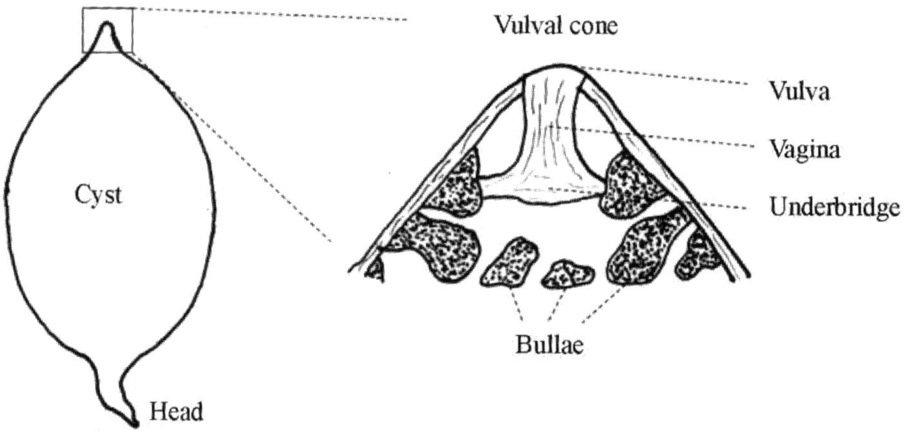

Fig. 7.3. Vulval cone of *Heterodera* cyst

Place a moist cyst on a Perspex slide and cut off the posterior end so that the perineal area is in the center of the cut piece. Trim the cut end so that it is slightly larger than the perineal area. Clean out the adhering body contents using very fine forceps and an eye brow handling needle (Fig. 7.4).

Wash the cleaned perineal end in distilled water and then pass it through 70, 95 and 100% ethanol to clove oil. Allow to clear in clove oil and mount in Canada balsam. Support the coverslip with pieces of glass wool of a suitable thickness to prevent crushing.

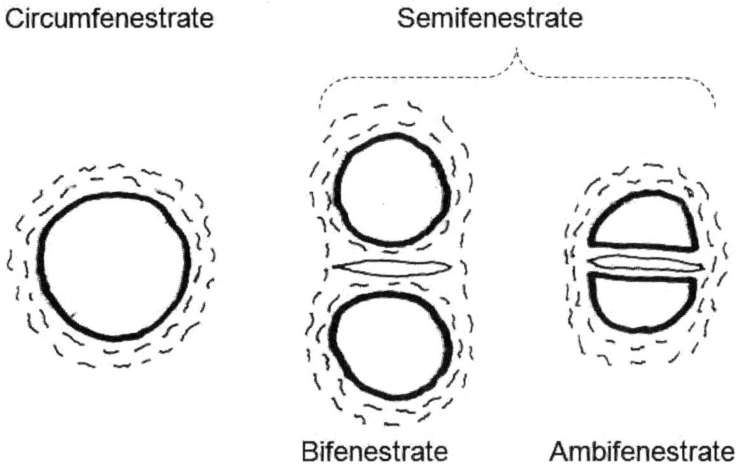

Fig. 7.4. Vulval fenestra of *Globodera* (left) and *Heterodera* (center, right) cysts

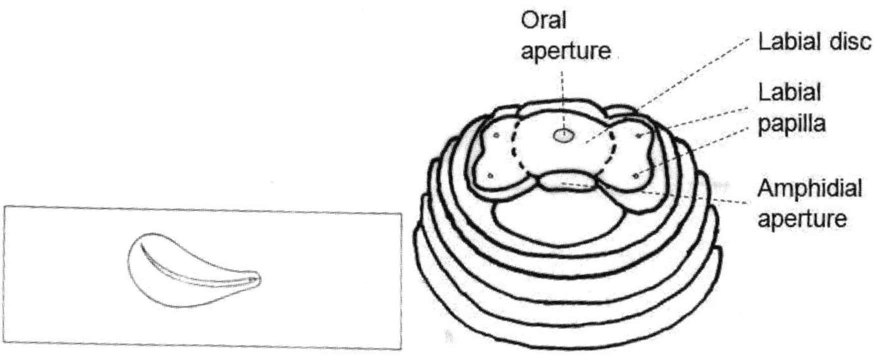

Fig. 7.5. Enface view of nematodes

7.8. PREPARATION OF ENFACE VIEW OF NEMATODES

Place a small bit of hard glycerin jelly (soak 9 g of gelatin in 40 ml of distilled water for 2 hours, then add 50 g of glycerin and 1 g of phenol; warm in a water bath for 10-15 minutes and stir till the mixture is homogeneous) on a glass slide and melt it over a small flame. Spread out a part of the liquid glycerin jelly drop at one point with a dissecting needle. Transfer a nematode which has already been processed to glycerin, to this melted glycerin jelly. Place the nematode with the head end pointing towards the spread out portion of the glycerin jelly drop (Fig. 7.5). Allow the glycerin jelly to become hard. Cut off the head end of the nematode usually about 1 head width from the anterior end. Transfer the head end onto the drop of melted glycerin jelly on another slide. Carefully turn the section with the handling needle so that the lip region faces up. Select 3 glass rods with the diameter less than the height of the section and arrange them radially. Allow the glycerin jelly to become hard. Apply a cover slip to the drop. Take care so that the lip is always facing up by adjusting the coverslip. Heat a dissecting needle and touch the coverslip above the head, melting the glycerin jelly until the glass rests near the face. Fix the coverslip down at 3 points with small drops of the ringing medium. Check once again if the lip is facing up. After half an hour seal the coverslip with a ringing medium on the turn table.

■■■■

Chapter 8

STAINING

8.1. INTRODUCTION

For the demonstration and study of the relationships of nematodes to plants, it is often desirable to stain nematodes in plant tissue.

Nematodes in plant tissues can be detected by staining. The advantage of staining is that it reveals the nematodes which do not actively emerge from the plant tissues (nematodes in some definite inactive larval stage, dying nematodes and eggs). Moreover, the staining is definitely advantageous in establishing the distribution and foci of the nematodes within the plant tissues.

8.2. ROOTS

8.2.1. Staining by Fleming's reagent

The washed roots are immersed in water at a temperature of 70 to 80°C for a period of 2 to 3 minutes to kill the nematodes within them. The roots are then placed in a Fleming's reagent for a period of 5 to 15 minutes. A closed container is used to protect the eyes from osmic acid vapor.

Fleming's reagent

1 % Chromic acid	-	15 parts
2% Osmic acid	-	4 parts
Glacial acetic acid	-	1 part

The nematodes are stained black while the plant tissue remains unstained. Finally, the roots must be thoroughly washed for a period of several hours in running water.

8.2.1.1. Advantages

- This method gives a sharp color differentiation which is suitable for microphotography.

8.2.1.2. Disadvantages

- The cost of osmic acid is very high. The fumes from osmic acid are harmful to the eyes.

8.2.2. Staining with picric acid and aniline blue for hardwood roots

The washed pieces of roots, not more than 1 cm in thickness, are placed for 48 hours in a fixing fluid of 100 ml of FAA. They are washed, cut into thin strips or slices and washed again for about 1½ hours in water. Subsequently, they are placed in a 1% solution of safranin for 1 to 2 minutes. The excess dye is washed off with water and the sections transferred to a solution of aniline blue in picric acid (100 ml of saturated aqueous solution of picric acid - 25 ml of saturated aqueous solution of aniline blue) and the staining solution heated almost up to boiling point. The sections are then washed once more with water. The nematodes stains blue and the plant tissues are reddish yellow.

8.2.3. Staining with bromphenol or bromthymol blue

The washed roots are placed in a solution of $NaOCl_2$ (0.5 to 1.0%) for a period of 4 hours. The roots are then thoroughly washed in water and immersed for several minutes in 50% alcohol. After drying, the roots are placed in a solution of bromphenol or bromthymol blue. A 1% solution of the dye in 50% alcohol and immersion for 4 hours is recommended if a stronger color differentiation is required, a 0.05 to 1.0% solution of dye in 50% alcohol and immersion for 12 to 16 hours is recommended when details are to be examined. The examination should be done in 50% alcohol acidified by acetic acid (0.3%). The nematodes and their eggs stain green to dark blue and the plant tissues remain unstained.

8.2.4. Staining by aceto-osmium method

This method is helpful for visualizing nematodes in more opaque roots (e.g. *Radopholus similis* in citrus roots). Fix and stain washed roots for 2 hours at 52°C in the following solution:

10% Acetic acid	-	10 parts
2% aqueous Osmic acid	-	2 parts
Distilled water	-	16 parts

Wash in running water for 1 hour, bleach in 10 - 30% H_2O_2 at 30°C till the darkened tissues lighten perceptibly, and then wash several times in water.

8.2.5. Staining by Lugol's solution

Iodine	-	1 part
Potassium iodine	-	2 parts
Distilled water	-	200 parts

Plunge the washed roots in the above solution. After some time the material is ready for observation.

8.2.6. NaOCl - acid fuchsine - glycerin method

Roots are washed, chopped into 1-2 cm segments and placed in a 150 ml beaker with 50 ml tap water. 20 ml of 5.255 NaOCl are added to give 1.5% NaOCl and the root tissue pieces are allowed to remain in this solution for 4 minutes with occasional agitation. Following the NaOCl treatment the root segments are rinsed in running water (30-45 seconds) and allowed to soak in tap water for 15 minutes to remove residual NaOCl. The material is then drained and transferred to a beaker containing 30 ml of water to which 1 ml of stain has been added (3.5 g acid fuchsine + 250 ml acetic acid + 750 ml distilled water). This solution is then heated to boiling for about 30 seconds.

After cooling to room temperature, excess stains are removed by rinsing in running water. The root material is then placed in 20-30 ml of glycerin acidified with a few drops of 5N HCl, heated to boiling and cooled. The root segments can then be pressed between microscopic slides for observations (Fig. 8.1).

Fig. 8.1. Staining of nematodes in acid fuchsine

8.2.7. Differential staining of endoparasitic nematodes (Alexander *et al.*, 1979)

Wash the nematode infested roots with tap water to remove the soil and adhering debris. Plunge the infested roots into the boiling stain (95% ethyl alcohol - 25 ml, malchite green - 25 mg, glycerol - 35 ml, acid fuchsine - 50 mg, phenol - 15 g, lactic acid - 1 ml and distilled water - 50 ml) for 30 to 40 seconds and allow to cool for 1 minute. Wash off the excess stain with running water and transfer the roots to a petri dish containing acidified water (3 drops of lactic acid in 100 ml water). After 2 minutes, place the roots in a solution containing 50 ml glycerol, 5 ml phenol, 3 drops of lactic acid and 45 ml water. Examine under a stereoscopic microscope. All stages of endoparasitic nematodes take on a crimson red color while the roots turn hyaline green.

8.3. SHOOTS

Unlike the staining of nematodes in the roots of plants, the staining of nematodes in the leaves and stems requires the removal of chlorophyll.

8.3.1. Staining with cotton blue or acid fuchsine in lacto-phenol

The infested tissue of the plant is immersed for a period of 3 to 5 min in a boiling solution of cotton blue (or acid fuchsine) in lacto-phenol and left for a period of approximately 30 minutes in the cooling dye. The excess color is then washed off with water and the plant material is immersed in 50% alcohol followed by liquid phenol to remove the chlorophyll. Tissues are allowed to differentiate for several hours to 2 to 3 days. Then examined in lacto-phenol. The nematodes stain blue (cotton blue) or red (acid fuchsine), whereas the plant tissue remains largely unstained.

8.3.2. Staining by Fleming's reagent

The infested plant tissue is immersed in boiling 80% acetone to remove the chlorophyll. The acetone is heated (boiling point, 63°C) in a water bath. The plant parts are left for several minutes in the boiling acetone for 3 to 4 hours, as it cools. After several washings in water, staining with Fleming's reagent is done (5 to 15 minutes).

8.3.3. Permanent mounts of nematodes within tissues stained with cotton blue or acid fuchsine

To make permanent mounts, cut out selected portions of the tissue containing well stained nematodes and further process them as follows:

8.3.3.1. Euparol mounts: Pass through 30, 50 and 70% ethanol, then to isobutanol and mount in 'Euparol'.

8.3.3.2. Canada balsam mounts: Pass through a graded ethanol series till absolutely clear in clove oil and mount in Canada balsam.

8.3.3.3. Glycerol mounts: Transfer to 5-10% glycerol in water and set aside in a desiccator until the glycerol is concentrated. Mount in pure glycerol and ring the coverslip with a ringing medium.

8.4. STAINING FOR DISTINGUISHING LIVE FROM DEAD NEMATODES

8.4.1. Staining by New Blue R

New Blue R is used to differentiate living and dead nematodes over a range of tylenchid genera and species. Nematodes or eggs are soaked in 0.05% aqueous solution of the dye for varying times (nematodes for up to 24 hour, *Heterodera* eggs up to 7 days). After the requisite time, dead specimens are darkly stained but live ones are unstained. This stain may be useful for estimating viability of nematodes subjected to heat or chemicals but not for populations that have died from natural causes.

8.4.2. Staining by Chrysoidin

Live nematodes can be distinguished from dead *Heterodera* larvae by staining with 50 ppm chrysoidin. Dead larvae stains uniformly yellow whereas live ones show yellow with orange contents.

8.4.3. Staining by Phloxine

Nematode mortality following heat treatment can be determined by adding a drop of 5% aqueous phloxine B to nematodes in a few ml of water. Dead specimens and *Meloidogyne* eggs stain immediately but moribund larvae are slow to accept a stain. This stain is unsatisfactory for determining viability of *Heterodera schachtii* eggs and larvae.

8.4.4. Staining by Eosin - Y

0.67% aqueous solution is used to stain 99% of heat killed, free living nematodes within 30 min, but not live ones. The dye also stains nematodes killed by freezing or by various chemicals.

8.4.5. Staining by Acridin Orange

100-200 ppm of Acridin Orange stains live nematodes green and dead ones red under the fluorescent microscope.

8.5. SPECIALIZED STAINING

8.5.1. Acetic Orcein

Kill nematodes by gentle heat and them fix in Carnoy's fluid (10-20 min).

Carnoy's fluid

Glacial acetic acid	-	1 part
Absolute ethanol	-	6 parts
Chloroform	-	3 parts

Stain the specimen in 1% Acetic Orcein for about 1 hour and mount in 45% acetic acid. This method is used for staining the reproductive system, general somatic nuclei and the cephalic framework.

8.5.2. Gold Chloride

Fixed nematodes are washed in distilled water and then placed in an aqueous solution of 0.15 $AuCl_3$ and 0.1% $HgCl_2$ until they are light yellow. After washing the specimens for 10 minutes in distilled water, they are placed in 1% formic acid and exposed to the light of a photoflood lamp. This causes a breakdown of the $AuCl_3$ to colloidal gold which stains the organs bright red. Nematodes can then be processed to glycerol and mounted permanently. This is an excellent staining method for showing the gonads, esophageal glands and nerve ring.

8.5.3. Silver Nitrate

Live nematodes are placed in a hypertonic solution of 10% $NaNO_3$ for 5 minutes which lessens their size by ex-osmosis then in 0.5% $AgNO_3$ for 15 seconds and finally in distilled water. Endosmosis occurs in the hypotonic $AgNO_3$ which gets flushed into the tissues. This treatment kills the specimens, but after illumination the peripheral nervous system becomes apparent due to the deposition of colloidal silver; it remains visible in specimens subsequently processed to glycerol. This method is successful only when used on live specimens.

0.5% $AgNO_3$ can also be used to deposit metallic silver on specimens so that surface structures such as body annulations and lateral incisures are more easily seen.

8.5.4. Vital Staining

Two rhabditid nematodes were seen to ingest methyl red and neutral red pH indicator dyes with their food *in vitro*. The dyes were absorbed in different concentrations by different regions of the intestinal wall. *Panagrellus redivius* sometimes ingests the redox indicator, 1-naphthol-2-sodium sulfonate-indole and the excretory system subsequently stains red.

■■■■

Chapter 9

MICROTOME SECTIONING

9.1. INTRODUCTON

Increasing attention is being given to the detailed structure of nematodes and examination of whole mounts and sections under the light microscope has become vital. To study fine structures in nematodes or in infected plants, sections are usually necessary. Hand sections are quick and adequate for studying gross structure in cross sections of nematodes and infected plants. But, for making histopathological observations of nematode infected roots or other plants, serial sections are required. To cut thinner sections that reveal more details, the specimens must be supported internally and externally. Therefore, they must be infiltrated with and embedded in a solidifiable matrix material. Of the embedding methods, those using paraffin wax are simple and rapid. Specimens can be oriented easily and good serial sections are possible because they come off the knife like a ribbon. With care, sections as thin as 5 μ can be cut.

9.2. KILLING AND FIXATION

Killing and fixing are the most critical operations in the processing of tissues. These are two distinct processes, but both are usually obtained by means of a single fluid which in turn is commonly a mixture of various chemical reagents.

The term killing means the sudden and permanent termination of the life processes within the cells. This should be accomplished with minimum structural disturbance within the cells and minimum distortion of the arrangement of cells in the tissues.

Killing invariably precedes fixation since the reagents which do the killing penetrates tissues faster than those upon which responsibility for fixation rests. Fixation means the preservation of all cellular and structural elements in as nearly the natural living conditions as possible. A good fixative is one that changes the cell chemistry the least and preserves the cell structure the best. Fixation is required in order that structures, which are obscured or entirely invisible when cells are observed in the living condition, may be seen more distinctly, and in order that soft structures may be hardened sufficiently for further treatment.

The smaller the piece of tissue and the larger the surface exposed to the direct action of the killing fluid, the better the killing and fixation.

FAA is one of the most useful types of killing and preserving fluid, which is represented by the following formula:

Ethyl alcohol (95%)	–	50 ml
Glacial acetic acid	–	5 ml
Formalin	–	10 ml
Water	–	35 ml

This fluid is stable, has good hardening action and material may be stored in it for years. The minimum time of fixation is 18 hours.

Carnoy's fluid is also suitable for fixing and storing infected plants for sectioning:

Ethyl alcohol (100%)	–	60 ml
Glacial acetic acid	–	10 ml
Chloroform	–	30 ml

This fluid kills protoplasm by rapid and probably violent dehydration. The material should not be left in this fluid for more than one hour. Wash the material in two changes of 70% ethyl alcohol and proceed to paraffin as quickly as possible.

9.3. DEHYDRATION

This operation removes water from the fixed and hardened tissues. Generally, replacement of water in tissues by other fluids should involve only gradual change of viscosity and decrease of polarity, to cause least shrinkage and distortion. The process consists of treating the tissues with a series of solutions containing progressively increasing concentrations of water. Two contrasting methods are used to dehydrate and prepare materials for infiltration. In the first method, the tissues are dehydrated in a non-solvent of paraffin and are then transferred to a solvent. In the second method, dehydrant is also a solvent of paraffin.

9.3.1. Dehydration by non-solvents of Paraffin

The most commonly used dehydrating agent in this category is ethyl alcohol. It is almost always necessary to wash out the killing fluid thoroughly before proceeding with the dehydration and infiltration. Alcoholic solutions should be washed out with plain alcohol of approximately the same percentage as that in the solution. For example, after FAA, begin dehydration in 50% alcohol. Solutions in the dehydrating series (60%, 70%, 80%, 90% and 100% alcohol) are changed by decanting the liquid from the tissues and promptly flooding the material with a generous volume of the solution next in the series. Leave the tissues in each grade for a period of about 3 hours. Make three changes of anhydrous alcohol.

The dehydrated tissues are transferred to a solvent. The most common solvent is xylene. As in the case of dehydration, a graded series is used for clearing. After dehydration in ethyl alcohol, the following absolute alcohol-xylene series is used (Table 9.1):

Table 9.1. Absolute alcohol-xylene series is used for clearing tissues

Grade number	Ethyl alcohol (%)	Xylene (%)
1	75	25
2	50	50
3	25	75
4	0	100

Leave the tissues in each grade for a period of about 3 hours.

9.3.2. Dehydration in solvents of paraffin

Normal butyl alcohol (Butanol) shows much promise as a dehydrating and infiltrating agent. Fixed material is washed in 50% ethyl alcohol and dehydrated as follows using ethyl alcohol-butyl alcohol grades leaving the material in each grade for a period of 3 hours (Table 9.2).

Table 9.2. Dehydrating and infiltrating tissue with different ethyl alcohol-butyl alcohol grades

Grade number	Ethyl alcohol	n-Butyl alcohol	Water
1	70	-	30
2	80	-	20
3	90	-	10
4	100	-	-
5	66	33	-
6	50	50	-
7	33	66	-
8	-	100	-
9	-	100	-

9.4. INFILTRATION AND EMBEDDING

Infiltration consists of dissolving the paraffin in the solvent containing the tissues, gradually increasing the concentration of paraffin and decreasing the concentration of solvent. The solvent is eliminated by decantation or evaporation or both. The paraffin matrix in which tissues are embedded serves to support the tissues against the impact of the knife and to hold the parts in proper relation to each other after the sections have been cut. These functions are best performed if all cavities within the tissues are filled with the matrix and if the matrix adheres firmly to the external and internal surfaces of the material.

Paraffin wax of 56-58°C melting point is used for infiltration and embedding. Small chips of paraffin wax are added successively to the medium of xylol or pure n-butanol containing dehydrated samples, until the medium reaches a saturation point at room temperature. Later they are kept under a table lamp (40 W) for 24 hours. The materials are then placed in an oven maintained at 60°C. Subsequently changes with fresh molten paraffin are administered at every 4 hour intervals to replace even the last traces of xylol or butanol with paraffin. This process has to be continued for 48 hours.

Heat one end of the casting plate with a small flame and place the empty paper boat at the melting zone. Pour the paraffin containing the tissues into the boat. Arrange the pieces with a bristle or warmed needle. When the pieces are suitably oriented, move the boat to the cold end of the plate. Sweep the Bunsen flame rapidly over the surface of the paraffin to prevent the formation of cavities. As soon as the paraffin is hardened enough to keep the pieces of tissue from moving, float the boat in a pan of cold water and brush with the Bunsen flame again. Allow the surface to solidify and submerge the boat in the water. When the paraffin is thoroughly cooled, peel off and discard the paper boat. Trim the edges of the cast block so that both surfaces are flat. Store the blocks in an envelope (bearing adequate identification data) at low temperature in order to minimize the damage to the tissues and to the texture of the paraffin.

9.5. MICROTOME SECTIONING

Trim the paraffin around the material so that the cutting plane is established. Lay the cutting face on a clean surface. Heat the mounting block, press it firmly on the back of the specimens and hold in contact until the wax cools. Build up a fillet of paraffin around the specimen to afford firm bracing for the tissues. Cool thoroughly and clamp the mounting block into a microtome before sectioning.

Material embedded in paraffin is almost invariably cut with a rotary microtome, in which the knife is stationary and the piece of tissue is moved up and down past the cutting edge. Cutting is accomplished by wedge action. After a section has been cut and the tissue carrier has passed the knife on the upstroke, an automatic feed mechanism advances the tissue carrier forward and the other section is cut. Successive slices remain attached to each other, forming a ribbon of paraffin. The successive sections cut from a piece of tissue are thus kept in serial order which helps in reconstruction of the external or internal structure of a tissue system. Sections of approximately 10μ thickness are optimum (Fig. 9.1).

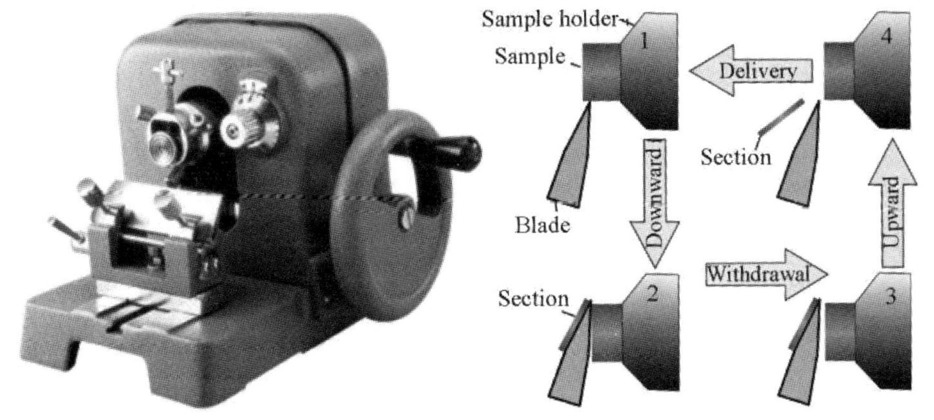

Fig. 9.1. *Left* – Rotary microtome. *Right* - Principle of sample movement for making a cut on a rotary microtome

Place a drop of egg albumen (white of an egg - 1 part, glycerin - 1 part, water - 1 part, sodium benzoate - 0.5% and homogenize in a blender) on a pre-cleaned slide, smear into a thin film with the ball of a finger, flood with water and place the paraffin sections in the form of a ribbon. Warm the slide over an alcohol lamp until the paraffin expands and lies flat on the slide. Allow the ribbon to cool and then blot with a filter paper. Wipe excess adhesive from around the edges of the ribbon. The slides are dried from 4 to 12 hours at a temperature (53°C oven) just under the melting point of the wax. Dried slides can be stored in a dust free place until they are stained.

9.6. STAINING (Fig. 9.2)

Before the sections can be stained the paraffin must be removed. Xylene is used for this purpose. Immerse the slide in a Colin jar of xylene for 5 minutes. Xylene is removed by immersing the slide in anhydrous alcohol. As outlined in the staining chart, progressive transfer of water is then made through the indicated grades of ethyl alcohol. The slide is then washed with water and is now ready to be stained in an aqueous safranin (Safranin - 2.25 g, 96% ethanol - 225 ml. This stock solution should be diluted with an equal volume of distilled water when required) for one hour. Rinse and wash the slide in water and proceed with dehydration as in the staining chart given below. The counter stain, fast green (add saturated solution of fast green in absolute ethanol to the mixture of 25 parts of absolute ethanol and 75 parts of clove oil) in 95% alcohol is now applied. Then the slide is passed through absolute alcohol giving two changes. The slide is cleared in carbol xylene (phenol - 1 part, xylene - 3 to 4 parts). Two changes of pure xylol are also given to the slide before mounting. Remove a slide from the last xylene and dry with blotting paper. Put a drop of DPX mountant on the tissues and lower a cover glass obliquely. Dry the slides in a 53°C oven for one or more days to harden the mountant.

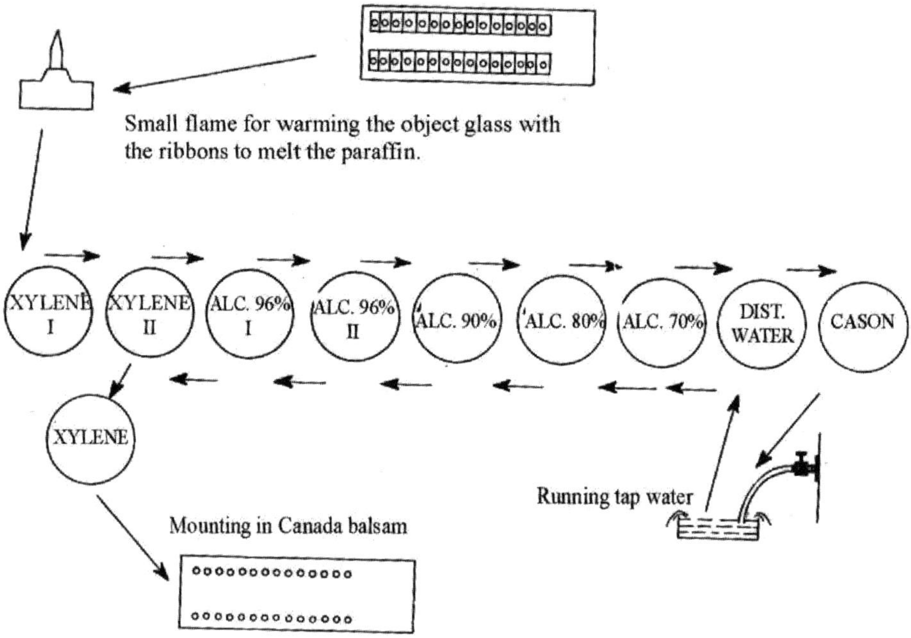

Fig. 9.2. Steps in staining of nematodes.

Safranin and fast green, is a basic acid dye combination often used on sections of nematodes in plant tissues, where cuticle, nucleoli, chromatin and lignin stains red and the remainder colors green.

■■■■

Chapter 10

CULTURING OF NEMATODES

Improving management strategies to limit crop losses attributed to plant parasitic nematodes; aseptic, monoxenic culture on excised plant tissue culture presents a possibility for production of larger numbers of nematodes than conventional greenhouse nematode-culture techniques.

10.1. *APHELENCHOIDES RITZEMABOSI*

10.1.1. Procedure

- Prepare a suspension of *Aphelenchoides ritzemabosi* in a solution of 2% methyl cellulose.
- Inoculate a 2-weeks old tobacco seedling with a droplet of the above mentioned nematode suspension.
- Place the seedling:
  - In a container having a relative humidity of 100%;
  - 3 days thereafter in the greenhouse at about 20°C;
  - 3 days thereafter in the container having a relative humidity of 100%;
  - 3 days thereafter again in the greenhouse until symptoms appear which takes about 14 days.

Sanwal (1959) set single leaves of chrysanthemum in wet sand in a moist, ventilated atmosphere. One or more nematodes were placed on each leaf in minute droplets of water which were then allowed to dry. At intervals, the leaves were sprayed with water from atomizer to maintain high humidity. Symptoms appeared on the leaves after 7-11 days and nematode populations increased greatly.

10.2. *DITYLENCHUS DESTRUCTOR*

10.2.1. On potato tubers

Seinhorst and Bels (1951) made holes about 5 mm deep and 2-3 mm in diameter on the surface of clean tubers, using cork borers. In each cavity, they placed a small quantity of thick suspension of *Ditylenchus destructor*. The cavity was sealed by smearing it with the melted surface of a candle. The wax set immediately and the heat was insufficient to damage either nematodes or potato tissue. The tubers were stored in wide-mouthed jars with damp filter paper to maintain a moist atmosphere.

10.2.2. On potato plugs

Dallimore (1966) embedded aseptic plugs of potato in potato-dextrose, corn-meal or water agar and inoculated them with a piece of infested tissue from young lesion of an infested potato. After 5 weeks, the plug was well infested and pieces could be used to inoculate further plugs or potato plants.

10.3. *GLOBODERA ROSTOCHIENSIS* AND *G. PALLIDA*

10.3.1. Procedure

- Wash potato tubers free from adhering soil and other debris. Store them for 3 weeks at 20°C in order to obtain sprouts of desired length.
- To obtain the hatching agent, let potato tubers sprout on a support above a beaker filled with tap water. The roots will grow into the water and excrete the desired root exudates.
- To induce larvae to hatch, place cysts in specially designed petri dishes and expose them for 3-14 days to the root exudates solution prepared earlier. The dishes with *G. pallida* are kept at 18°C, and those with *G. rostochiensis* at 22°C.
- For culturing, a 3% water agar is used. After boiling, the water agar is poured into an enamel jar, and after having removed the foam, the agar poured into autoclaved petri dishes in a sterile culture room or under a laminar flow. Pour in each dish about 20 ml agar; as soon as the agar has solidified the dishes are closed with the lid and stored at room temperature for a few days.
- A wedge shaped potato piece containing the sprout is cut from the well cleaned potato with a scalpel. Sterilize the scalpel regularly with 96% ethanol or flame the scalpel. The cut is placed on the agar near the edge of the petri dish with the sprout pointing towards the center of the dish.
- The dishes are stored in a horizontal position in an incubator at 20°C. When the roots are 2 cm long, the second stage larva is added to the root one by one with a handling needle using a dissecting microscope with overhead light. Inoculate these second stage larvae near root hair zone (on the root). After about 18 days, regular observation of the dishes is necessary.

Williams (1963) surface sterilized small potato tubers, removed eye tissue to prevent sprouting, and cut them in half. A concentrated suspension of *Globodera rostochiensis* larvae was added to the cut surface, and pieces were buried in washed sand kept moist by the addition of 20 ppm $CuSO_4$ solution. Females developed on the cut surfaces and they could be counted after about 6 weeks.

10.4. *MELOIDOGYNE* SP. ON EXCISED TOMATO ROOTS

10.4.1. Surface sterilization of tomato seed

The tomato seeds in a beaker are stirred in a 4.5% commercial 'Clorox' (HClO) solution (commercial 'Clorox' = 10g chlorine in 100 ml water) on a magnetic stirrer, using low speed, for about 10 minutes. As soon as the skin of tomato seed turns yellow in color, the excess of 'Clorox' solution is separated from the seed by pouring it into another beaker. The seed settled to the bottom of the beaker is transferred to an autoclaved petri dish filled with sterile water using flamed forceps.

The tomato seeds can also be surface sterilized using sodium hypochlorite (NaOCl):

- Pour 100 to 200 tomato seeds into the sterile beaker. Fill beaker with a 1: 1 mixture of 95% ethanol and 5.25% NaOCl. Stir gently – allow to soak for 8 minutes.
- Drain alcohol - NaOCl mixture and rinse 4 to 6 times with sterile water.
- Drain on the sterile paper in the petri dish.
- Place 8 to 12 seeds per dish on the germination plates.
- Allow seeds to grow for 3 days before use.

10.4.2. Transferring of the seed

Transfer the seeds with flamed forceps to a petri dish filled with 2% water agar under a laminar flow. Germinate in an incubator at 20°C.

10.4.3. Planting of the root tips in a petri dish filled with the nutrient solution

Modified Skoog, Tsui and White (1954) nutrient solution (Netscher, 1973)

Solution A1: Containing the macro-nutrients (purest grade commercially available chemicals) consisting of:

$Ca(NO_3)_2 . 4H_2O$ -------------------1.44 g

KNO_3 -------------------------------- 0.8 g

KCl -------------------------------- 0.65 g

KH_2PO_4 ---------------------------- 0.38 g

NH_4NO_3 -------------------------- 4.00 g

Dissolve in 800 ml demineralized water

Solution A2: Mg SO_4. $7H_2O$ ------------------- 0.72 g

Dissolve in 200 ml demineralized water

Combine the solution A1 and A2 thus obtaining 1000 ml of stock solution A.

Solution B: Fe SO_4. $7H_2O$ --- 10.19 g

Na_2 EDTA. $2H_2O$ (disodium salt of ethylene diaminotetra acetic acid) --- 8.25 g

Dissolve in 1000 ml demineralized water.

Aerate this solution for 1 day and keep the light yellow filtered solution in a brown bottle (stock solution B).

Solution C: Containing the trace elements (purest grade commercially available chemicals) consisting of:

Zn SO_4. $7H_2O$ --------------------------- 270 mg

Mn SO_4. $1H_2O$ -------------------------- 372 mg

HB_3O_3 ------------------------------------- 160 mg

KI --- 75 mg

Dissolve in 1000 ml demineralized water (= stock solution C)

Stock A, B and C should be stored at ± 5°C.

Solution D: Containing the vitamins and amino acids (purest grade commercially available chemicals) combining in one solution:

Nicotinic acid --------------------------------- 50 mg

Pyridoxine ------------------------------------ 75 mg

Thiamine HCl --------------------------------- 10 mg

Glycine --------------------------------------- 200 mg

Dissolve in 1000 ml demineralized water (= stock solution D).

Stock solution D should be made up just before preparing the medium. For obtaining 1 liter of the Skoog medium, take 884 ml of demineralized water, and dissolving therein 20 g saccharose and 15 g agar and adding 100 ml stock A, 5 ml stock B, 10 ml stock C and 1 ml stock D. Autoclave for 20 min at 110°C. Prevent overheating, because vitamins are unstable at high temperatures.

- Sterilize the table, on which this work is carried out, with 70% ethanol.
- Pour the autoclaved and melted nutrient solution (20 ml) into the sterilized petri dishes (this work should be carried out under a laminar-flow). Allow the plates to cool down to room temperature.
- Flame the lids of the petri dishes (never flame a polystyrene dish).

- When the roots are about 1 cm, they are cut off from the seed, using a flamed scalpel. Transfer the root onto Skoog, Tsui and White nutrient agar medium.
- Seal the dishes with adhesive tape (Para film).
- Place the petri dishes in a plastic box and transfer them to an incubator at 28°C.
- Allow the roots to grow for 2 weeks before inoculation with nematodes.
- Tissues under operation, tools, petri dishes, etc. must, at all times be protected from contamination.

10.4.4. Inoculation of excised tomato roots with the juveniles

- Pour the surface sterilized juveniles (for 2-3 min in 0.5% hibitane or 5-10 min in 0.1% hibitane) into a sterile petri dish.
- Suck nematodes from the water with a sterile micropipette and inoculate them on the growing tip of the excised tomato roots.
- Seal the petri dishes with a Para film and label them (date, nematode species, and origin of population).
- Place the petri dishes back in the plastic box and place the box in the incubator at 28°C.

10.4.5. Inoculation of excised tomato roots with the eggs

With a scalpel, remove an egg mass containing eggs and/or juveniles from an old root and transfer to a new growing root tip on a new petri dish.

10.5. *RADOPHOLUS SIMILIS* AND *PRATYLENCHUS* SP. ON CARROT DISCS

10.5.1. Preparation of carrot discs

It will never be possible to completely sterilize the carrots, but they should be cleaned, surface sterilized and peeled to reduce the risk that fungi and bacteria will develop in the carrot tissue. Then, the carrots are cut into discs and placed into sterile petri dishes.

10.5.1.1. Procedure: Under the Laminar-flow:

- Dip a pair of tongs in ethanol and flame them.
- Hold the carrot with the tongs, spray ethanol on the carrot and flame it. Repeat this step 3 times.
- Dip a knife in ethanol, flame it and cut off the tip of the carrot. Work on a glass plate.
- Dip a potato peeler in ethanol, flame it and peel the carrot. Repeat this step 3 times.

- Dip a knife in ethanol, flame it and cut the carrot into discs, 0.5 cm thick. Work on a glass plate.
- Dip a pair of tweezers in ethanol, flame them and put 1-2 discs together in a petri dish.
- Seal the petri dishes well with para film.
- Place the petri dishes in a plastic box and store in a refrigerator.

10.5.2. Selection and surface sterilization of nematodes

Actively moving, gravid female nematodes are preferred to initiate culture. After selection, the nematodes are transferred to sterile water. From now on, it is important to work under a laminar-flow cabinet. The nematodes should be sterilized twice: first in 0.01% mercuric chloride ($HgCl_2$) for 2 min and afterwards in 2000 ppm streptomycin sulfate overnight. The next morning, the nematodes are rinsed several times with sterile water.

10.5.2.1. Procedure: Under the Laminar-flow using the microscope:

- Pour the nematodes into a sterile counting dish.
- Select female nematodes, especially gravid females that are actively moving.
- Transfer the nematodes using a needle to sterile water in a sterile watch glass.
- Transfer the nematodes with a sterile pipette to a small sterile 5, 10 or 20 µm sieve.
- Place the small sieve in a sterile petri dish with 0.01% $HgCl_2$ for 2 minutes
- Rinse the nematodes in the small sieve two times with sterile water.
- Place the sieve in a sterile petri dish with sterile water and transfer the nematodes with a sterile pipette to sterile water in a sterile test tube.
- Add with a sterile pipette 1 ml of 6000 ppm streptomycin sulfate solution per 2 ml nematode suspension (thus obtaining a final concentration of 2000 ppm streptomycin sulfate).
- Leave the solution in a sealed test tube overnight.

Next morning

- The nematodes will have settled on the bottom of the test tube, visible as a pellet. With a sterile pipette, remove the streptomycin sulfate solution from above the nematodes.
- Add fresh sterile water to the nematodes with a sterile pipette.
- Wait until the nematodes have settled again to the bottom of the test tube, remove the water from above the nematodes and add again fresh sterile water with a sterile pipette. Repeat this for 2 to 3 times.

10.5.3. Inoculation of the carrot discs with nematodes

The nematodes are transferred to the carrot discs with a sterile micropipette. Small drops of nematode suspension are placed on the margin of the carrot discs. For culture maintenance, low inoculum levels (25 nematodes or less) are used, while for mass multiplication, high inoculum levels (more than 25 nematodes) are used.

The petri dishes are sealed well with para film, labelled (date, nematode species, origin of population, number of females) and placed in a plastic box in an incubator at 28°C.

Depending on the initial inoculum and on the nematode species, the nutrients will be depleted after some weeks and the nematodes will leave the carrot discs. It becomes then necessary to subculture the nematodes onto fresh carrot discs. The nematodes are extracted from the old carrot discs using the modified Baermann-funnel method. For sterilization of the nematodes, sterilize-ation in $HgCl_2$ is skipped to avoid a too high pressure on the nematodes. The sterilization in streptomycin sulfate is however done as described before. The inoculation of the carrot discs with nematodes is also done in a similar way.

10.6. ENTOMOPATHOGENIC NEMATODES

10.6.1. *In vivo* production

In vivo production yields vary greatly among insect hosts and nematode species. Due to high susceptibility to most nematodes, wide availability, ease in rearing, and the ability to produce high yields, the most common insect host used for laboratory and commercial EPN culture is the greater wax moth, *Galleria mellonella*. Significant research (and commercial application) has also been achieved for production of EPNs in the yellow mealworm, *Tenebrio molitor*.

In vivo can be accomplished by pipetting or spraying nematodes onto a substrate, immersion of insects in a nematode suspension, or (in some cases) applying the nematodes to the insect's food. Immersion of hosts is generally more time efficient but requires more nematodes than other procedures.

10.6.2. *In vitro* production

10.6.2.1. Solid culture: *In vitro* solid culture advanced considerably with the invention of a three-dimensional rearing system involving nematode culture on crumbled polyether polyurethane foam. A liquid medium is mixed with foam, autoclaved, and then inoculated with bacteria followed by the nematodes. Nematodes are then harvested within 2-5 weeks by placing the foam onto sieves immersed in water. Media for this approach was initially animal product based (e.g., pork kidney or chicken offal) but was later improved and may include various ingredients including peptone, yeast extract, eggs, soy flour, and lard.

10.6.2.2. Liquid culture: Generally, in liquid *in vitro* culture, symbiotic bacteria are first introduced followed by the nematodes. Various ingredients for liquid culture media have been reported including soya flour, yeast extract, canola oil, corn oil, thistle oil, egg yolk, casein peptone, milk powder, liver extract and cholesterol. Culture times vary depending on media and species, and may be as long as three weeks though many species can reach maximum IJ production in two weeks or less. Once the culture is completed, nematodes can be harvested from media via centrifugation.

In vitro liquid culture is considered to be the most cost efficient process for producing entomopathogenic nematodes and thus accounts for the bulk of the world market in EPNs. Liquid culture of entomopathogenic nematodes has been accomplished in bioreactors of up to 80,000 liters capacity (Fig. 10.1).

Fig. 10.1. *Left* – Culturing *Heterorhabditis bacteriophora* in liquid media: (Left) Fermenter prior to *Photorhabdus* inoculation. Right - After 21 days of rearing nematodes.

Chapter 11

MISCELLANEOUS TECHNIQUES

11.1. PROOF OF PATHOGENICITY

Many of us have learned and even memorized Koch's postulates as being the acceptable scientific method for demonstrating pathogenicity. Essentially, a pathogen must be isolated from host tissue, grown in pure culture on artificial media, a susceptible host must be inoculated, and the pathogen reisolated following the development of symptoms.

Most plant parasitic nematodes have never been cultured on artificial media and so cannot be subjected to Koch's postulates to prove pathogenicity. Even if this were possible, the question can be raised as to how closely sterile laboratory conditions compare with field situations.

Because of the problems with Koch's postulates, nematode pathogenicity often consists of an "educated" guess and this is where experience, art, and intuition come into play. It is also a reason why some diagnostic laboratories will indicate which nematodes are present but hesitate to provide management advice.

If plant parasitic nematodes are known to be present in a field, they are unlikely to disappear from that field whether or not they are found in a particular sample. Each plant parasitic nematode will have a particular crop or group of crops on which they have been observed to cause problems. There is extensive literature on nematode-host associations. In some cases, these are just what the word association indicates and pathogenicity has not been proven even to the extent of conducting greenhouse or micro plot trials or extracting nematodes from plant tissue. In other cases, various types of research have been conducted to provide more positive evidence that symptoms observed are actually the result of nematode infestation. Several texts listing nematode-host associations have been written and may be found in public libraries or private collections of nematologists.

11.2. METHODS OF ESTIMATING YIELD LOSSES

Pot studies can be used to determine some of the basic information on yield-loss relationships, but because of environmental differences and interactions, field studies are also needed. There are two approaches—one is to use nematicides at relatively uniformly infested sites; the other is to work at sites with a range of population densities, but which are uniform in other respects. A combination of both approaches is often a happy compromise. The former gives practical information on the effectiveness and potential value of a particular treatment, but tells little about the nature of the relationship. It also suffers from the criticism that nematicides have a range of side effects. The latter has the benefit of producing information on the relationship between initial population (Pi) and yield, but it requires experimental errors to be minimized. Because Pi estimates have large errors, accuracy is improved by reducing plot size and by taking and processing multiple samples from each plot. However, plot size must be large enough to obtain a realistic yield, and adequate guard plants are essential.

Another option is to establish many small plots in large but otherwise uniform fields. These can be at random, in a grid pattern, or along known trends in Pi. The plots can be split and a nematicide applied to one half. For each plot, the Pi and yield are determined. The results will produce a scatter of points, hopefully with yield decreasing as Pi increases. Much of the scatter is due to errors in estimating Pi and yield, and it can be minimized by taking the average of all the results within each error band. Such an approach needs:

- A wide range of initial populations
- A uniform field
- A large number of plots (100 or more)
- The plots to be part of an otherwise uniform crop.

11.3. SCREENING OF GERMPLASM

11.3.1. Screening of banana germplasm against migratory endoparasitic nematodes (*Pratylenchus coffeae* and *Radopholus similis*) under greenhouse conditions

11.3.1.1. Procedure

- Start with rooted *in-vitro* plantlets, ready for transfer to *in-vivo* conditions. For suckers, use suckers of uniform size, which are peeled immediately after detaching from the mother plant and are immersed in hot water maintained at 55°C for 15 min.
- Sieve the soil through a 2 mm sieve and sterilize it.
- Fill pots with the sterilized soil and place them on greenhouse bench.
- Water the soil up to field capacity.
- Make a hole in the middle of each pot.
- Place an *in-vitro* plantlet (or sucker) in each hole and fill the hole with soil.

- Place the pots under a plastic or Styrofoam cover for adaptation to the new environment (not necessary for suckers). Gradually open the cover after 2 weeks.
- Four weeks after planting, inoculate 8-10 plants of each genotype with nematodes (see next step), while keeping another 8-10 plants as control plants (nematode-free).
- Make 3-4 holes, about 3 cm deep, into the soil adjacent to the base of each plant.
- Inoculate about 1000 vermiform nematodes in a few ml of water into the holes.
- Fill the holes with soil.
- Water the soil lightly.
- During the whole experiment, water the plants as needed and fertilize them with a nutrient solution every 2-4 weeks.
- Label all the pots (genotype, inoculated or control) and arrange them in a randomized complete block design.

For experiments with *R. similis*, collect data at 8 weeks after inoculation; for experiments with *P. coffeae*, collect data at 10 weeks after inoculation.

- Measure the plant height, the girth of the pseudo stem at the base, the shoot weight, the weight of the root system and the number of healthy leaves (> 70% of the leaf surface healthy).
- Select at random 5 functional primary roots, at least 10 cm long, cut them into 10 cm pieces and slice the roots lengthwise. Score one half of each of the 5 roots for the percentage of the root cortex showing necrosis. The maximum root necrosis per half root can be 20%, giving a maximum root necrosis of 100% for five root halves together.
- Estimate the nematode population (males, females, juveniles and eggs).

Section III
MORPHOLOGY AND TAXONOMY

Chapter 12

MORPHOLOGY

12.1. INTRODUCTION

Nematodes exhibits considerable variation in their external and internal structure. Despite this structural complexity certain basic principles are common to all nematodes.

Nematodes are triploblastic, bilaterally symmetrical, unsegmented, pseudocoelomate and vermiform animals. Typically, the nematode body is elongate, spindle shaped or fusiform, tapering toward both ends and circular in cross section. The length may vary from 0.2 mm (*Paratylenchus*) to about 11.0 mm (*Paralongidorus maximus*) but averaging about 1.0 mm in length. The body width may vary from 0.01 to 0.5 mm. In few plant parasitic genera, the body may be pear, lemon, reniform or irregularly saccate shaped as in the females of *Meloidogyne, Heterodera, Nacobbus, Rotylenchulus* and *Tylenchulus* (Fig. 12.1). The body width may be less than one per cent of its length, as in *Ecphyadophora,* or as much as 15 per cent, as in *Criconema.* The nematode body is generally considered to be bilaterally symmetrical. Radially symmetric traits (triradiate, tetraradiate ad hexaradiate) occur in the anterior region, whereas intestine excretory and reproductive systems show tendencies towards asymmetry. They have one or two tubular gonads which open separately in the female and into the rectum in the male which also has copulatory spicules. The body cavity of nematodes constitutes a pseudocoelom (false body cavity) since, it is not lined by epithelium. Respiratory and circulatory systems are lacking in nematodes.

Gross morphology of typical female and male plant parasitic nematodes is presented in Fig. 12.2.

Longitudinally, the body is divided into four regions: dorsal, right and left lateral, and ventral. For purposes of orientation, we generally determine the ventral region by the fact that the bodies of a majority of the species are naturally bent ventrally to some degree. But should the body be straight, the location of the anal opening is always a definite character on which we can depend, even in the larval stages. Mature females possess a vulva, a prominent organ, which also is always placed ventrally. Usually, the vulva is near the middle of the body, although it may be far back near the anus as in *Criconema* or forward toward the base of the esophagus like that of *Tylencholaimellus*. If an excretory pore is present, it too, is located ventrally, somewhere between the nerve ring and the base of the esophagus, except in rare instances as in *Tylenchulus* where it is far backward.

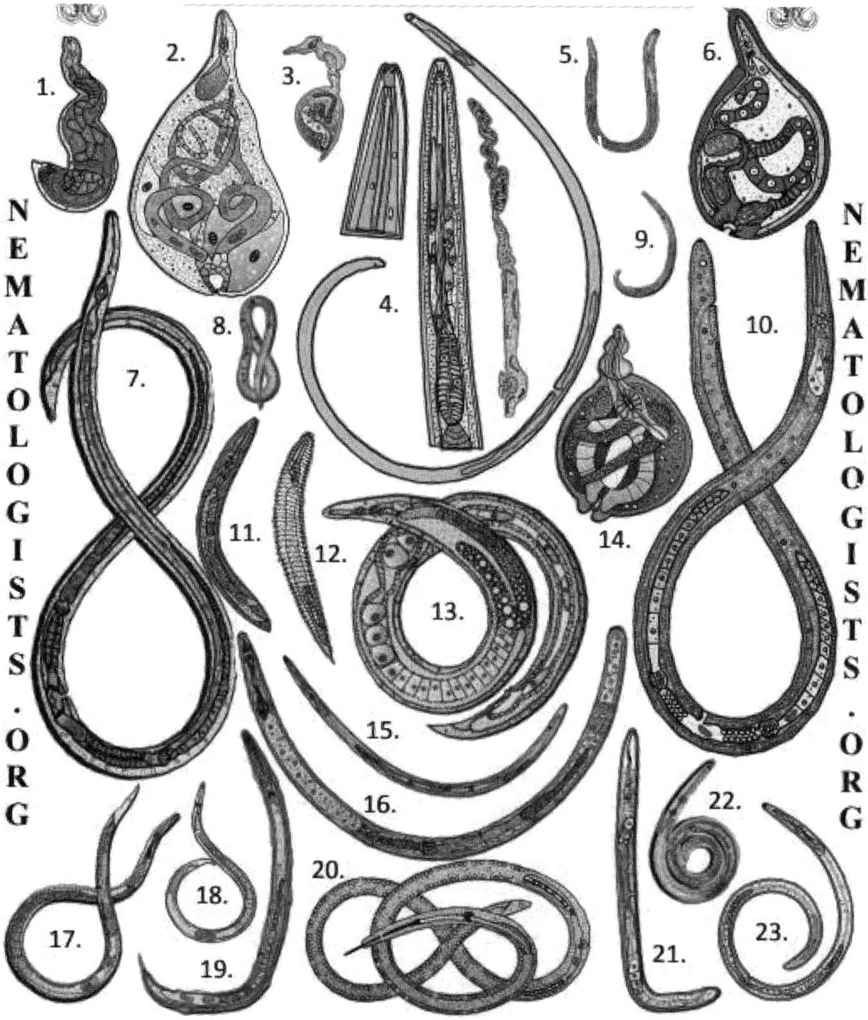

Fig. 12.1. Common genera of plant parasitic nematodes. 1 - Reniform (*Rotylenchulus* spp.), 2 - Root-knot (*Meloidogyne* sp.), 3 - Citrus (*Tylenchulus* sp.), 4 - Needle (*Longidorus* sp.), 5 - Lesion (*Pratylenchus* sp.), 6 - Cyst (*Heterodera* sp.), 7 - Awl nematodes (*Dolichodorus* sp.), 8 - Burrowing (*Radopholus* sp.), 9 - Pin nematodes (*Paratylenchus* sp.), 10 - Sting (*Belonolaimus* sp.), 11 - Ring (*Criconemoides* sp.), 12 - Ring (*Ogma* sp.), 13- Seed gall (*Anguina* sp.), 14 - Sphaeroid (*Sphaeronema* sp.), 15 - Stunt (*Tylenchorhynchus* sp.), 16 - Lance (*Hoplolaimus* sp.), 17 - Bulb and stem (*Ditylenchus* sp.), 18 - Foliar (*Aphelenchoides* sp.), 19 - Sheath nematodes (*Hemicycliophora* sp.), 20 - Dagger (*Xiphinema* sp.), 21 - Stubby root (*Trichodorus* sp.), 22 - Spiral (*Helicotylenchus* sp.), 23 - Spiral (*Rotylenchus* sp.) (Photo - Society of Nematologists)

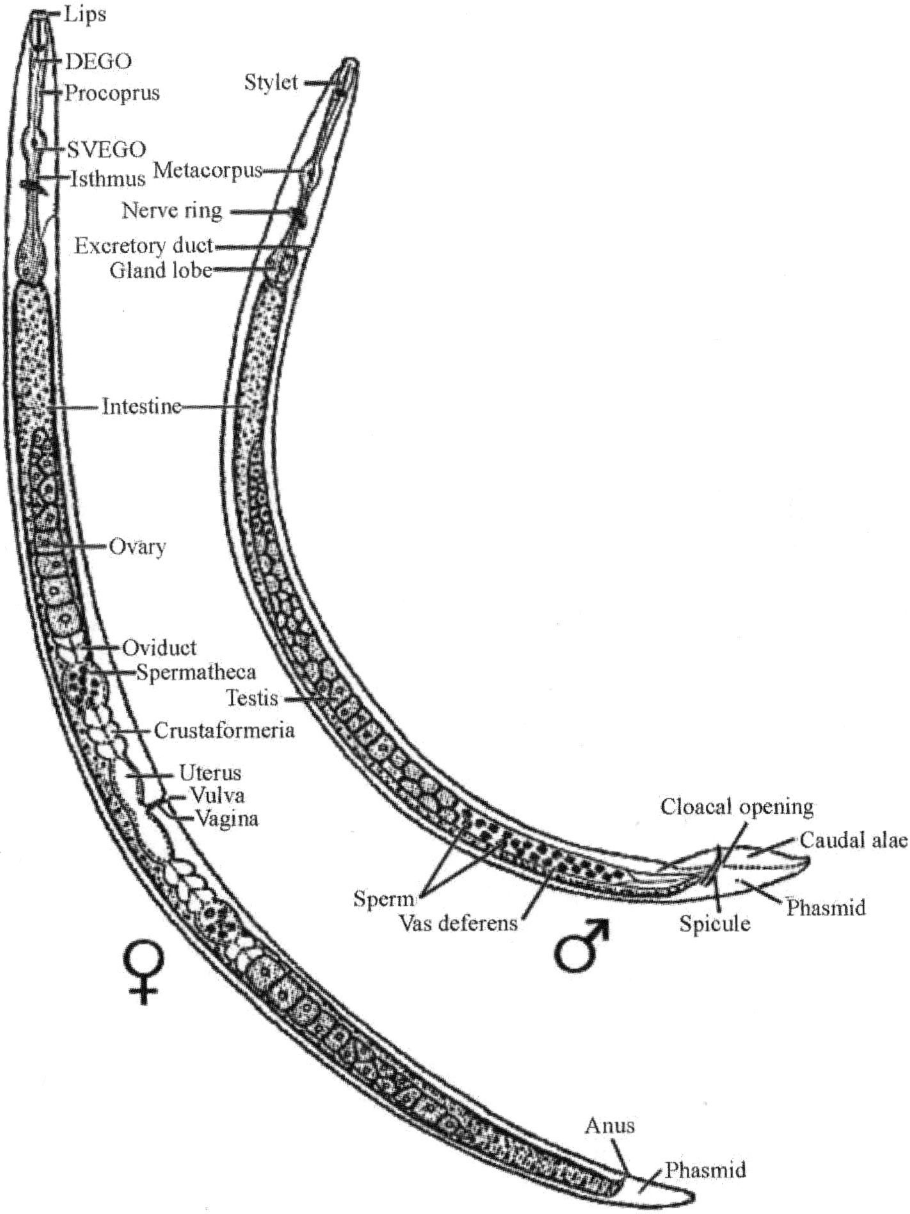

Fig. 12.2. Gross morphology of a typical vermiform female (*left*) and male (*right*) plant-parasitic nematodes (DEGO-Dorsal esophageal gland opening, SVEGO-Sub-ventral esophageal gland opening)

The nematode body is divided into three regions - outer body tube or body wall (ectoderm and mesoderm), inner body tube (endoderm) and body cavity (pseudocoelom) (Fig. 12.3).

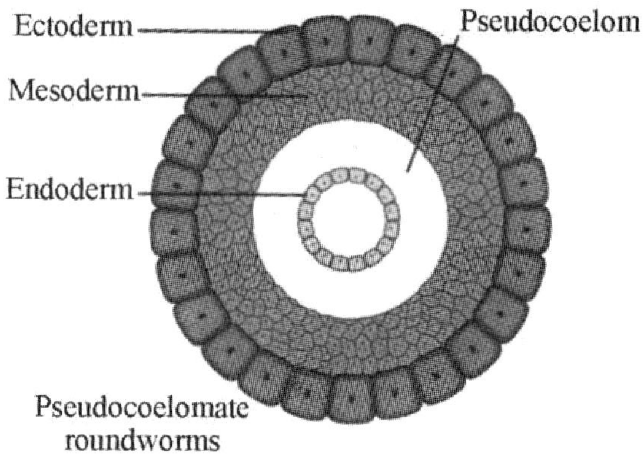

Fig. 12.3. Three regions of nematode body

12.2. OUTER BODY TUBE

The outer body tube comprises of exoskeleton or cuticle, hypodermis and muscle layer (Fig. 12.4).

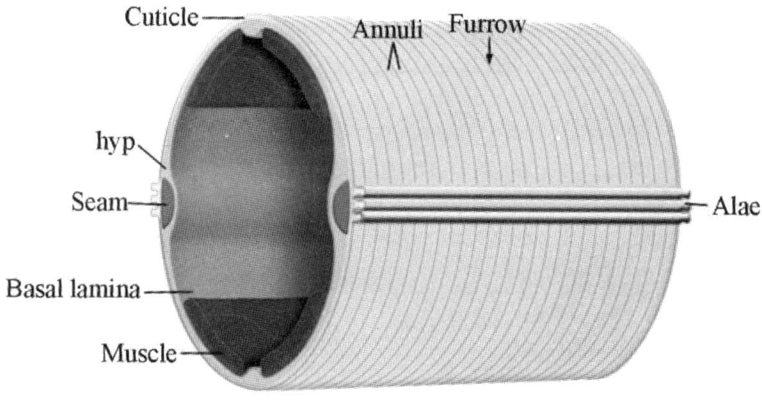

Fig. 12.4. Three layers of outer body tube (cuticle, hypodermis and muscle layer)

12.2.1. Exoskeleton or cuticle

The exoskeleton of nematodes may be defined as the outer covering or cuticle together with its invaginations and outgrowths. It comprises a cortical layer, median layer and basal layer. The exoskeleton is a non - cellular, multilayered structure that is secreted by the hypodermis, which functions both as barrier to undesirable elements in the environment and as a flexible skeleton. It invaginates at the mouth, rectum, cloaca, vagina, excretory pore, amphids and phasmids.

12.2.1.1. External structures: The markings on the surface of the cuticle are varied and complex and have been used extensively by taxonomists to assist in the identification of various species.

(a) Punctuations: They commonly appear as minutes, round areas which are often arranged in a pattern (Fig. 12.5). Punctuations act as structures for strengthening the cuticle rather than as pore canals through which cuticular proteins may be transported.

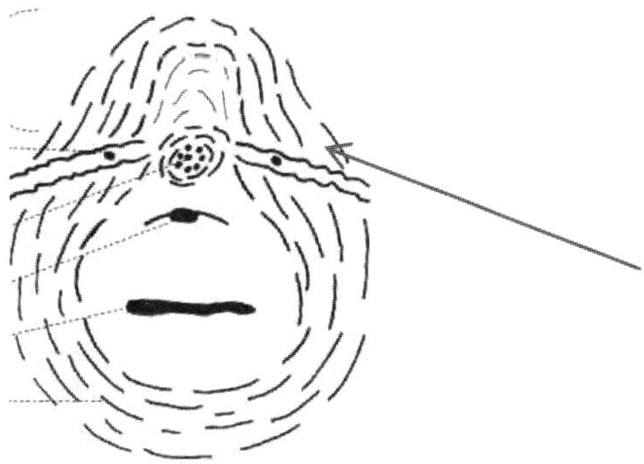

Fig. 12.5. Punctations (arrow) on tail terminus of the root-knot nematode

(b) Transverse markings: These take the form of transverse striations or annulations and are found in most nematodes giving them a segmented appearance (Fig. 12.6). These annulations are very pronounced in some families, such as Tylenchidae, Heteroderidae and particularly the Criconematidae where the annulations are so pronounced that they become outgrowths and are known as scales. The perineal pattern in the posterior body region of *Meloidogyne* females as well as the rugose wall pattern of *Heterodera* cysts is believed to be modification of transverse markings.

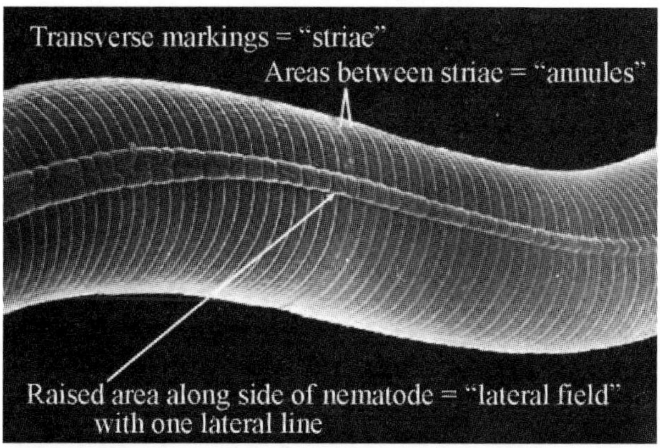

Fig. 12.6. Annulations on root-knot nematode larva (Photo J.D. Eisenback)

Electron microscope examination of members of the Dorylaimidae showed that they do have a cuticle with transverse annulations, although these are not shown in the literature in various drawings done with the aid of a light microscope. Their presence may be necessary to allow dorsoventral undulatory movement.

(c) Longitudinal markings: These usually take the form of longitudinal ridges or alae.

(i) Ridges: These are raised areas which run the length of the body and occur on the submedian as well as the lateral surface. The arrangement of ridges is used by taxonomists as an aid to species identification.

(ii) Alae: These are thickenings or projections of the cuticle which occur in the lateral or sublateral region.

Caudal alae: These are found in the posterior region and are restricted to males as copulatory bursae.

Cervical alae: These are confined to the anterior part of the nematode body.

Longitudinal alae: These delimit the lateral fields and are known as lateral alae. Their form varies considerably in different species and is characteristic for a particular species. They are traversed by striations or furrows varying in number from 1 to 12. Functionally, they probably assist locomotion and may permit slight changes in the width of nematodes.

(d) Cuticular layering: The nematode cuticle is basically a 3-layered structure and consists of an outer layer (cortical), a middle layer (median or matrix) and an inner layer (basal or fiber) (Fig. 12.7).

In some nematodes, there are only 2 layers as in the adult females of the family Heteroderidae. These nematodes are parasitic and have an initial rapid period of growth of about 3 weeks duration in the roots of host plants, where movement is not required. This may give rise to a modification of the basic 3-layered structure which is found in pre-parasitic larvae and males of the Heteroderidae.

(i) Cortical layer: In the larger nematodes the cortical layer is often divided into an external cortical layer and an internal cortical layer. The surface of the external cortical layer is exposed to the nematodes' environment. This layer is very thin (about 25 to 40 mμ) and can be subdivided into an outer membrane which is about 3-5 mμ thick and may correspond with a plasma membrane (triple layered membrane); a middle, relatively non osmiophilic layer of about 7 to 8 mμ thick and an inner osmiophilic layer which varies in thickness from 12 to 25 mμ.

The external cortical layer has been considered to be a Keratin chemically. The disulfide group present in this layer may be partly responsible for its resistant properties. Quinone tanning, which would also make this a more resistant layer, has been shown to occur in this layer in *Globodera rostochiensis.* This is a rather specialized case, as the cuticle of this nematode becomes a cyst wall on the death of the female and acts as a protective cover over the eggs. It is thought to aid in slowing down the rate of desiccation of the eggs under dry conditions.

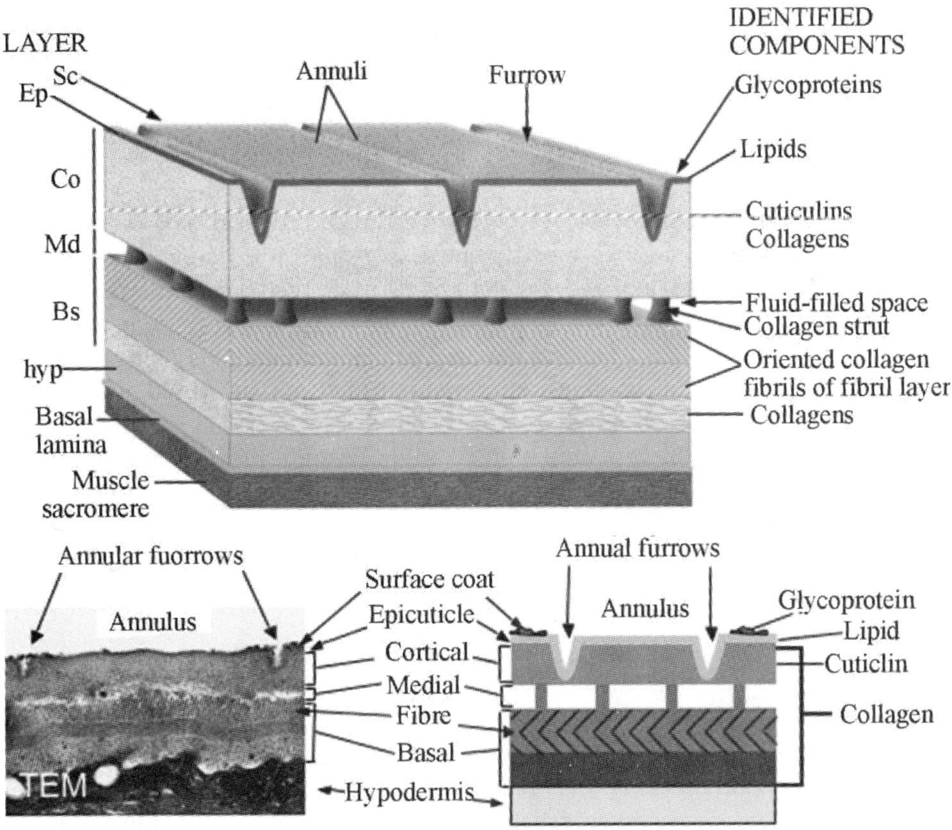

Fig. 12.7. Schematic showing layers of the adult cuticle and identified molecular components in each. (Sc) Surface coat; (Ep) epicuticle; (Co) cortical zone; (Md) medial zone; (Bs) basal zone; (hyp) - hypodermis.

The internal cortical layer varies considerably in thickness in different nematodes. In pre-parasitic larval forms, it may be only about 0.15 to 0.25 μ thick. It has a fibrous structure. There is no clear line of demarcation separating the internal cortical layer and the median layer. This layer appears to be biochemically active in some nematodes, as both enzymes and RNA have been detected in it.

(ii) Median layer: This layer in pre-parasitic larvae belonging to the family Heteroderidae is largely electron transparent and contains globular electron dense bodies. It is traversed by thin columns of material which connect the cortical layer to the basal layer. The median layer is fluid in a number of different species of nematodes.

The average thickness of the median layer is 0.1 μ in the larvae of *Meloidogyne* and *Heterodera*. This layer undergoes marked increases in width in *Hemicycliophora arenaria* as it grows through the larval stages to become adult nematode. The median layer in this nematode is about 0.2 μ in width in the second stage larvae and about 0.7 μ in width in the adult female.

Chemically, the median layer consists of proteins which resemble collagen. A non-specific esterase, acid mucopolysaccharide and some lipid have also been detected in this layer. It does not appear to be as metabolically active as the cortical layer.

(iii) Basal layer: It consists of regularly arranged vertical rods or striations. The depth of this layer varies in different species and in different stages of the same species, e.g. it is about 125 mμ thick in *Meloidogyne javanica* larvae, but is 250 mμ thick in *Globodera rostochiensis* larvae and 500 mμ thick in the adult male of *M. javanica.* This layer is composed of a protein with very close linkages between the molecules, resulting in a resistant layer that would protect the nematode from a fluctuating environment. The basal layer and the hypodermis are separated by a highly convoluted, triple-layered plasma membrane or basement membrane.

The presence of enzymes, RNA, ascorbic acid and hemoglobin in some cuticles, together with their observed growth after molting, supports the hypothesis that the cuticle is an active living structure and not just an inert covering.

12.2.2. Hypodermis

The hypodermis lies between the cuticle and the somatic muscle layer. It is particularly important and the metabolically active part of a nematode, responsible among other things, for the secretion and maintenance of the cuticle.

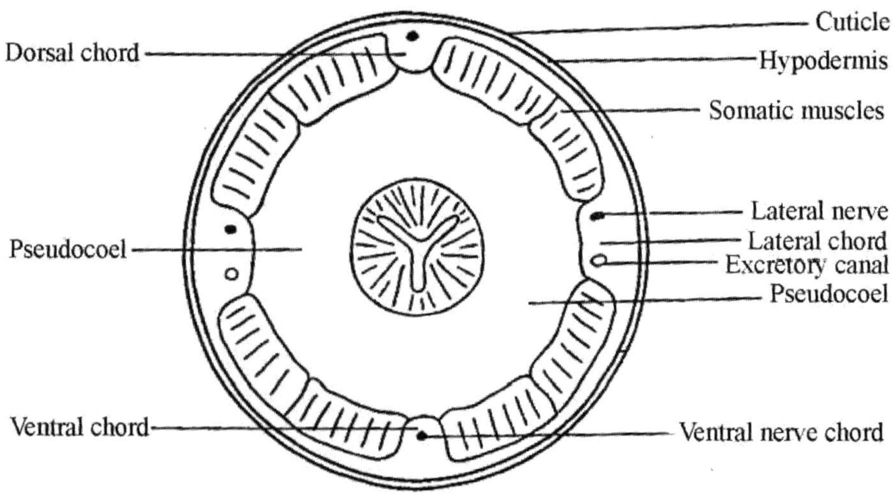

Fig. 12.8. Cross section of nematode in the region of esophageal glands.

The hypodermis, which is occasionally called the epidermis of sub-cuticle, is a thin layer that is characteristically thickened in the dorsal, ventral and lateral positions to form 4 hypodermal cords (Fig. 12.8). These protrude into the pseudocoelomic cavity between the somatic muscles and divide them into 4 quadrants. The 2 largest projections are the lateral cords which commonly run the whole length of the body. Their position is often demarcated externally by cuticular structures known as lateral alae. Nuclei are always found in the lateral cords of nematodes. In some groups, lateral canals of the excretory system lie embedded in the lateral hypodermal cords.

The hypodermis in some nematodes is cellular and presumably both cuticle and hypodermis, and muscle and hypodermis are separated by cell membranes as observed in *Xiphinema index*.

12.2.2.1. Hypodermal glands: Glands occur in the hypodermis of many different species of nematodes. It has been suggested that they act as either osmotic or ionic regulators. The function of the caudal glands of the hypodermis is more obvious as they produce an adhesive secretion which serves to anchor the nematode. Hypodermal glands are also associated with certain sensory organs (amphids, phasmids and deirids).

12.2.2.2. Composition: The hypodermis is at all times rich in glycogen, lipid and a number of enzymes. At times, such as just before the onset of molting, there is an accumulation of protein and nucleic acids.

Hypodermis is clearly one of the most metabolically active regions in the nematode body.

12.2.3. Muscle layer

The somatic muscles of nematodes are grouped into rows between the hypodermal cords. These may not, in some cases, be any rows or only 2 (holomyarian) or, more commonly, there may be 2-5 rows (meromyarian) or there may be a large number of rows (polymyarian).

There are 3 types of muscles cells, a flat type of cell (platymyraian), a U - shaped type of cell (coelomyarian) in which the fibers extend up to the side of the cell and the sarcoplasm bulges into the pseudocoelom, and finally, cells in which the fibers encircle the sarcoplasm (circomyarian).

The most common configuration is platymyrian cells grouped in meromyarian arrangement a condition also found in the larval stages of some the more complex coelomyarian polymyarian types.

Muscle cells are connected to each other by means of cytoplasmic bridges in many nematodes and may have more than one nervous connection. These structures obviously assist greatly in co-ordinated muscular contraction.

In addition to somatic muscle, there are a number of specialized muscles which are associated with feeding, food movement and defecation (cephalic, esophageal, intestinal and anal muscles) as well as with reproduction (vulvar, spicular, gubernacular, copulatory and bursal muscles).

The myofilaments of the contractile part of the nematode muscle cell are remarkably similar. The myofilaments are of 2 kinds, thin and thick. The thick filaments are thought to contain myosin and the thin filaments actin. These actin filaments slide past the myosin filaments during contraction.

The coelomyraian muscle cell is virtually a storage depot for glycogen.

The nematode muscles are unusual in that the muscle sends branches to the nerve rather than the nerve sending branches to the muscle.

12.3. INNER BODY TUBE

The nematode digestive system is essentially the inner body tube into which some glands open. It starts at the mouth as a cuticular inpushing known as stomodeum, which incorporates the buccal cavity and esophagus, and ends at the anus as another cuticular invagination that is known as the proctodeum, which includes the rectum in females and the cloaca in males. On molting, the linings of the stomodeum and proctodeum are shed with the cuticle. The middle part of the digestive system is the intestine or midgut, which is the only part that is endodermal in origin. The ducts of the esophageal glands enter the stomodeum and the ducts of the rectal glands enter the proctodeum.

12.3.1. Stomodeum

The Stomodeum includes the mouth and lips, the buccal cavity and the pharynx or esophagus.

12.3.1.1. Mouth and lips: The general pattern is a head with 6 lips (Fig. 12.9). This may, however, be reduced by partial fusion to 3 or by complete fusion to form a united ring around the mouth.

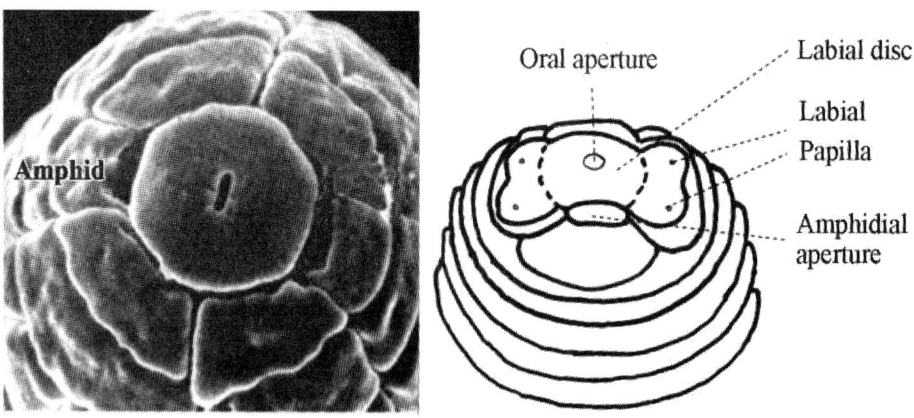

Fig. 12.9. Lip region of a nematode head.

12.3.1.2. Buccal cavity: The buccal cavity or stoma lies between the mouth and the pharynx. Plant parasitic nematodes have a hollow buccal spear or stylet whose lumen is usually less than 1 μ in diameter. The stylet can, therefore, act as a bacterial filter as well as a type of micro-hypdermic needle through which material may be injected into the plant cell and through which nutrients from the plant are ingested by the nematode. A buccal spear is also found in other parasitic nematodes as well as in predatory forms. The latter may either pierce their prey and suck out their contents using a buccal stylet, as in *Dorylaimus* and *Seinura* or they may puncture their prey by means of teeth, as in *Mylonchulus.*

The buccal stylet of nematodes may also be either odontial or onchial in origin. Thus, in forms such as the Tylenchida, the stylet, which is a hollow, tripartite, cuticular structure usually with basal knobs or swellings for the attachment of muscles, is developed, at least in part, from a fusion of the walls of the buccal cavity or stoma and is known as stomatostyl or stomatostylet (Fig. 12.10). In forms such as the Dorylaimida, the anterior part of the buccal stylet originates from the same region as do onchia and spears in these nematodes are referred to as odontostyl or odontostylet (Fig. 12.10).

ESOPHAGEAL TYPES OF COMMON SOIL NEMATODES

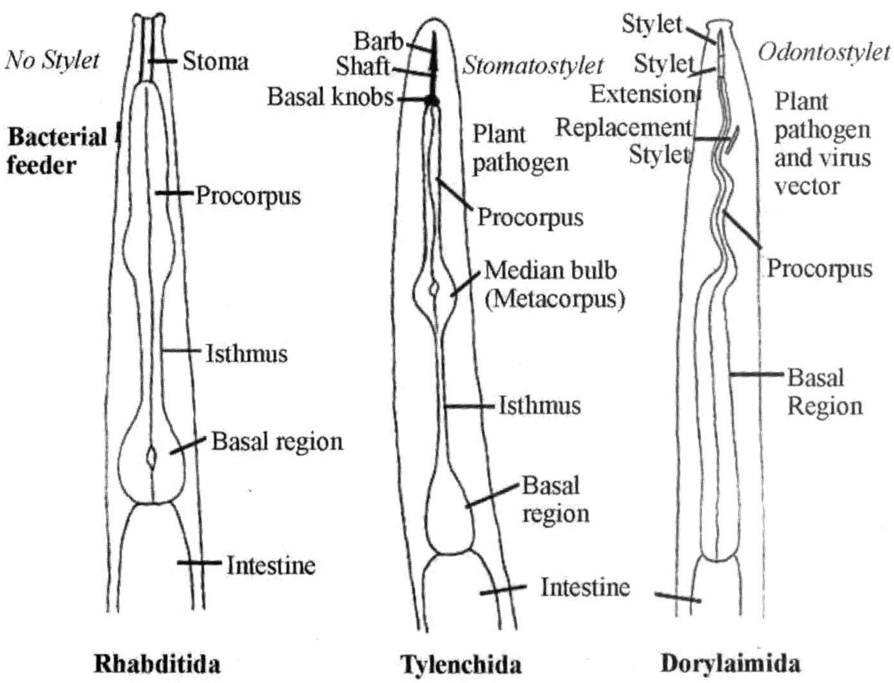

Fig. 12.10. Esophageal types of common soil nematodes

12.3.1.3. Pharynx or esophagus: The pharynx is a muscular pumping organ with a triradiate, cuticle lined lumen of which one ray normally projects ventrally and two project laterodorsally. The pharynx contains radial muscles, esophageal glands and valves which prevent the regurgitation of food.

In some nematodes, both the median and posterior parts of the pharynx are swollen to form muscular bulbs. The three part cylindrical pharynx contains three, well defined regions, namely, the corpus, the isthmus and the bulb (Fig. 10.5). The corpus may be further modified to form a procorpus and a swollen posterior part, the metacorpus. The pharynx is a cellular structure with distinct muscle cells, supporting cells, nerve cells and gland cells.

12.3.2. Intestine

The nematode intestine is a simple tube consisting of a single layer of epithelial cells. The intestine is generally divided into three regions, which merge into each other without any perceptible boundaries. These are the anterior or ventricular region, the midintestinal region and the posterior or pre-rectal region. In most forms the intestine is a straight tube.

The microvilli are finger like projections of the plasma membrane projecting into the intestinal region. They considerably increase the surface area of the intestine and are thought to be associated with absorption, although in some cases, it appears that they are concerned with secretion (Fig. 12.11).

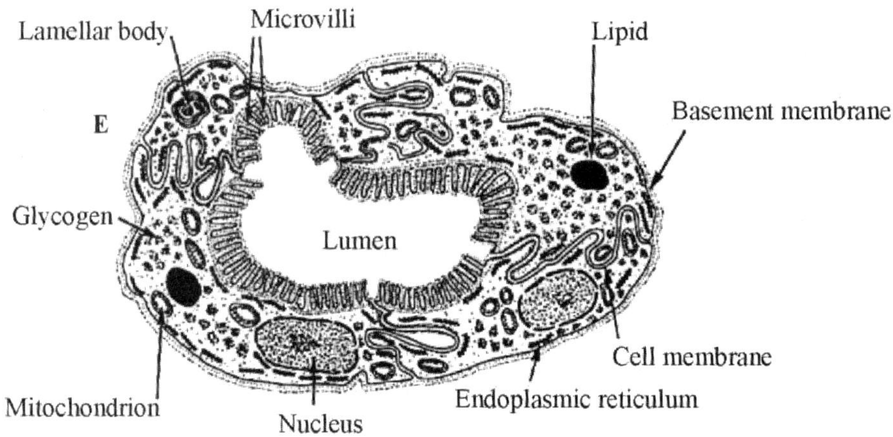

Fig. 12.11. Cross section of nematode intestine

Each intestinal cell is surrounded by a plasma membrane and the whole intestine is separated from the pseudocoelom by a basement membrane.

In addition to glycogen and lamellar bodies, the intestinal cells of nematodes often contain large protein and lipid or fat bodies. These make up the principal storage material in the infective larvae of many parasitic nematodes.

In most nematodes, food is moved along the intestine by the ingestion of more food and by the locomotory activity of the worm.

12.3.3. Proctodeum

The proctodeum or rectum is a cuticle lined invagination into which rectal glands open in many nematodes. In the female, the proctodeum normally consists of a simple tube leading to the anus, whereas in the male, the reproductive system opens into it to form a cloaca which contains the spicules and other copulatory structures.

The intestine is modified to form an intestine-rectal valve at its junction with the rectum. It is a simple structure and is closed by means of a unicellular sphincter muscle.

In dorylaims, a distinct region known as the pre-rectum occurs between the intestine and the rectum.

The anus consists of a slit-like structure on the ventral surface, whose opening is controlled by the unicellular, H-shaped depressor animuscle, which acts by raising the dorsal wall of the rectum and pulling on the posterior lip of the anus to open it.

12.3.4. Glands

The glands fall into two main classes namely, the esophageal or pharyngeal glands which enter the stomodeum and the rectal glands which enter the proctodeum.

12.3.4.1. Esophageal glands: Typically there are three, uninucleate esophageal glands, one dorsal and two ventrolateral or sub-ventral. These glands connect with the lumen of the esophagus by means of cuticularized ducts, often by way of a terminal swelling or ampulla. The dorsal gland usually opens much further forward than do the subventrals and may open into or close to the stylet (Fig. 12.12).

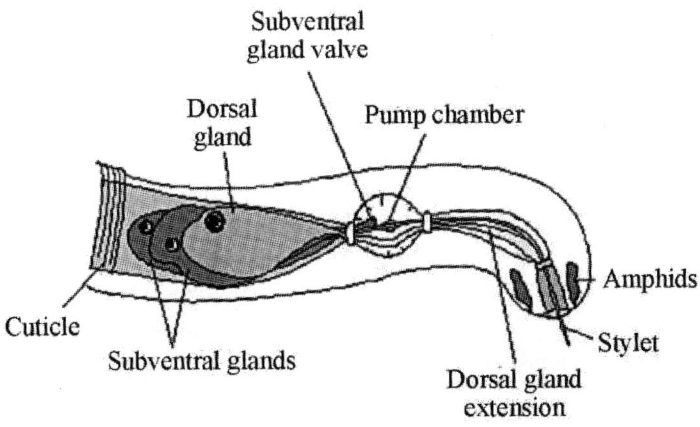

Fig. 12.12. Esophageal glands in a typical sedentary plant parasitic nematode

The esophageal glands have been studied most thoroughly in the genus *Meloidogyne* in which they appear to have important roles in hatching, host penetration and the establishment of a complex host-parasite relationship (development, control and maintenance of the tumor-like giant cells in plants).

12.3.4.2. Rectal glands: The number varies in different species and between the sexes in the same species. They are responsible for the production of copious amounts of a gelatinous mucopolysaccharide matrix which serves to protect the eggs.

12.3.5. Function

12.3.5.1. Feeding: Digestive juices which originate from the dorsal esophageal gland, which is filled with small, refractive, granule-like bodies, are injected into the host plant cell by means of the buccal stylet. During feeding, a distinct zone develops around the stylet tip in the host cell. There are two feeding phases. The first is the injection or salivation phase, during which the flow of juices into the host cell may be enhanced by contractions of the lateral muscle of the median bulb. The second phase is the ingestion phase, during which rhythmical contraction of the posterior part of the esophagus associated with the median bulb and, in some forms, the esophago-intestinal valve or cardia, results in the ingestion of material from the host.

12.3.5.2. Secretion: Various glands associated with the digestive system play an important role in secretion. The cells of these glands are associated with active protein and mucopolysaccharide synthesis and their products are shed through cuticle lined ducts either into the stomodeum or proctodeum or directly to the exterior and function in either intra or extra oral digestion or as external protection or adhesive structures. The intestinal cells of nematodes also have a secretory function.

12.3.5.3. Excretion: The intestine acts as an excretory organ.

12.3.5.4. Defecation: Defecation in nematodes is mechanically controlled and is an intense process.

12.4. BODY CAVITY

The body cavity or pseudocoelom of nematodes differs from that of animals with a true coelomic cavity in that it is not completely lined with tissue of mesodermal origin. It is lined externally by somatic muscle cells, which are mesodermal in origin, and internally by the cells of the alimentary canal, which are ectodermal in origin. The pseudocoelom is filled with fluid which bathes all the internal organs and functions as part of the turgor pressure system on the exoskeleton.

The pseudocoelomic fluid contains protein, glucose, sodium, phosphorus, chloride, potassium, magnesium and small amounts of copper, zinc, iron, hematin and ascorbic acid with a neutral pH.

The pseudocoelom is thought to be lined with a network of fenestrated membranes which help support the visceral organs and yet permit continuity of the pseudocoelomic fluid.

Associated with the pseudocoelomic membranes are large, fixed cells known as coelomocytes. These structures are usually two, four or six in number and are ovoid or many branched. There is some evidence accumulating which suggests that they may have no excretory function.

Nervous, excretory and reproductive systems are present in the pseudocoelom.

12.4.1. Nervous system

The cephalic sense organs of nematodes consist basically of 16 papillae (6 inner labial, 6 outer labial and 4 cephalic papillae) arranged in three circles and a pair of laterally placed amphids (Fig.12.13). The amphids, which occur in all nematodes, occupy a relatively large area of the head and it is thought that they must have a very important sensory function. It has now been established that nematodes do have cilia which are found in the sense organs. Only two or three cilia are found in the papillae compared with as many as 23 in the amphids. These cilia appear to be derived from the nerve endings of the amphidial nerve. They become packed together and lose their cilia-like configuration as they approach the opening of the amphid.

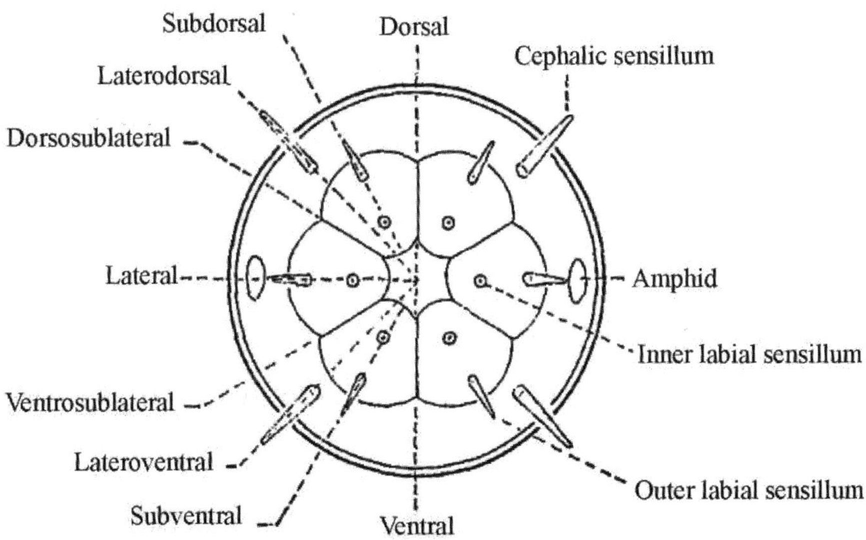

Fig. 12.13. Arrangement of anterior sensilla in nematodes

The nematodes are abundantly supplied with free nerve endings, have a peripheral nervous system, and are sensitive to changes in light and heat in parts of their anatomy which are lacking in complicated sensory structures.

Hemizonids, hemizonions, cephalids, caudalids and phasmids are thought to be sensory structures.

More obvious sensory structures, found in some nematodes are paired pigment spots and ocelli which are shown to have a photoreceptor function.

The caudal sensory organs are associated with reproduction and exhibit sexual dimorphism.

The peripheral nervous system consists of a lattice network of nerves connecting the papillae and setae with each other and with the central nervous system. There are about 250 nerve cells in nematodes. The main nerve centers are around the esophagus and rectum as groups of ganglia (Fig. 12.14). The esophageal nerve ring itself is composed of packed fibers. These two heavily ganglionated areas are connected by longitudinal nerves, the largest being the ventral which really consists of a chain of ganglion; the other principal longitudinal nerves, the dorsal and two laterals, are much smaller.

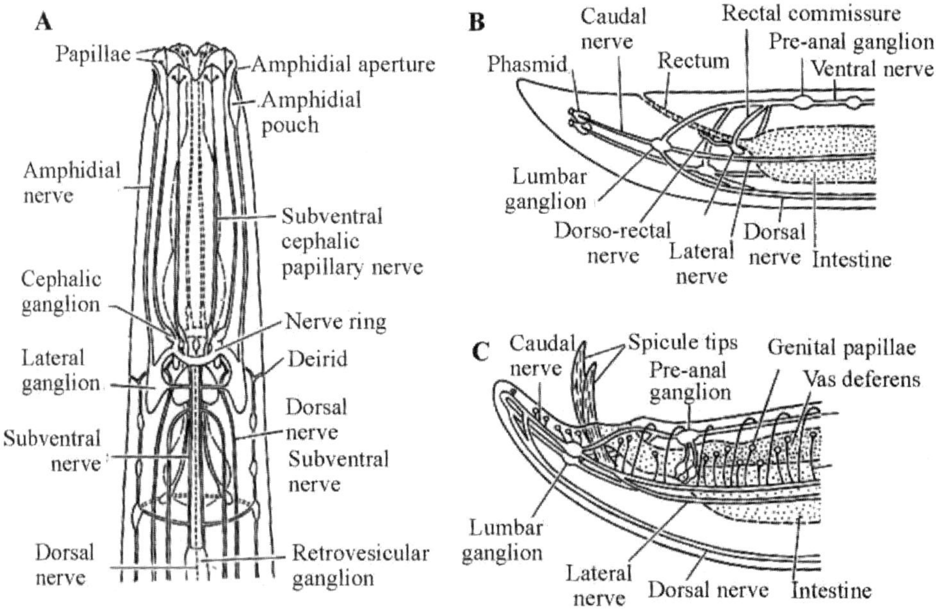

Fig. 12.14. The nervous system. A - Anterior part, B - Posterior part of female nematode. C - Posterior part of male nematode.

Some nerve cells appear to have a neurosecretory function. Studies of sections through neuromuscular connections show that, in certain areas, the muscle and nerve are separated by a distance of only about 50 mμ. These areas have a specialized cytoplasm which is thought to function in impulse transfer.

The responses of nematodes to various stimuli indicate very sensitive receptor systems which must be coupled with a high degree of nervous coordination.

12.4.2. Excretory system

The nematode excretory system has been assigned an excretory function on morphological grounds. It is much reduced in forms such as the Dorylaimoidea. It also appears to lack structures such as flame cells and cilia.

The excretory duct opens to the exterior through an excretory pore that is commonly situated in the anterior mid-ventral line close to the nerve ring. However, its position may vary considerably, even in different stages of the same species. Thus, in *Meloidogyne javanica* the excretory pore is located near the basal region of the median bulb in the infective larva and is found about 25 μ in front of the anterior part of the median bulb in adult females. In *Tylenchulus semipenetrans,* the excretory pore is located in the posterior region but this is a typical and, indeed, its function in this particular instance is secretory rather than excretory (secretes gelatinous matrix).

There are two basic types: a glandular type of the excretory system and a tubular type. The former consists of a large cell in the pseudocoelomic cavity known as the ventral gland, which is connected to the excretory pore by a duct that terminates in an ampulla (Fig. 12.15).

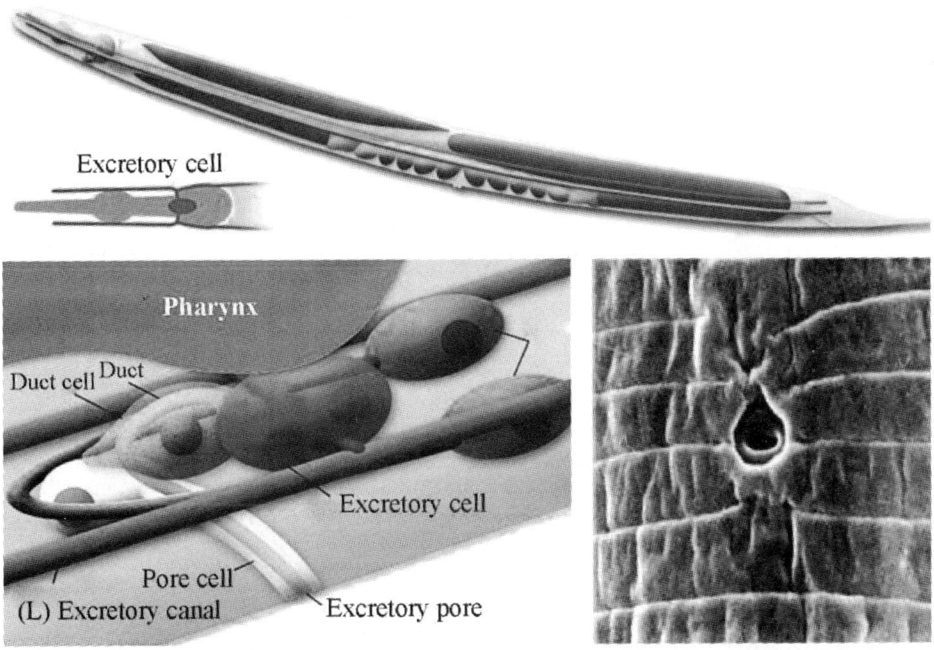

Fig. 12.15. *Top* - Schematic showing the full length of the H-shaped excretory cell as seen from lateral side. In an adult nematode, the excretory canals reach from the nose of the nematode to the tail region. *Bottom left* – Expanded view of excretory pore region. *Bottom right* - Scanning electron micrograph of the external morphology of the excretory pore of a second stage juvenile of a *Meloidogyne* spp.

The excretory system consists of the fused pair of gland cells (blue), the excretory (canal) cell (red), the duct cell (brown) and the pore cell (yellow). Somata of all of these cells are located in the head region. The excretory cell is the largest cell in the nematode and is located juxtaposed to the terminal bulb of the pharynx on the ventral side.

There are several different types of tubular excretory systems. It consists of two long canals in the lateral hypodermal cords connected anteriorly to each other by means of a transverse canal that is connected to the excretory pore by a median duct. This system is known as an H- system for obvious reasons. In other forms there is a reduction of the anterior lateral canals and the system is then known as an inverted U - system. In forms with only one lateral canal the system is asymmetric (Fig. 12.16).

Tylenchid Type:

A single tube runs throughout the body length.

It is found in either of the lateral hypodermal chords.

Excretory sinus connected with canal.

No gland cell.

Excretory duct which opens via excretory pore.

Example : Tylenchids

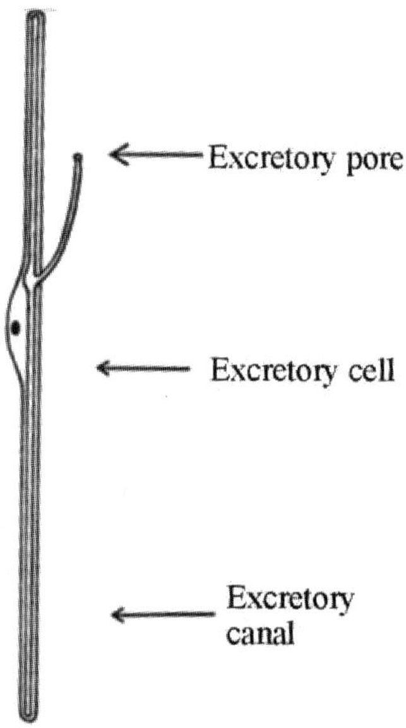

Fig. 12.16. Tubular type of excretory system (Tylenchoid type).

It is clear that a wide range of chemical compounds passes out of the excretory pore, including some of the relatively high molecular weight such as polypeptides and proteins. During molting, these substances may be of very great physiological importance.

12.4.3. Reproductive system

The majority of nematodes are dioecious in which the males are smaller than the females and are equipped with various copulatory organs such as bursae, papillae and spicules. Both hermaphroditism and parthenogenesis have been shown to occur in nematodes. In hermaphroditic forms, the gonad usually produces sperms first and these are stored until they fertilize the eggs which are produced from the same gonad after sperm formation is completed. Parthenogenesis is not uncommon in plant parasitic nematodes and has been reported in various genera including *Meloidogyne, Heterodera, Tylenchulus* and *Hemicycliophora* (van Gundy, 1958). Intersexes are found in some genera, such as *Ditylenchus* and *Meloidogyne*. These are individuals that have a blending of male and female characters. Usually, they are females which show secondary male characters.

12.4.3.1. Male reproductive system: The male reproductive system arises from genital primorida which arise in the larval stages. The stage of development of the genital primordium has proved to be a useful characteristic for differentiating various larval stages. The manner in which the male and female genital primordia develop is often the only reliable way to distinguish males from females before the fourth molt.

Hirschmann (1962) observed the genital primordium in *Ditylenchus triformis* even before the embryo assumed the vermiform shape. The genital primordium consists of a large, central germinal nucleus bordered by two small somatic nuclei. The genital primordium increases in size slightly in the second molt. The anterior somatic nucleus divides several times to give rise to several nuclei which eventually, in turn, give rise to the male gonoduct. The posterior somatic nucleus divides once to give a cap cell nucleus and a somatic nucleus which migrates anteriorly to the germinal nucleus and ultimately to the epithelial cells of the testis. In the rectal region, a large group of small nuclei appear and eventually give rise to the spicules and gubernaculum. At the third molt, the anterior group of cells differentiates into two rows of small spherical cells which have reflexed and grow posteriorly along the ventral side, whereas the germinal nucleus, cap nucleus and epithelial nuclei have moved anteriorly. During the fourth molt, the male gonad is clearly differentiated into testis and vas deferens, the latter increases in length and its posterior end opens into the rectum at the completion of the fourth molt; the spicules and gubernaculum become differentiated.

Adult nematodes usually have a single testis (monorchic), although there may be two (diorchic). The male reproductive system usually consists of three primordial parts: the testis, seminal vesicles and vas deferens (Fig. 12.17). The testis normally consists of two zones, a germinal zone, in which spermtogonial divisions take place; and a growth zone, in which the spermatocytes increase in size. Sperms are formed at the posterior end of the testis and are usually stored in the seminal vesicle. This structure merges into the vas deferens which consists of an anterior glandular region and a posterior muscular region which opens ventrally into the cloaca.

Male nematodes have a pair of copulatory spicules which lie in an invagination of the cloaca and are thought to function in assisting copulation. They are aided in this function by gubernaculum and telamon which guide and deflect the spicules. At copulation, sperms are deposited in the vagina and make their way up the uterus.

(a) Spermatogenesis: Spermatogenesis starts in the germinal zone of the testis and the closely packed spermatogonia enlarges in the growth zone to become spermatocytes and then spermatids. This process involves two maturation divisions and ends with the haploid spermatozoa.

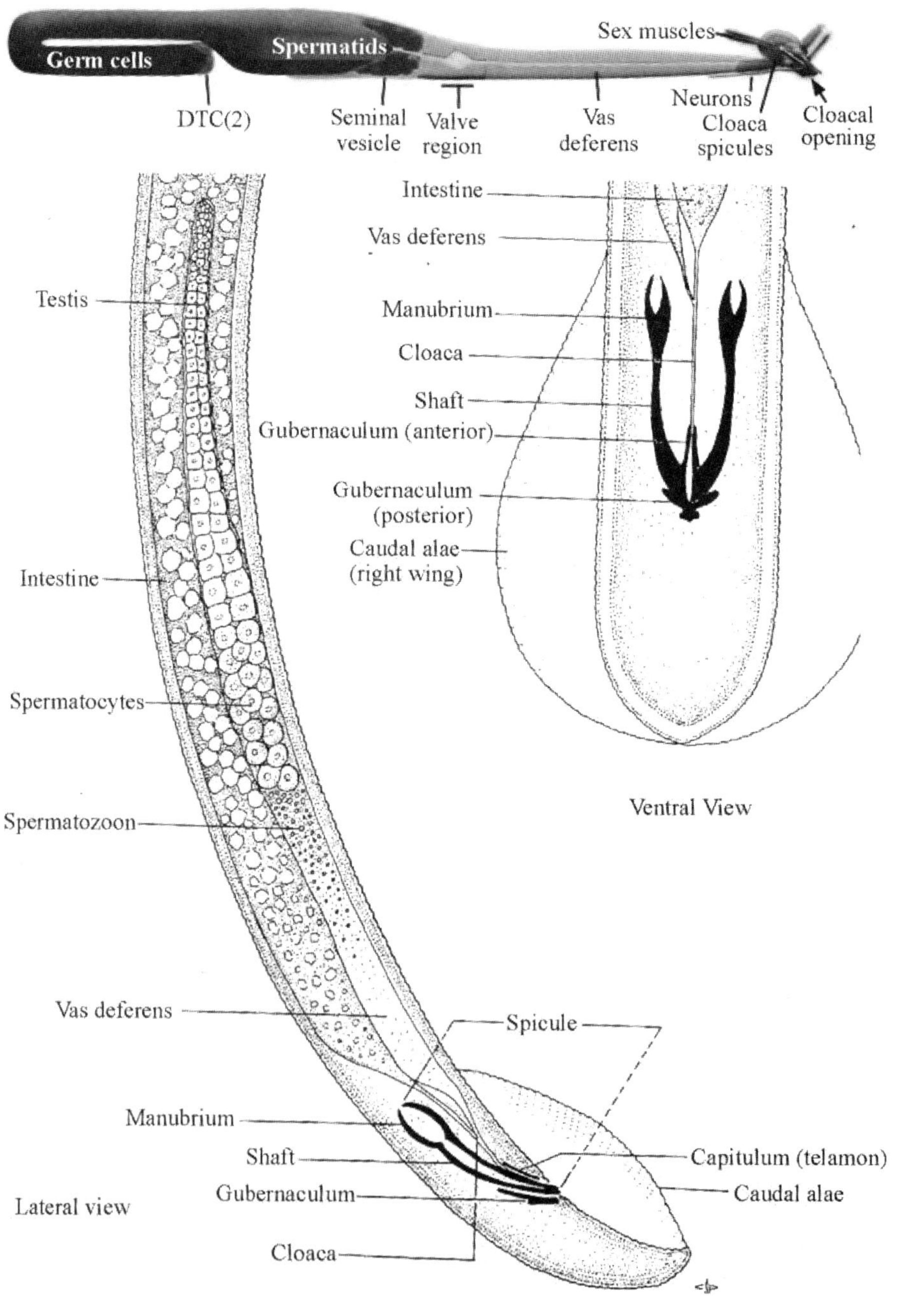

Fig. 12.17. The male reproductive system – Lateral and ventral views

12.4.3.2. Female reproductive system: The nematodes may have a single ovary (monodelphic) or two ovaries (didelphic) (Fig. 12.18). In some forms, the ovaries lie at opposite sides of the vulva, one anterior and the other posterior (amphidelphic). When both ovaries are anteriorly directed they are said to be prodelphic; when both are posteriorly directed they are said to be opisthodelphic.

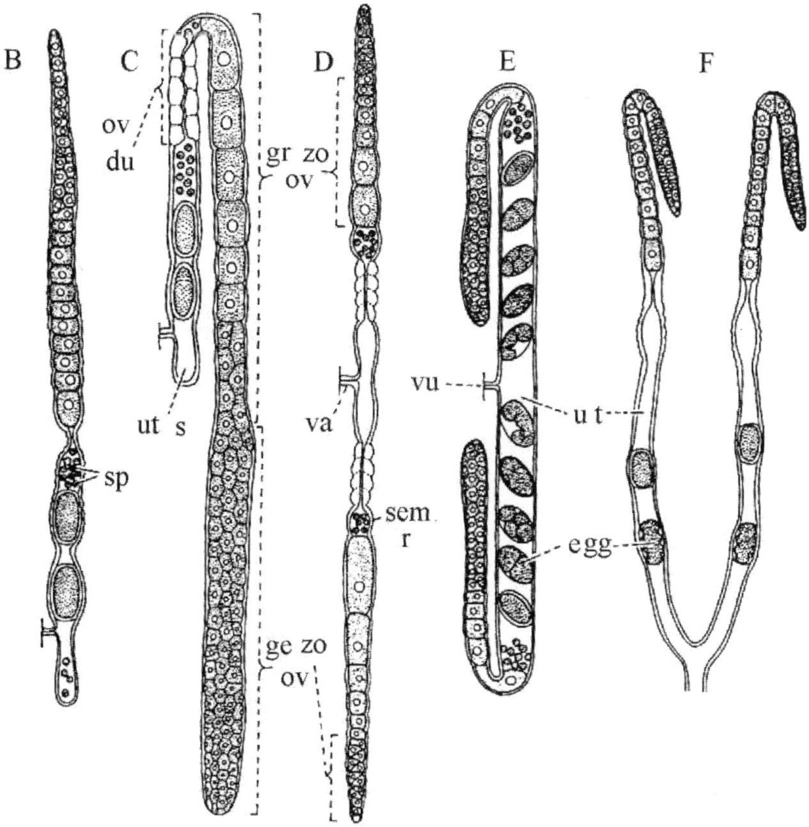

Fig. 12.18. Different types female reproductive system. B - Monodelphic, prodelphic outstretched ovary of *Ditylenchus* sp. C - Monodelphic, prodelphic reflexed ovary of *Panagrolaimus* sp. D - Didelphic, amphidelphic branches opposed ovaries of *Tylenchorhynchus* sp. E - Didelphic, amphidelphic branches opposed reflexed ovaries of *Rhabditis* sp. F - Didelphic, amphidelphic, prodelphic reflexed ovaries of *Meloidogyne* sp. (after Hirschmann, 1962)

Hirschmann (1962) gave a detailed description of the development of a typical monodelphic prodelphic type of female reproductive system in *Ditylenchus triformis* (Fig. 12.19). Sexual differentiation first occurs in the second stage larvae because of the presence of a pair of ventral cord nuclei which divides once to give rise to a group of four nuclei which lie slightly posterior to the genital primordium and eventually give rise to the vagina.

These nuclei are shown at the second molt. At the second molt, the posterior somatic nucleus of the genital primordium divides several times to form nuclei which eventually give rise to the female gonoduct, while the anterior nucleus gives rise to the cap cell nucleus and epithelial nuclei of the ovary. The female third stage larva has a group of small cells at the posterior end of the gonad and there are only a few nuclei in the rectal area. The four ventral cord nuclei enlarges and lie close to the developing gonad. By the third molt, the posterior cell group had increased in size and the ventral cord nuclei have begun to divide. The oviduct starts to form and consists of two rows of six columnar epithelial cells followed by double rows of six cells, which ultimately form the spermatheca. This is followed by a wider part of 16 large cells which form the quadricolumella and this is followed by the uterus and the post vulvar uterine sac which is both formed from a layer of flattened epithelial cells.

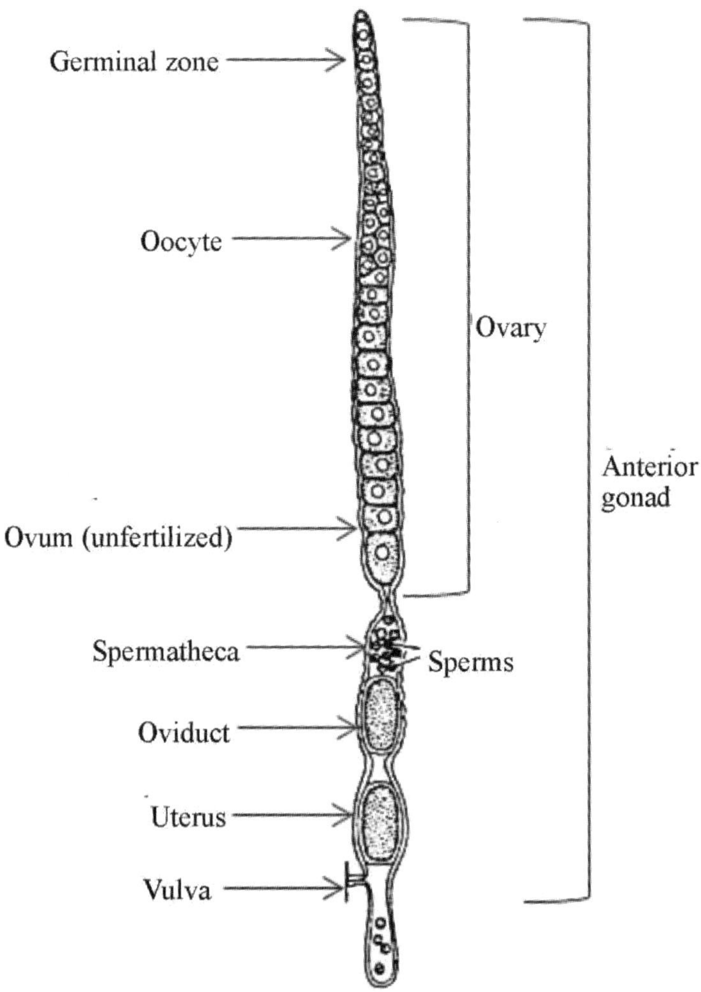

Fig. 12.19. Female reproductive system.

Usually, the ovaries are filled with oocytes attached radially about a central rachis by cytoplasmic bridges. The rachis is anucleate but contains various cytoplasmic inclusions. The oocytes break away from the rachis and come to lie in the oviduct.

When present, spermatheca or sperm storage organ occurs at the junction of the oviduct and the uterus. It appears to function in keeping the sperm in a viable state.

The uterus is often divided into an anterior glandular portion, whose cells are rich in cytoplasmic components, which is indicative of high metabolic and synthetic activity and a posterior muscular portion that is associated with movement of eggs into the vagina, which is also muscular and is lined with cuticle. The eggs are expelled through a vulva that is normally located in the middle of the body.

In parasitic nematodes, there is often extreme sexual dimorphism which results in considerable increases in egg laying capacity associated with an enlarged female reproductive system.

(a) Oogenesis: Oogenesis starts in the germinal zone at the tip of the ovary, involves growth, maturation and fertilization. This takes place in the spermatheca, if present, and is followed by egg development and shell formation.

The sperm enters the egg cell and initiates the formation of a fertilization membrane, contraction of the protoplasm and the start of shell formation. While the egg membranes are being formed, development within the egg takes place. This involves divisions of the egg nucleus by two maturation divisions, a mitotic and then a meiotic division.

12.4.4. Chromosome numbers

The chromosome number observed in plant parasitic nematodes are presented in Table 10.1.

Table 12.1. Chromosome numbers of selected plant parasitic nematodes

Nematode species	Chromosome number
Anguina tritici	$n = 19$
Aphelenchoides composticola	$n = 3$
Ditylenchus dipsaci	$n = 12$
Globodera rostochiensis	$n = 9$
Helicotylenchus erythrinae	$n = 5$
Hemicriconemoides sp.	$n = 5$
Heterodera avenae	$n = 9$
H. oryzae	$n = 9$

Nematode species	Chromosome number
Hoplolaimus galeatus	$n = 10$
Meloidogyne arenaria	$n = 27$
M. graminicola	$n = 18$
Pratylenchus coffeae	$n = 7$
Rotylenchulus reniformis	$n = 9$
Rotylenchus buxophilus	$n = 8$
Tylenchorhynchus claytoni	$n = 8$
Tylenchulus semipenetrans	$n = 5$
Xiphinema index	$n = 10$

Chapter 13

TAXONOMY

13.1. INTRODUCTION

The classification of nematodes is done for convenience, and to enable us to name and catalogue specimens easily; it is, therefore, largely artificial. The classification, however, is difficult and is constantly being changed.

In Tylenchida, taxa are recognized and classified mainly on morphological characters. It is only recently that ecological, biochemical, cytological and embryological characters have been introduced, but still in a very limited manner.

13.2. TAXONOMY OF NEMATODES

Nematodes are placed under the kingdom Animalia. Chitwood and Chitwood (1974) from their studies and review concluded that Nematoda is a sound phylum, properly associated with Rotatoria, Gastotricha and Echinodera (Kinorhynchs). Hyman (1951) revised the phylum Aschelminthes and included 6 classes; Rotifera, Gastotricha, Echinodera, Priapulida, Nematoda and Nematomorpha. Borradiale *et al.* (1961) placed nematodes in the category of a separate phylum Nematoda. Paramonov (1963) proposed the phylum Schizocoeliides to include 4 classes: Gastotricha, Kinorhyncha, Rotatoria and Nematoda. Barnes (1968) in his textbook treats nematodes as a phylum Nematoda.

Thorne (1961) considered the group as a phylum Nemata, divided into two classes, the Secernentea and Adenophorea. Goodey (1963) considered the group to have the status of class, the Nematodea, and he rejects the initial binary grouping.

The apparently relevant distinguishing characteristics of Aschelminthes (Nemathelminthes) are the presence of a pseudocoelom, bilateral symmetry, a protonephridial excretory system or a derivation of this system, cell constancy and adhesive glands. Nematodes lack photonephridial excretory system. Gastotricha, Kinorhyncha and Priapulida lack cell constancy.

There are no profound characters or even a single character that can be used at present to unite all the diverse groups in one phylum. From the diversified form and origin of the many types of body cavities in pseudocoelomate bilateria and the lack of proof of cell constancy or of homology of the excretory system, it is concluded that nematodes are a separate phylum Nematoda.

13.2.1. Types of specimens

While describing a species of a nematode, the following terms are in vogue connected with the specimens:

13.2.2.1. Type specimen: The original specimen from which a description was drawn up. It is the basis for species description and comparison.

13.2.2.2. Holotype: The one specimen possessed by the describer of a species and forming the basis of original diagnosis.

13.2.2.3. Paratype: A specimen belonging to the original series, but not the type selected by the author. It must be erected by the original author.

13.2.2.4. Allotype: A paratype specimen designated in the original description of a species as the type of sex other than that of the holotype.

13.2.2.5. Neotype: A specimen from the original locality where the true type was obtained, which has been lost or destroyed.

13.2.2.6. Lectotype: Specimens of the original series, chosen after the original description to be the type.

13.3. TYPES OF TAXONOMY

13.3.1. Morphological taxonomy

The morphological characters most useful in the classification and identification of plant-parasitic groups includes the cuticle, alimentary canal and an array of mostly sensory cuticular organs that have many diagnostic features useful to taxonomists. One such feature is the stylet, which is found in plant-parasitic and fungal-feeding species, and in some predators. The size and shape of the stylet are related to feeding habit and therefore are also useful for ecologists when assigning trophic groups. The portion of the alimentary canal between the stoma and the intestine is known as the oesophagus (or pharynx) and is of major diagnostic value in many nematode groups. Also, descriptions are given of the numerous features within the reproductive systems of male and female nematodes that are of taxonomic value.

Morphological characters provide more than 90% of the data used in taxonomy and classification of plant and soil nematodes. These characters show considerable variation and care must be taken in using them. Morphometric and allometric measurements often vary greatly under the influence of geographical and ecological conditions. Techniques, types of equipment, methods of observation and personal skills also result in variations. Some characters used to differentiate species of *Pratylenchus*, e.g. stylet knob shape, length of outer margins of cephalic framework, the shape of the spermatheca and tail tip, tend to vary through host influence.

13.3.1.1. Nematode measurements useful in morphological taxonomy: Descriptions of new species should be as detailed as possible and contain all measurements likely to be of taxonomic value. Only a few measurements may be necessary for identification within a particular group. Before making routine measurements, recent appropriate publications should be consulted to see which measurements are of value. The following are commonly used:

n = number of specimens

L = total body length in mm or μ

a = body length ÷ greatest body width

b = body length ÷ distance from anterior end to junction of esophagus and intestine

b' = body length ÷ distance from anterior end to posterior end of esophageal glands (used when glands overlap intestine)

c = body length ÷ tail length (anus or cloaca to tail terminus)

c' - tail length ÷ body width at anus or cloaca

V = distance of vulva from anterior end x 100 ÷ body length

T = distance from cloaca to anterior most part of testis x 100 ÷ body length

Length of stoma or spear in μ

Spicule length and gubernaculum length

Other values, which are sometimes used include:

b_1 = body length ÷ distance from anterior end to base of median esophageal bulb

G_i = Overall length of anterior gonad x 100 ÷ body length

H or h = length of hyaline area in the tail between body contents and cuticle at tail terminus

M = length of anterior part (conical) of spear (in Tylenchida) x 100 ÷ total spear length

O = distance from spear base (in Tylenchoidea) to dorsal esophageal gland outlet x 100 ÷ total spear length

P = distance of phasmid from anus x 100 ÷ tail length (+ = anterior to anus, - = posterior to anus)

Pa = distance of anterior phasmid from anterior end x 100 ÷ body length

Pp = distance of posterior phasmid from anterior end x 100 ÷ body length

S = Spear length ÷ body width at base of spear

The following are useful for some nematodes:

L' = distance from anterior end to anus

V' = distance from anterior end to vulva x 100 ÷ L'

MB = distance to median bulb from anterior end x 100 ÷ total esophageal length

The following are used when describing criconematid nematodes:

R = number of body annules

RB = breadth of one body annule

R St = annules between labial disc and base of stylet knobs

R Oes = annules between labial disc and esophago-intestinal valve

R hem = annules between labial disc and first annule after hemizonid

R ex = annules between labial disc and first annule after excretory pore

RV = number of annules from tail terminus to vulva

R an = number of annules from tail terminus to anus

R Van = annules between vulva and anus

13.3.2. Molecular/ biochemical taxonomy

The approaches to taxonomy have shifted from conventional description of species primarily based on morphological characters to biochemical and molecular parameters. Most of the reputed journals accept taxonomic papers provided with traditional descriptions of species supplemented with biological, biochemical and molecular information (DNA barcoding, 4D microscopy, video vouchering). Access to new technologies provides some new opportunities for nematode identification and quantification and there is an increasing reliance on developing molecular-based diagnostic protocols to identify plant parasitic nematodes. The development of molecular diagnostics can quantify specific plant-parasitic nematode species.

Gel electrophoresis has been widely used as a technique in the identification of plant-parasitic nematode species. Evans (1971) was able to distinguish between *Ditylenchus destructor* and *D. myceliophagus* populations through the study of their esterase patterns as revealed by gel electrophoresis. Different protein band profiles were obtained by using disc electrophoretic separation of soluble proteins and enzymes of some species of *Ditylenchus, Heterodera* and *Meloidogyne* (Dickson *et al.*, 1970). However, these techniques have been used only in a limited manner, although the scope is considerable. Immunological techniques involving DNA hybridization and isoenzyme analysis have been applied to certain aspects of plant nematode taxonomy, although specialized biochemical knowledge, skills and instrumentation are required to use such techniques. The current way forward seems to be via molecular techniques, particularly those employing polymerase chain reaction (PCR) and amplification of DNA.

13.3.3. Cytogenetic taxonomy

Cytological analysis can be utilized in the differentiation of species and pathotypes. Chromosome number has been used in the differentiation between some species of *Meloidogyne* and between *Radopholus similis* and *R. citrophilus*, although in the latter case chromosome number has been shown to be more variable than previously thought and therefore useless as a race marker. Chromosome number also differs between nematode groups, for example, Criconematoidea and Hoplolaimidae have a basic chromosome number of $n = 5$; Dolichodoridae including Belonolaiminae have $n = 8$.

13.4. MORPHOLOGICAL TAXONOMY OF PHYLUM NEMATODA

The outline classification of Phylum Nematoda is presented in Table 13.1.

Table 13.1. Outline classification of Phylum Nematoda

Phylum	Class	Order	Sub-order	Super-family
Nematoda	Secernentea	Tylenchida	Tylenchina	Tylenchoidea
				Heteroderoidea
				Neotylenchoidea
				Atylenchoidea
				Criconematoidea
			Aphelenchina	Aphelenchoidea
	Adenophorea	Dorylaimida	Dorylaimina	Dorylaimodea
				Trichodoroidea

13.4.1. Class Secernentea and Class Adenophorea

The distinguishing characters between class Secernentea and class Adenophorea are presented in Table 13.2.

Table 13.2. Distinguishing characters between Secernentea (Tylenchida) and Adenophorea (Dorylaimida)

Class : Secernentea	Class : Adenophorea
1. Lateral canals opening into the excretory duct	1. Lack lateral canals and excretory duct ends in a cell
2. Amphids labial in position	2. Amphids post labial in position
3. Caudal glands absent	3. Caudal glands present

13.4.2. Order Tylenchida and Order Dorylaimida

The classification of the order Tylenchida proposed by Golden (1971) has been followed. It includes 2 sub-orders, 6 super-families, 23 families, 46 sub-families and 136 genera.

The distinguishing characters between Order Tylenchida and Order Dorylaimida are presented in Table 13.3.

Table 13.3. Distinguishing characters between Order Tylenchida and Order Dorylaimida

Order : Tylenchida	Order : Dorylaimida
1. Three part esophagus	1. Two part esophagus
2. Stomatostylet present	2. Odontostylet present
3. Cuticle has distinct annulations	3. Annulations of the cuticle are not usually visible

13.4.3. Sub-order Tylenchina and Sub-order Aphelenchina

The distinguishing characters between Sub-order Tylenchina and Sub-order Aphelenchina are presented in Table 13.4.

Table 13.4. Distinguishing characters between Sub-order Tylenchina and Sub-order Aphelenchina

Sub-order: Tylenchina	Sub-order: Aphelenchina
1. Dorsal esophageal gland outlet located in procorpus of esophagus, often near base of stylet	1. Dorsal esophageal gland outlet situated in median bulb (Metacorpus) of esophagus
2. Metacorpus less than 3/4 body width	2. Metacorpus more than 3/4 body width

13.5. SUB-ORDER: TYLENCHINA

Key to super-families of Tylenchina

1. Head with setae .... Atylenchoidea
   Head without setae 2
2. Valvular median esophageal bulb absent .... Neotylenchoidea
   Valvular median esophageal bulb present 3
3. Procorpus fused with median bulb, terminal bulb small .... Criconematoidea
   Procorpus separated from median bulb, terminal bulb well developed........... 4

4. Both sexes active, vermiform, usually with only reduced or no sexual dimorphism; eggs deposited outside body not in gelatinous matrix; bursa present, often large .... Tylenchoidea

   Females inactive, swollen, sessile attached to or within plant roots, with pronounced sexual dimorphism; eggs may be retained in body or often deposited in gelatinous matrix; males vermiform, with small bursa or morc oftcn without .... Heteroderoıdea

13.5.1. Superfamily: Tylenchoidea (Table 13.5)

Table 13.5. Families, sub-families and genera of superfamily Tylenchoidea

Family	Sub-family	Genera
Tylenchidae	Dactylotylenchinae Tylodorinae Psilenchinae Sychnotylenchinae Anguininae Tylenchinae Ditylenchinae	*Tylodorus, Psilenchus, Neopsilenchus, Clavilenchus, Basiria, Basiroides, Sychnotylenchus, Neoditylenchus, Paranguina, Auguina, Chitinotylenchus, Aglenchus, Cephalenchus, Malcnchus, Miculenchus, Ditylenchus, Pseudhalenchus*
Tylenchorhynchidae	Tylenchorhynchinae	*Geocenamus, Tylenchorhynchus, Nagelus, Tetylenchus*
	Trophurinae	*Trophurus, Macrotrophurus*
Doilichodoridae		*Dolichodorus* *Brachydorus*
Belonolaimidae	Belonolaiminae	*Carphodorus, Belonolaimus, Morulaimus*
	Telotylenchinae	*Trichotylenchus, Telotylenchus*
Pratylenchidae	Pratylenchinae Radopholinae	*Pratylenchus, Hoplotylus, Hirschmanniella, Zygotylenchus, Radopholus, Pratylenchoides*
Hoplolaimidae	Hoplolaiminae Rotylenchoidinae Aphasmatylenchinae Rotylenchinae	*Rotylenchoides, Acontylus, Aphasmatylenchus, Rotylenchus, Helicotylenchus, Scutellonema, Peltamigratus, Hoplolaimus, Aorolaimus*

Key to families of Tylenchoidea

1. Esophageal glands enclosed in a basal bulb (except *Pseudhalenchus* and in certain *Ditylenchus* species with 1 ovary and small stylet) 2

   Esophageal glands not enclosed in bulb but lobed, overlapping anterior end of intestine 4

2. Head with massive sclerotization; stylet long and well developed; ovaries 2; bursa large, terminal, trilobed .... *Dolichodoridae*

   Head without or with light to moderate sclerotization; stylet weakly to moderately developed; ovaries 1 to 2; bursa adanal or enveloping tail but then not trilobed 3

3. Vulva in posterior part of body (about 60-90%); ovary 1 (except 2 and with median vulva in *Psilenchus* with slit-like amphid apertures); female tail attenuated, fusiform, clavate or elongate conoid, very rarely short and stout .... *Tylenchidae*

   Vulva always median; ovaries 2 (except 1 in *Trophurus* and then tail cuticle thick and swollen); female tails cylindrical, conoid, with rounded, swollen or acute terminus .... *Tylenchorhynchidae*

4. Female head low, broad at its base, and rounded or flattened anteriorly; about 1/2 - 3/5 as wide as length of stylet .... *Pratylenchidae*

   Female head high or elevated, not so broad in relation to its length; convex conoid, hemi-spherical to broadly rounded; sometimes being knob-like, with or without lobes 5

5. Female tail not more than 2 anal body widths in length, generally less; phasmids may be small, pore-like and in anal region or large and variable in position on body or absent......................................*Hoplolaimidae*

   Female tail at least 2 anal body widths in length, usually more; phasmids always small, pore like and located on tail .... *Belonolaimidae*

13.5.1.1. *Family: Dolichodoridae (Only subfamily: Dolichodorina)*

Key to genera

Stylet over 60 μ female tail rounded to conically pointed ....*Dolichodorus*

Stylet very strong, much less than 60μ female tail attenuated ....*Brachydorus*

13.5.1.2. *Family: Tylenchidae*

Key to Subfamilies of Family Tylenchidae

1. Body cuticle exceptionally thick; distinct cephalic framework digitate ....*Dactylotylenchinae*

   Body cuticle thin; cephalic framework normal shape if present 2

2. Stylet very elongate and over 75 μ in length .... *Tylodorinae*

   Stylet small, less than 30 μ and usually about 10-20μ 3

3. Amphid apertures silt like post-labial, easily visible; stylet with or without basal knobs; ovaries 1 or 2 ....*Psilenchinae*

   Amphid aperture pore like, labial, difficult to see; stylet always with basal knobs; only 1 ovary 4

4. Female tail short and stout, rounded; vulva at about 90% bursa enveloping tail (insect associates) .... *Sychnotylenchinae*

   Female tail elongate, conoid, attenuated, filliform; vulva generally at about 60-85%; bursa adanal...5

5. Species about 1.5-4.0 mm, stout or obese; ovary often extending to basal bulb or beyond, with 1 or 2 flextures; oocytes generally in multiple rows arranged about a rachis, seldom only in 2 rows .... *Anguininae*

   Species about 0.5 - 1.5 mm, not stout; ovary usually straight, without flexture; oocytes in single row, rarely in 2; rachis absent 6

6. Esophageal glands in well-defined basal bulb, pyriform to clavate, with distinct cardia; Vulva usually about 60-70%; tail long tapering, often filliform .... *Tylenchinae*

   Esophageal glands may be either in basal bulb, sometimes expanded and with small lobe extending over intestine or not enclosed, overlapping intestine; cardia absent or reduced; vulva about 75-85% tails elongate conoid .... *Ditylenchinae*

Key to genera of Family Tylenchidae

1. Head in form of high lobes, with annulation .... *Dactylotylenchus*

   Head normal and continuous, without annulation .... *Tylodorus*

2. Ovaries 2; stylet basal knobs absent .... *Psilenchus*

   Ovary 1; stylet with or without basal knobs 3

3. Basal knobs of stylet absent .... *Neopsilenchus*

   Basal knobs of stylet present 4

4. Terminally tail clavate .... *Clavilenchus*

   Terminally tail finely rounded or pointed 5

5. Median bulb posterior to or at middle of esophagus .... *Basiria*

   Median bulb well anterior to middle of esophagus .... *Basiroides*

6. Excretory pore at level of or anterior to median bulb .... *Sychnotylenchus*

   Excretory pore posterior to median bulb, usually near nerve ring or basal bulb .... *Neoditylenchus*

7. Esophageal glands 4, appearing forked basally and lying about halfway between median bulb and base of esophagus, forming odd-shaped basal bulb .... *Paranguina*

   Esophageal glands 3, enclosed in somewhat swollen but well defined bulb basally abutting intestine 8

8. Oocytes in multiple rows arranged about a rachis; occurring in seeds or galls on above-ground plants parts .... *Auguina*

   Oocytes in 2 rows, not on rachis; occurring in root galls of graminae .... *Subanguina*

9. Base of stylet furcate (inverted Y) .... *Chitinotylenchus*
   Base of stylet normal, not furcate 10
10. Lateral vulvar membranes present .... *Aglenchus*
   Lateral vulvar membranes absent 11
11. Head prominent, set off by constriction; lateral field with 6 incisures .... *Cephalenchus*
   Head continuous or slightly offset; lateral field with 4 or less incisures, sometimes as plain band 12
12. Body narrowing markedly immediately posterior to vulva .... *Malenchus*
   Body only tapering normally posterior to vulva 13
13. Female tail usually curved or hooked ventrally; bursa present .... *Tylenchus*
   Female tail straight; bursa absent .... *Miculenchus*
14. Esophageal glands as distinct basal bulb, sometimes variously expanded, with a small lobe projecting over anterior end of intestine in some species...*Ditylenchus*
   Esophageal glands as a large, elongate lobe overlapping anterior end of intestine ....*Pseudhalenchus*

13.5.1.3. *Family: Tylenchorhynchidae*

Key to Subfamilies of Family Tylenchorhynchidae

Cuticle strongly annulated, typically not thickened or swollen on tail; amphid apertures labial; always 2 ovaries .... *Tylenchorhynchinae*
Cuticle weakly annulated and thickened or swollen on tail; amphid apertures post-labial; 1 ovary (if 2, the stylet over 90 μ) .... *Trophurinae*

Key to Genera of Family: Tylenchorhynchidae

1. Stylet guide tubular throughout its length, capped anteriorly with perioral disc; stylet exceedingly slender, fragile .... *Geocenamus*
   Stylet guide lyre shaped; perioral disc absent; stylet well developed............... 2
2. Cephalic framework present, often with sclerotization; bursa terminal, enveloping tail .... *Tylenchorhynchus*
   Cephalic framework absent; bursa sub-terminal 3
3. Stylet muscles attached to inner cephalic wall anterior to base of head; female tail cylindroid; tapering to rounded ....*Nagelus*
   Stylet muscles attached to band at base of head; female tails conoid, with acute, subacute or finely rounded terminus ....*Tetylenchus*
4. Ovary 1 (but vulva still near median); stylet small, about 15 - 30 μ ... *Trophurus*
   Ovaries 2; stylet very slender, around 90 μ or more .... *Macrotrophurus*

13.5.1.4. *Family: Pratylenchidae*

Key to Subfamilies of Family Pratylenchidae

Ovary 1 .... *Pratylenchinae*

Ovaries 2 .... *Radopholinae*

Key to Genera of Family Pratylenchidae

1. Lobe of esophageal glands overlapping intestine on ventral side; males not degenerate .... *Pratylenchus*

   Lobe of esophageal glands overlapping intestine on dorsal side; male degenerate 2

2. Stylet less than 20 μ; female head very low and flattened .... *Radopholoides*

   Stylet about 21-26μ; head not as low, conoid to the flattened labial surface .... *Hoplotylus*

3. Large nematodes, about 1-4 mm in length; phasmids in posterior third of tail, tail terminus pointed or with mucro .... *Hirschmanniella*

   Small nematodes; about 0.5-1.0 mm; phasmids in middle or anterior portion of tail; tail terminus rounded 4

4. Lobe of esophageal glands overlapping intestine on ventral side.. *Zygotylenchus*

   Lobe of esophageal glands overlapping intestine on dorsal side...................... 5

5. Esophageal gland lobe large, elongate; deirids absent; phasmids in anterior part of tail; Males degenerate .... *Radopholus*

   Esophageal gland lobe short and oblique basally; deirids very distinct; males not degenerate.. *Pratylenchoides*

13.5.1.5. *Family: Hoplolaimidae*

Key to Subfamilies of Family Hoplolaimidae

1. Ovary 1; vulva in posterior fourth of body .... *Rotylenchoidinae*

   Ovaries 2; vulva about median or nearly so 2

2. Phasmids absent; esophageal lobe narrow and on ventral side .... *Aphasmatylenchinae*

   Phasmids present; esophageal lobe dorsal or lateral 3

3. Phasmids small, pore like, in anal region.................................... *Rotylenchinae*

   Phasmids large, variable in position on body .... *Hoplolaiminae*

Key to Genera of Family Hoplolaimidae

1. Outlet of dorsal esophageal gland less than 1/2 stylet length from stylet base; females without any swelling in posterior portion.. *Rotylenchoides*

Outlet of dorsal esophageal gland at least 1/2 stylet length from base of stylet; female with slight swelling posteriorly .... *Acontylus*

2. Amphid apertures elongate, labial; stylet rather slender .... *Aphasmatylenchus*

3. Outlet of dorsal esophageal gland near base of stylet (1/2 or less of stylet length); large portion of esophageal lobe overlapping intestine usually on dorsal side .... *Rotylenchus*

   Outlet of dorsal esophageal gland further back, usually 1/2 or more of stylet length from base of stylet; largest portion of esophageal lobe overlapping intestine generally on ventral side .... *Helicotylenchus*

4. Large phasmids opposite each other or nearly so and located on tail or in immediate anal area .... *Scutellonema*

   Large phasmids not opposite each other and not in immediate and region 5

5. Both large phasmids located in posterior half of body but closest one still well anterior to anal area; head without striation .... *Peltamigratus*

   Large phasmids commonly in different halves of body; head with transverse and often longitudinal striae 6

6. Moderate to large size nematodes (1-2 mm), with very strong stylet, basal knobs with anterior projections, cephalic framework massive .... *Hoplolaimus*

   Small to moderate size forms (0.7-1.2 mm), with moderate stylet, basal knobs without Anterior projections; cephalic framework strong but not massive .... *Aorolaimus*

13.5.1.6. *Family: Belonolaimidae*

Key to Subfamilies of Family Belonolaimidae

Head usually prominent, often deeply offset and with labial disc; stylet over 60 μ .... *Belonolaiminae*

Head smaller, well offset or continuous, without labial disc; stylet less than 50 μ .... *Telotylenchinae*

Key to Genera of Family Belonolaimidae

1. 2 incisures in lateral field; hemizonid posterior to excretory pore .... *Carphodorus*

   Always more or less than 2 incisures in lateral field; hemizonid usually anterior to excretory pore 2

2. Lateral field represented by only 1 incisure ....*Belonolaimus*

   Lateral field with 3 incisures... .... *Morulaimus*

3. Base of stylet diverging or forked, with 3 rounded knobs; lateral field with only 3 incisures .... *Trichotylenchus*

Base of stylet normal not diverging; and with basal knobs; lateral field with at least 4 incisures .... *Telotylenchus*

13.5.2. Superfamily: Heteroderoidea (Table 13.6)

Table 13.6. Families, sub-families and genera of superfamily Heteroderoidea.

Family	Sub-family	Genera
Heteroderidae	Heterodorinae Meloidoderinae Meloidogyninae	*Globodera, Heterodera, Meloidoderita, Meloidodera, Cryphodera, Meloidogyne, Hypsoperine*
Nacobbidae	Nacobbinae	*Nacobbus*
	Rotylenchulinae	*Rotylenchulus*

Key to families of Heteroderoidea

Females becoming cysts or remaining soft-bodies, pyriform saccate, spheroid or lemon shaped; cephalic framework of females not sclerotized; males cylindroid, well developed and without bursa .... *Heteroderidae*

Female bodies soft, saccate, saccate elongate or kidney shaped; cephalic framework of female's sclerotized; males vermiform, sometimes degenerate anteriorly and with small but distinct bursa .... *Nacobbidae*

13.5.2.1. Family: Heteroderidae

Key to Sub-families of Family Heteroderidae

1. Female body becoming a hard, tough cyst .... *Heteroderinae*
   Female body not becoming hard but remaining relatively soft 2
2. Body cuticle very thick, with sub-cuticular punctuations or needle like spines on cuticle; vulva well removed from anus and located subequatorial to terminal; some or all eggs retained in body .. *Meloidoderinae*
   Body cuticle very thin to moderately thick, without consistent punctuation or spines; vulva and anus close together, being terminal or nearly so; eggs in gelatinous matrix outside body ....*Meloidogyninae*

Key to Genera of Family Heteroderidae

1. Cysts round to globular; cyst wall variously patterned, often being zig zag or wavy lined or network .... *Globodera*
   Cysts broadly pyriform or lemon-shaped .... *Heterodera*
2. Body cuticle with small, needlelike spines ....*Meloidoderita*

Body cuticle without spines, only with sub-cuticular punctuation 3

3. Vulva sub-equatorial to sub-terminal; vulva/anus distance about 100-300 μ; larvae with 4 lines in lateral field .... *Meloidodera*

   Vulva terminal or nearly so; vulva/anus distance about 45-70 μ; larvae with 3 lines in Lateral field .... *Cryphodera*

4. Adult females typically pyriform with elongate, tapering neck basically in line with length-wise median body axis; female cuticle moderately thin *Meloidogyne*

   Adult females elongate ovoid to lemon-shaped or oval, with small well defined neck usually situated to one side; female cuticle generally very thick, but sometimes extremely thin .... *Hypsoperine*

13.5.2.2. Family: Nacobbidae

Key to Subfamilies of Family Nacobbidae

Vulva sub-terminals; 1 ovary .... *Nacobbinae*

Vulva post-equatorial; 2 ovaries ....*Rotylenchulinae*

Key to Genera of Family Nacobbidae

Outlet of dorsal esophageal gland close to base of stylet; male not degenerate in anterior portion, with strong stylet and with bursa enveloping tail ....*Nacobbus*

Outlet of dorsal esophageal gland over 1/2 stylet length from base of stylet; males degenerate in anterior portion, with weak stylet and with bursa adanal .... *Rotylenchulus*

13.5.3. Superfamily: Neotylencoidea (Table 13.7)

Table 13.7. Families, sub-families and genera of superfamily Neotylencoidea

Family	Sub-family	Genera
Ecphyadophoridae	Ecphyadophorinae	*Ecphyadophora, Ecphyadophoroides*
Neotylenchidae	Neotylenchinae	*Deladenus, Gymnotylenchus, Hexatylus*
Paurodontidae	Paurodontinae	*Paurodontus, Stictylus*
Nothotylenchidae		*Nothanguina, Nothotylenchus, Boleodorus, Boleodoroides, Dorsalla, Halenchus, Thada*

Key to families of Neotylenchoidea with plant nematodes

1. Extremely slender forms, with a = 75 or more; body of male narrowing markedly and abruptly at cloaca......................... *Ecphyadophoridae*

Vermiform nematodes with a = much less than 75; males not narrowing posteriorly as above 2

2. Cephalic framework in 3 sectors .... *Neotylenchidae*

Cephalic framework in 6 sectors 3

3. Esophageal basal bulb with stem-like extension posteriorly........ *Paurodontidae*

Basal bulb without stem-like extension, joining intestine by obscure valvular apparatus .... *Nothotylenchidae*

13.5.3.1. Family: Ecphyadophoridae (Only Sub-family Ecphyadophorinae)

Key to genera of Sub-family Ecphyadophorinae

Head dorso-ventrally flattened; female body not abruptly narrowed at vulva .... *Ecphyadophoroides*

Head not flattened; female body narrowed abruptly at vulva .... *Ecphyadophora*

13.5.3.2. Family: Neotylenchidae (Only Sub-family Neotylenchinae)

Key to genera of Sub-family Neotylenchinae

1. Esophageal glands forming long overlapping lobe; basal bulb with valveless chamber; male with gubernaculum and bursa .... *Deladenus*
2. Esophageal glands forming long overlapping lobe; basal bulb without valveless chamber, male without gubernaculum and bursa .... *Gymnotylenchus*
3. Spear knobs bifid; no distinct separation between esophagus and intestine........ *Hexatylus*

13.5.3.3. Family: Paurodontidae (Sub-family: Paurodontinae)

Key to genera of Sub-family: Paurodontinae

Tail long; spear knobs symmetrical; bursa adanal .... *Paurodontus*

Tail short, spear knobs asymmetrical; bursa nearly wholly enveloping tail .... *Stictylus*

13.5.3.4. Family: Nothotylenchidae

Key to genera of Family Nothotylenchidae

1. Female obese, resembling *Anguina*; gall forming; bursa not enveloping tail tip .... *Nothanguina*
2. Female vermiform; body straight in death; basal bulb distinct and swollen; spear knobs round; tail short, subacute; bursa extending to about middle of tail........ *Nothotylenchus*

3. As 2, but spear knobs flange like; base of basal bulb behind middle of esophagus; lateral lips much smaller than submedian ones; tail bent to ventral side, tip round or subacuate; post-uterine sac present .... *Boleodorus*
4. As 3, but base of basal bulb anterior to middle of esophagus; tail outstretched, tip clavate .... *Boleodoroides*
5. As 3, but body bent dorsally in death, lateral lips not smaller than submedian ones; post uterine sac absent; vulva heavily cuticularized.. *Dorsalla*
6. Basal bulb with valveless chamber, tail tip hooked; bursa adanal ......*Halenchus*
7. Cuticle very thick, with prominent striae; bursa adanal; body small, stout .... *Thada*

13.5.4. Superfamily: Atylenchoidea (Table 13.8)

Table 13.8. Families, sub-families and genera of superfamily Atylenchoidea

Family	Sub-family	Genera
Atylenchidae	Atylenchinae	*Atylenchus*, *Eutylenchus*

13.5.4.1. Family: Atylenchidae (only Sub-family Atylenchinae)

Key to genera of Sub-family Atylenchinae

Lip region continuous, annulated, males without bursa ....*Atylenchus*
Lip region offset, not annulated, males with bursa ....*Eutylenchus*

13.5.5. Superfamily: Criconematoidea (Table 13.9)

Table 13.9. Families, sub-families and genera of superfamily Criconematoidea

Family	Sub-family	Genera
Criconematidae	Criconematinae	*Criconemoides, Hemicriconemoidies*
	Hemicyliophorinae	*Criconema, Bakernema, Hemicycliophora, Caloosia*
Paratylenchidae		*Cacopaurus, Paratylenchus, Gracilacus*
Tylenchulidae	Tylenchulinae	*Tylenchulus, Triphotylenchulus*
	Sphaeronematinae	*Sphaeronema, Trophonema*

Key to families of Criconematoidea

1. Female body cuticle heavily annulated, the large annules sometimes with spines, scale like or membranous projections extending posterior; a cuticular sheath present in some forms ....*Criconematidae*

Female body cuticle finely annulated, the small annules without spines or projections; cuticular sheath never present..2

2. Stylet of female well developed and elongate; female body vermiform, slender and less frequently plump to obese .... *Paratylenchidae*

Stylet of female well developed but very short, averaging around 15 μ; female body subspherical, saccate or elongate saccate ..... *Tylenchulidae*

13.5.5.1. Family: Criconematidae

(i) Key to Subfamilies

Female rather stout, often fusiform, usually with less than 200 annules; stylet knobs typically anchor-shaped .... *Criconematinae*

Female more elongate, with annules numbering about 200-400; stylet knobs spheroid, sloping posteriorly ....*Hemicyliophorinae*

(ii) Key to Genera

1. Adult female with prominent cuticular sheath .... *Hemicriconemoidies*

Prominent cuticular sheath absent 2

2. Annules without prominent projection posteriorly .... *Criconemoides*

Annules with projections extending posteriorly 3

3. Very coarse annules with cuticular spines or scale like projections extending posteriorly.. *Criconema*

Very coarse annules with rather large, hyaline, membranous structures on their posterior margins tending to fade or disappear in glycerin .... *Bakernema*

4. Adult female with cuticular sheath ....*Hemicycliophora*

Adult female without cuticular sheath .... *Caloosia*

13.5.5.2. Family: Paratylenchidae (Only subfamily: Paratylenchinae)

Key to genera of Sub-family: Paratylenchinae

1. Female obese, with cuticular ornamentations and very short, blunt tail.. *Cacopaurus*

Female slender to plump or obese, without cuticular ornamentations; tail short and conical to elongate 2

2. Female stylet 36 μ or less; excretory pore always near nerve ring or posterior to it ....*Paratylenchus*

Female stylet 45-120 μ; excretory pore typically anterior to nerve ring, often near valves of median bulb ....*Gracilacus*

13.5.5.3. Family: Tylenchulidae

Key to Subfamilies

Excretory pore well posterior to nerve ring; males with weak stylet .... *Tylenchulinae*

Excretory pore near nerve ring; males without stylet ....*Sphaeronematinae*

Key to Genera

1. Excretory pore subequatorial; outlet of dorsal esophageal gland about a stylet length from base of stylet .... *Tylenchulus*

   Excretory pore anterior to midbody, being 33-44% from anterior end; outlet of dorsal esophageal gland about 1/2 stylet length from base of stylet ....*Trohotylenchulus*

2. Females subspherical ....*Sphaeronema*

   Females much more elongate, obese .... *Trophonema*

13.6. SUB-ORDER: APHELENCHINA

13.6.1. Superfamily: Aphelenchoidea (Table 13.10)

Table 13.10. Families, sub-families and genera of superfamily Aphelenchoidea

Family	Sub-family	Genera
Aphelenchidae		*Aphelenchus*
Aphelenchoididae	Aphelenchoidinae	*Aphelenchoides, Bursaphelenchus, Megadorus*
	Rhadinaphelenchinae	*Rhadinaphelenchus*
	Seinurinae	*Seinura, Paraseinura*
Paraphelenchidae		*Paraphelenchus, Metaphelenchus*
Anomyctidae		*Anomyctus*

Key to families of Aphelenchoidea

1. Head with prominent, bowl-shaped frontal disc ....*Anomyctidae*

   Head normal, without frontal disc 2

2. Esophageal glands enclosed in a basal bulb .... *Paraphelenchidae*

   Esophageal glands lobed, overlapping intestine 3

3. Lateral field with 6-14 striae at midbody; male with costate bursa and gubernaculum; spicules slender, arcuate; ....*Aphelenchidae*

Lateral field with usually less than 6 striae; male commonly without, but may have a non- costate bursa; gubernaculum absent; spicules typically thorn-shaped .... *Aphelenchoididae*

Key to Genera

1. Body extremely slender (a = about 100); male with small terminal bursal flap .... *Rhadinaphelenchus*
2. Body less slender (a = 60 or less); tail short, conoid, often mucronate; male tail without bursal flap .... *Aphelenchoides*
3. As 2, but male tail with bursal flap, insect associates .... *Bursaphelenchus*
4. As 2, but tail long, filliform, without mucro .... *Seinura*
5. As 3, but anterior conoid part of spear jointed ....*Paraseinura*
6. Spear heavy, with large round basal knobs .... *Megadorus*

13.7. SUB-ORDER: DORYLAIMINA

13.7.1. Superfamily: Dorylaimoidea (Table 13.11)

The classification of Ferris (1971) is followed.

Table 13.11. Families, sub-families and genera of superfamily Dorylaimoidea

Family	Sub-family	Genera
Longidoridae		*Longidorus, Xiphinema, Paralongidorus*

13.7.1.1. Family: Longidoridae

All species have a greatly attenuated spear with long extensions, the posterior enlarged part of the esophagus reduced in length, and similar tails in male and female.

(i) *Longidorus:* Large pouch like amphids, opening minute slit; spear extensions are not flanged; guiding ring located near the spear tip.

(ii) *Xiphinema:* Amphids funnel shaped, opening wide; spear extensions bearing bursal flanges; guiding ring located near the spear base.

(iii) *Paralongidorus*: Amphids funnel-shaped, opening wide; spear extensions are not flanged; guiding ring located near the spear tip.

13.7.2. Superfamily: Trichodoroidea (Table 13.12)

The classification of Loof (1975) is followed.

Table 13.12. Families, sub-families and genera of superfamily Trichodoroidea

Family	Sub-family	Genera
Trichodoridae		*Trichodorus, Paratrichodorus*

13.7.2.1. Family: Trichodoridae

The onchiostyle is long, curved, without basal knobs of distinct apical sclerotizations; it is embedded in muscular sheath. The male supplements are well developed. The female rectum runs almost parallel to the longitudinal body axis and the anus lies sub-terminally.

(i) *Trichodorus*: Caudal alae absent, male tail in death curved ventral, suspensor muscle conspicuous (oval to circular), vaginal sclerotization strong, cuticle not swollen after fixation, lateral pores present near vulva.

(ii) *Paratrichodorus*: Caudal alae present, male tail in death straight, suspensor muscle inconspicuous, vaginal sclerotization weak, cuticle swollen after fixation, lateral pores near vulva absent.

13.8. MOLECULAR/BIOCHEMICAL TAXONOMY

Numerous molecular techniques have been developed that are capable of identifying and quantifying nematodes at the species level and below. Techniques such as isoenzyme pattern analysis, restriction fragment length polymorphism (RFLP) analysis, random amplified polymorphic DNA (RAPD) analysis, polymerase chain reaction (PCR), quantitative polymerase chain reaction (qPCR), and sequencing of diagnostic rDNA regions have all been used successfully to identify and quantify several agriculturally important plant-parasitic nematodes. These methods have their own advantages and limitations (Table 13.13).

Table 13.13. Advantages, limitations, and applications of molecular methods in nematode identifications.

Method	Use	Limitations	Application example
Allozymes	- First biochemical method in nematode taxonomy - Provides comparable data between geographical isolates	- Lack of confirmed genetic basis - Cannot be used for single nematode studies - Life stage specific - Lacks data on specific characters - Requires non-denatured proteins	- Nematode genera identified using this method includes: *Meloidogyne, Heterodera, Pratylenchus, Ditylenchus* and *Aphelenchus*. (Esbenshade & Triantaphyllou 1990; Navas *et al*., 2001).

Method	Use	Limitations	Application example
Restriction fragment length polymorphism (RFLP)	- Suited to differentiate between closely related taxa based on presence/ absence of restriction fragment bands	- Lacks homology of characters - Requires large amounts of PCR products to use for different restriction enzymes.	-Detection of species or populations within species (Castagnone-Sereno *et al.*, 1993; Fargette *et al.*, 1996).
Amplified fragment length polymorphism (AFLP)	- Suited to assess variation among individuals of the same species	- Lengthy procedure	- Detection of species in *Heterodera avenae* group (Subbotin *et al.*, 1999), molecular characterization of *Pratylenchus* species (Waeyenberge *et al.*, 2000) and the study of intraspecific variation in *Radopholus similis* (Elbadri *et al.*, 2002).
Random amplified polymorphic DNA (RAPD)	- Unlike AFLP this method doesn't require a restriction step - Simple and rapid	- Not suitable to all organisms and lacks reproducibility - Sensitive to variations in primer and DNA concentration	- Detection of *Globodera rostochiensis, G. pallida, Meloidogyne incognita, M. javanica*, and *M. arenaria*. (Fullaondo *et al.*, 1999; Zijlstra *et al.*, 2000)
Real-time PCR	- Qualitative and quantitative detection of species - Rapid and precise	- Requires species specific primers - Multiple species detection requires lengthy optimization	- Detection of potato cyst nematode (*G. pallida*) and beet cyst nematode (*Heterodera schachtii*), quantification of *Meloidogyne* spp. from tomatoes, *Pratylenchus thornei* and *Pratylenchus neglectus*, detection of *Bursaphelenchus xylophilus* (Madani *et al.*, 2005, Stirling *et al.*, 2004, Hollaway *et al.*, 2004).
Multiplex PCR	- Detects more than one species at a time - Cost effective	- Involves optimization of reagent concentrations,	- Identification of *Globodera pallida* and *G. rostochiensis*

Method	Use	Limitations	Application example
	- Rapid	variation in primer annealing temperatures, primer interactions, and amplification of non-specific products - Requires species-specific primers	(Fullaondo *et al.*, 1999), *Pratylenchus penetrans* and *P. scribneri* (Setterquist *et al.*, 1996), *Meloidogyne chitwoodi* and *M. fallax* (Petersen *et al.*, 1997), and *P. coffeae* and *P. loosi* (Uehara *et al.*, 1998).
DNA sequencing	- Preferred from the above mentioned methods as factors that affect patterns in these methods (e.g. variation in primer and DNA concentration, DNA template quality, gel electrophoresis, and the type of DNA polymerase) can be controlled and the sequencing step can be optimized. - Fast and accurate	- Difficult in finding an ideal gene for taxonomic identification as well as phylogenetic inference, that works in all nematode groups - Choice of marker is still an open issue	Widely used in recent years; in various groups including family Hoplolaimidae, (Subbotin *et al.*, 2007), order Tylenchida (Subbotin *et al.*, 2006), suborder Criconematina (Subbotin *et al.*, 2005), *Pratylenchus* (Al-Banna *et al.*, 1997), *Steinernema* (Stock *et al.*, 2001), *Meloidogyne* (Castillo *et al.*, 2003), *Longidorus* (Mekete *et al.*, 2009).
DNA sequencing	- Preferred from the above mentioned methods as factors that affect patterns in these methods (e.g. variation in primer and DNA concentration, DNA template quality, gel electrophoresis, and the type of DNA polymerase) can be controlled and the sequencing step can be optimized. - Fast and accurate	- Difficult in finding an ideal gene for taxonomic identification as well as phylogenetic inference, that works in all nematode groups - Choice of marker is still an open issue	Widely used in recent years; in various groups including family Hoplolaimidae, (Subbotin *et al.*, 2007), order Tylenchida (Subbotin *et al.*, 2006), suborder Criconematina (Subbotin *et al.*, 2005), *Pratylenchus* (Al-Banna *et al.*, 1997), *Steinernema* (Stock *et al.*, 2001), *Meloidogyne* (Castillo *et al.*, 2003), *Longidorus* (Mekete *et al.*, 2009).

Method	Use	Limitations	Application example
Denaturing gradient gel electrophoresis (DGGE)	- Easy and straight forward approach for nematode community analysis - Used to differentiate multiple species in a sample	- DNA extraction for DGGE problematic	– Community analysis in different ecosyst-ems (Griffiths *et al.*, 2006; Shi-Bin *et al.*, 2008)

13.9. MONOGRAPHIC REFERENCES (References to nematological keys for identification of species)

13.9.1. *Anguina*

Brzeski, M.W. 1981. The genera of the Anguinidae (Nematoda: Tylenchida). *Revue de Nématologie* 4: 23-34.

Fortuner, R. and Maggenti, A.R. 1987. A reappraisal of the Tylenchina (Nemata). 4. The family Anguinidae Nicoll, 1935 (1926). *Revue de Nématologie* 10: 163-176.

Chizhov, V.N., and Subbotin, S.A. 1990. Phytoparasitic nematodes of subfamily Anguininae (Nematoda, Tylenchida). Morphology, trophic specialization, taxa (In Russian). *Zoologicheskii Zhurnal* 69:15-26.

13.9.2. *Aphelenchoides*

Nickle, W.R. and Hooper, D.J. 1991. The Aphelenchinae bud, leaf and insect nematodes. In: W.R. Nickle (ed.) *Manual of Agricultural Nematology*. Marcel Dekker, New York, p. 465-507.

Hunt, D.J. 1993. *Aphelenchida, Longidoridae, and Trichodoridae: Their systematics and Bionomics*. CAB International, Wallingford, UK.

Shahina, F. 1996. A diagnostic compendium of the genus *Aphelenchoides* Fischer, 1894 (Nematoda: Aphelenchida) with some new records of the group from Pakistan. *Pakistan Journal of Nematology* 14:1-32.

13.9.3. *Belonolaimus*

Fortuner, R. and Luc, M. 1987. A reappraisal of Tylenchina (Nemata). 6. The family Belonolaimidae Whitehead, 1960. *Revue de Nématologie* 10: 183-202.

13.9.4. *Bursaphelenchus*

Tarjan, A.C. and Aragon, C.B. 1982. An analysis of the genus *Bursaphelenchus* Fuchs, 1937. *Nematropica* 12:121-144.

Yin, K., Fang, Y. and Tarjan, A.C. 1988. A key to the species in the genus *Bursaphelenchus* with a description of *Bursaphelenchus hunanensis* sp. n. (Nematoda: Aphelenchoididae) found in pine wood in Hunan province, China. *Proc. Helminthol. Soc. Wash.* 55: 1-11.

13.9.5. *Coloosia*

Mathur, V.K., Khan, E., Nand, S. and Prasad, S.K. 1969. Two new species of *Caloosia* Siddiqi and Goodey (Nematoda: Hemicycliophoridae) from India. *Ent. Soc. India Bull.* 10: 27-31.

13.9.6. *Criconema* and *Neocriconema*

Diab, K.A. and Jenkins, W.R. 1965. Description of *Neocriconema adamsi*, n gen., n. sp. (Criconematidae: Nematoda) with a key to the species of *Neocriconema. Proc Helminth Soc. Wash.* 32: 193-197.

Mehta, U.K. and Raski, D.J. 1971. Revision of the genus *Criconema* Hofmanner and Menzel. 1914 and other related genera (Criconematidae: Nematoda). *Indian J. Nematol.* 1: 145-198.

Yeates, G.W., Loof, P.A.A. and Wouts, W.M. 1997. Criconematidae (Nematoda: Tylenchida) from the New Zealand region: analysis of and key to *Criconema* (*Nothocriconemella*) species. *New Zealand Journal of Zoology* 24: 153-162.

13.9.7. *Criconemella* and *Discocriconemella*

Ebsary, B.A. 1982. *Bakernema yukonense* n. sp. (Nematoda: Criconematidae) with keys to the species of *Criconemella* and *Discocriconemella. Canadian Journal of Zoology* 60: 3033-3047.

Hashim, Z. 1984. Re-diagnosis and a key to the species of *Neolobocriconema* Mehta & Raski, 1971 (Nematoda: Tylenchida), with a description of *N. orlarum* n. sp. from Jordan. *Systematic Parasitology* 6:69-73.

Loof, P.A.A. and De Grisse. 1989. Taxonomic and nomenclatorial observation of the genus *Criconemella* De Grisse & Loof, 1965 *sensu* Luc & Raski (Criconematidae). *Mededlingen Landbouwhogeschool Wageningen, Rijksuniversiteit Gent* 54: 53-74.

13.9.8. *Ditylenchus*

Fortuner, R. 1982. On the genus *Ditylenchus* Filipjev, 1936 (Nematoda: Tylenchida). *Revue de Nematol* 5: 17-38.

Brzeski, M.W. 1991. Review of the genus *Ditylenchus* Filipjev, 1936 (Nematoda: Anguinidae). *Revue de Nématologie* 14:9-59.

13.9.9. *Dolichodorus*

Smart, G.C. and Khuong, N.B. 1985. *Dolichodorus miradvulus* n sp. (Nematoda: Tylenchida) with a key to species. *J. Nematol.* 17: 29-37.

Siddiqi, M. R. 1986. *Tylenchida: Parasites of Plants and Insects*. CAB International, Wallingford, UK, 645 pp.

Luc, M. and Fortuner, R. 1987. A reappraisal of Tylenchina (Nemata). 5. The family Dolichodoridae Chitwood, 1950. *Revue de Nématologie* 10: 177-181.

13.9.10. *Helicotylenchus*

Anderson, R.V. 1979. A supplemental key to species of *Helicotylenchus* Steiner, 1945 (Nematoda: Hoplolaimidae) described since 1972 and a description of *H. ascephalus* n. sp. *Can J. Zool.* 57: 337-342.

Boag, B. and Jairajpuri, M.S. 1985. *Helicotylenchus scoticus* n. sp. and a conspectus of the genus *Helicotylenchus* Steiner, 1945 (Tylenchida: Nematoda). *Systematic Parasitology* 7: 47-58. (Compendium of 154 species).

Fotedar, D.N. and Kaul, V. 1985. A revised key to the species of genus *Helicotylenchus* Steiner, 1945 (Nematoda: Rotylenchoidinae). *Indian J. Nematol.* 15: 138-147.

Firoza, K. and Maqbool, M.A. 1994. A diagnostic compendium of the genus *Helicotylenchus* Steiner, 1945 (Nematoda: Hoplolaimidae). *Pakistan Journal of Nematology* 12: 11-50.

13.9.11. *Hemicriconemoides*

Edward, J.C. Misra, S.L. and Singh, G.R. 1965. *Hemicriconemoides birchifieldi* n sp (Nematoda: Criconematidae) from Allahabad, Uttar Pradesh, India: with a revisions of the key to species of *Hemicriconemoides. Nematologica* 11: 157-161.

Dasgupta, D.R., Raski, D.J. and van Gundy S.D. 1969. Revision of the genus *Hemicriconemoides* Chitwood and Birchfield, 1957 (Nematoda: Criconematidae). *J. Nematol.* 1: 126 - 145.

Esser, P.R. and Volvas, N. 1990. A diagnostic compendium to the genus *Hemicriconemoides* (Tylenchida: Criconematidae). *Proceedings of the Soil and Crop Science Society of Florida* 49: 211-219.

13.9.12. *Hemicycliophora*

Brzeski, M.W. 1974. Taxonomy of Hemicycliophorinae (Nematoda, Tylenchida). *Zess Problem Postep Nauk Roln* 121: 235 - 244.

Siddiqi, M.R. 1980. Taxonomy of the plant nematode superfamily Hemicycliophoroidea, with a proposal for Criconematina, a new suborder. *Revue de Nématologie* 3: 179-199.

Van Den Berg, E. 1987. *Hemicycliophora* species from the Western Cape Province with a key to the South African species (Hemicycliophoridae: Nematoda). *Phytophylactica* 19: 303-307.

13.9.13. *Heterodera* and *Globodera*

Sharma, S.B. and Swarup, G. 1980. Identification scheme for *Heterodera* species in India. *Indian J. Nematol.* 13: 171-180.

Hesling, J.J. 1982. Cyst nematodes: morphology and identification of *Heterodera, Globodera* and *Punctodera.* In J.F. Southey (ed.) *Plant Nematology.* HMSO, London, pp. 125 -155.

Mulvey, R.H. and Golden, A.M. 1983. An illustrated key to the cyst -forming genera and species of Heteroderidae in the Western Hemisphere with species morphometrics and distribution. *J. Nematol.* 15: 1-59.

Wouts, W.M. 1984. *Globodera zelandica* n. sp. (Nematoda: Heteroderidae) from New Zealand, with a key to the species of *Globodera*. *New Zealand Journal of Zoology* 11: 129-135.

Golden, A.M. 1986. Morphology and identification of cyst nematodes. In F. Lamberti and C.E. Taylor (eds.) *Cyst Nematodes*. NATO ASI Series, Series A: Life Sciences Vol. 121. Plenum Press, New York, pp. 23-45.

Wouts, W.M., Schoemaker, A., Sturhan, D. and Burrows, P.R. 1995. *Heterodera spinicauda* sp. n. (Nematoda: Heteroderidae) from mud flats in the Netherlands, with a key to the species of the *H. avenae* group. *Nematologica* 41: 575-583.

13.9.14. *Hirschmanniella*

Ebsary, B.A. and Anderson, R.V. 1982. Two new species of *Hirschmaniella* Luc and Goody, 1963 (Nematoda: Pratylenchidae) with a key to the nominal species. *Canadian Journal of Zoology* 60: 530-534.

Sivakumar, C.V. and Khan, E. 1982. Description of *Hirschmanniella kaverii* sp. n. (Radopholidae: Nematoda) with a key for identification of *Hirschmanniella* spp. *Indian J. Nematol* 12: 86-90.

13.9.15. *Hoplolaimus*

Handoo, Z.A. and Golden, A.M. 1992. A key and diagnostic compendium to the species of the genus *Hoplolaimus* Daday, 1905 (Nematoda: Hoplolaimidae). *Journal of Nematology* 15: 45-53.

13.9.16. *Longidorus* and *Paralongidorus*

Chen, Q.W., Hooper, D.J., Loof, P.A.A. and Xu, J.H. 1997. A revised polytomous key for the identification of species of the genus *Longidorus* Micoletzky, 1922 (Nematoda: Dorylaimoidea). *Fundamental and Applied Nematology* 20: 15-28.

Escuer, M. and Arias, M. 1997. *Paralongidorus iberis* sp. n. from Spain with a polytomous key to the species of the genus *Paralongidorus* Siddiqi, Hooper & Khan, 1963 (Nematoda: Longidoridae). *Fundamental and Applied Nematology* 20: 135-148.

Loof, P.A.A. and Chen, Q.W. 1999. A revised polytomous key for the identification of species of the genus *Longidorus* Micoletzky, 1922 (Nematoda: Dorylaimoidea). *Supplement I. Nematology* 1: 55-59.

Taylor, C.E. and Brown, D.J.F. 1997. *Nematode Vectors of Plant Viruses*. CAB International, Wallingford, UK, p. 286.

13.9.17. *Macroposthonia* (*Criconemoides*)

Ganguly, S. and Khan, E 1983. Revision of the genus *Caloosia* Siddiqi and Goodey, 1963 (Caloosiidae : Nematoda). *Indian J Nematol* 13: 181-198.

Khan, E., Chawla, M.L. and Saha, M. 1975. Criconematoidea (Nematoda: Tylenchida) from India, with descriptions of nine new species, two new genera and a family. *Indian J. Nematol* 5: 70-100.

Loof, P.A.A. and de Grisse, A 1973. Interrelationships of the genera of Criconematidae (Nematoda: Tylenchida). *Meded Fak Landb Welensch Gent* 38: 1303-1328.

Luc, M. and Raski, D.J. 1981. Status of the genera *Macroposthonia, Criconemoides, Criconemella* and *Xenocriconemella* (Criconematidae: Nematoda). *Revue de Nematol* 4: 3 -21.

13.9.18. *Meloidogyne*

Eisenback, J.D. and Triantaphyllou, H.H. 1991. Root-knot nematodes: *Meloidogyne* species and races. In W.R. Nickel (ed.) *Manual of Agricultural Nematology*. Marcel Dekker, New York, p. 191-274.

Jepson, S.B. 1987. *Identification of Root-Knot Nematodes* (Meloidogyne *Species).* CABI, Wallingford, UK, p. 265.

Sasser, J.N. and Carter, C.C. 1986. *An Advanced Treatise on* Meloidogyne. *Vol. I. Biology and Control.* North Carolina State Univ. Graphics, Raleigh, NC, USA, p. 422.

Taylor, A.L. 1987. Identification and Estimation of Root-Knot Nematode Species in Mixed Populations. Florida Department of Agriculture and Consumer Services.

13.9.19. *Morulaimus*

Sauer, M.R. 1966. *Morulaimus,* a new genus of the Belonolaiminae. *Nematologica* 11: 609-618.

13.9.20. *Nacobbus*

Sher, S.A. 1970. Revision of the genus *Nacobbus* Thorne and Allen, 1944. (Nematoda: Tylenchoidea). *J. Nematol* 2: 278-235.

13.9.21. *Paratylenchus*

Esser, R.P. 1992. A diagnostic compendium to species included in Paratylenchinae Thorne, 1949 and Tylenchocriconematinae Raski & Siddiqui, 1975 (Nematoda: Criconematoidea). *Nematologica* 38: 146-163.

Raski, D.J. 1975a. Revision of the genus *Paratylenchus* Micoletzky. 1922 and descriptions of new species. Part I of 3 parts. *J. Nematol.* 7: 15-34.

Raski, D.J. 1975b. Revision of the genus *Paratylenchus* Micoletzky, 1922 and descriptions of new species. Part II of 3 parts. *J. Nematol* 7: 274-295.

Raski, D.J. 1976. Revision of the genus *Paratylenchus* Micoletzky, 1922 and descriptions of new species. Part III of 3 parts. *J. Nematol* 8: 97-115.

Sologeva, G.I. 1975. *Parasitic Nematodes of Woody and Herbaceous Plants. A Review of the Genus* Paratylenchus *Micoletzky, 1922.* Amerind Publishing Co Pvt Ltd., New Delhi, Bombay, Calcutta and New York, p. 134.

13.9.22. *Peltamingratus* (*Peltamigratus* was placed in synonymy with *Aorolaimus*)

Baujard, P., Castillo, P., Doucet, M.E., Martiny, B. and Mounport, D. 1994. Taxonomic studies on the genus *Aorolaimus* Sher, 1963 (Nemata: Hoplolaimidae) I.

Bibliographic analysis and tentative key to species. *Fundamental and Applied Nematology* 17:103-115.

Bitencourt, C. and Huang, C.S. 1986. Brazilian *Peltamigratus* Sher, 1964 (Nematoda: Hoplolaimidae), with description of six new species. *Revue de Nématologie* 9: 3-24.

Fortuner, R. 1987. A reappraisal of Tylenchina (Nemata). 8. The family Hoplolaimidae Filip' ev, 1934. *Revue de Nématologie* 10: 219-232.

Mulk, M.M. and Siddiqi, M.R. 1982. Three new species of Hoplolaimid nematodes from South America. *Indian Journal of Nematology* 12: 124-131.

Rashid, F, Geraert, E. and Sharma, R.D. 1987. The genus *Peltamigratus* Sher, 1964 with description of two new species (Nematoda: Hoplolaimidae). *Revue de Nématologie* 10: 3-21.

13.9.23. *Pratylenchoides*

Baldwin, J.G. *et al.* 1983. Contribution to the study of the genus *Pratylenchoides* Winslow (Nematoda: Tylenchidae). *Revue de Nématologie* 6:111-125.

Ryss, A.Y. 1980. *Pratylenchoides ivanovae* sp. n. (Nematoda: Pratylenchidae) and key to species of the genus *Pratylenchoides*. *Parazitologiya* 14:516-520.

Sher, S.A. 1970. Revision of the genus *Pratylenchoides* Winslow. 1958. (Nematoda: Tylenchoidea). *Proc Helminth Soc Wash* 37: 154-166.

13.9.24. *Pratylenchus*

Café Filho, A.C. and Huang, C.S. 1989. Description of *Pratylenchus pseudofallax* n. sp. with a key to species of the genus *Pratylenchus* Filipjev, 1936 (Nematoda: Pratylenchidae). *Revue de Nématologie* 12: 7-15.

Frederick, J.J. and Tarjan, A.C. 1989. A compendium of the genus *Pratylenchus* Filipjev, 1936 (Nemata: Pratylenchidae). *Revue de Nématologie* 12:243-256.

Handoo, Z.A. and Golden, A.M. 1989. A key and diagnostic compendium to the genus *Pratylenchus* Filipjev, 1936 (lesion nematodes). *Journal of Nematology* 21:202-218.

Loof, P.A.A. 1978. *The Genus Pratylenchus Filipjev, 1936, (Nematoda: Pratylenchidae): A Review of its Anatomy, Morphology, Distribution, Systematics and Identification.* Swedish Univ Agri Sci, Res Information Centre, Uppasala, Sweden, 50 pp.

13.9.25. *Radopholus* and *Radopholoides*

Colbran, R.C. 1971. Studies of plant and soil nematodes. 15. Eleven new species of *Radopholus* Thorne and a new species of *Radopholoides* de Gurian (Nematoda: Tylenchoidea) from Australia. *Qd J. Agric Anim. Sci* 27: 437- 460.

Sher, S.A. 1968. Revision of the genus *Radopholus* Thorne, 1949 (Nematoda: Tylenchoidea). *Proc Helminth Soc Wash* 35: 219-237.

13.9.26. *Rotylenchulus*

Dasgupta, D.R., Raski, D.J. and Sher, S.A. 1968. A revision of the genus *Rotylenchulus* Linford and Oliveira, 1940 (Nematoda: Tylenchida). *Proc. Hilminth Soc Wash* 35: 169-192.

Robinson, A.F., Inserra, R.N., Caswell-Chen, E.P., Vovlas, N. and Troccoli, A. 1997. *Rotylenchulus* species: identification, distribution, host ranges, and resistance. *Nematropica* 27:128-180.

13.9.27. *Rotylenchus*

Boag, B. and Hooper, D.J. 1981. *Rotylenchus ouensensis* n. sp. (Nematoda: Hoplolaimidae) from the British Isles. *Systematic Parasitology* 3: 119-125.

Castillo, P., Vovlas, N., Gomez-Barcina, A. and Lamberti, F. 1993. The plant parasitic nematode *Rotylenchus*, (a monograph). *Supplement to Nematologia Mediterranea* 21: 200.

Geraert, E. and Barooti, S. 1996. Four *Rotylenchus* from Iran, with a key to the species. *Nematologica* 42: 503-520.

13.9.28. *Scutellonema*

Germani, G., Baldwin, J.G., Bell, A.H. and Wu, X.Y. 1986. Revision of the genus *Scutellonema* Andrássy, 1958 (Nematoda: Tylenchida). *Revue de Nématologie* 8:289-320.

Van Den Berg, E. and Heyns, J. 1973. South African Hoplolaiminae 2. The genus *Scutellonema* Andrassy, 1958. *Phytophylactica* 5: 23-40.

13.9.29. *Trichodorus* and *Paratrichodorus*

Decraemer, W. 1991. Stubby root and virus vector nematodes: *Trichodorus, Paratrichodorus, Allotrichodorus* and *Monotrichodorus*. In W.R. Nickle (ed.) *Manual of Agricultural Helminthology*. Marcel Dekker, New York, p. 587-625.

Decraemer, W. 1995. *The Family Trichodoridae: Stubby Root and Virus Vector Nematodes.* Kluwer Academic Publishers, p. 36.

Decraemer, W. and Baujard, P. 1998. A polytomous key for the identification of species of the family Trichodoridae Thorne, 1935 (Nematoda: Triplonchida). *Fundamental and Applied Nematology* 21: 37-62.

Taylor, C.E. and Brown, D.J.F. 1997. *Nematode Vectors of Plant Viruses*. CAB International, Wallingford, UK, p. 286.

13.9.30. *Tylenchorhynchus* and *Tetylenchus*

Brzeski, M.W. and Dolinski, C.M. 1998. Compendium of the genus *Tylenchorhynchus* Cobb, 1913 *sensu lato* (Nematoda : Belonolaimidae). *Russian Journal of Nematology* 6: 189-199. (Synonymy of six genera with *Tylenchorhynchus*).

Fortuner, R. and Luc, M. 1987. A reappraisal of Tylenchina (Nemata). 6. The family Belonolaimidae Whitehead, 1960. *Revue de Nématologie* 10: 183-202. (Five genera synonymized with *Tylenchorhynchus*).

Jairajpuri, M.S. and Hunt, D.J. 1984. The taxonomy of Tylenchorhynchinae (Nematoda: Tylenchida) with longitudinal lines and ridges. *Systematic Parasitology* 6: 261-268. (Keys to *Dolichorhynchus*, *Neodolichorhynchus*, *Tessellus*, and to genera of subfamily).

Tarjan, A.C. 1973. A synopsis of the genera and species in the Tylenchorhynchinae (Tylenchoidea: Nematoda). *Proc. Helminth Soc Wash* 40: 123-144.

13.9.31. *Tylenchulus*

Inserra, R.N., Vovlas, N., O' Bannon, J.H. and Esser, R.P. 1988. *Tylenchulus graminis* n. sp. and *T. palustris* n.sp. (Tylenchulidae), from native flora of Florida, with notes on *T. semipenetrans* and *T. furcus*. *Journal of Nematology* 20: 266-287.

13.9.32. *Xiphinema*

Loof, P.A.A. and Luc, M. 1990. A revised polytomous key for the identification of species of the genus *Xiphinema* Cobb, 1913 (Nematoda: Longidoridae) with exclusion of the *X. americanum*-group. *Systematic Parasitology* 16: 35-66.

Lamberti, F. and Carone, M. 1991. A dichotomous key for the identification of species of *Xiphinema* (Nematoda: Dorylaimida) with the *X. americanum*–group. *Nematologia Mediterranea* 19: 341-348.

Loof, P.A.A. and Luc, M. 1993. A revised polytomous key for the identification of species of the genus *Xiphinema* Cobb, 1913 (Nematoda: Longidoridae) with exclusion of the *X. americanum*-group: Supplement 1. *Systematic Parasitology* 24: 185-189.

Loof, P.A.A., Luc, M. and Baujard, P. 1996. A revised polytomous key for the identification of species of the genus *Xiphinema* Cobb, 1913 (Nematoda: Longidoridae) with exclusion of the *X. americanum*-group: Supplement 2. *Systematic Parasitology* 33: 23-29.

Robbins, R.T., Brown, D.J.F., Halbrendt, J.M. and Vrain, T.C. 1996. Compendium of juvenile stages of *Xiphinema* species (Nematoda: Longidoridae). *Russian Journal of Nematology* 4: 163-171.

Section IV
BIOLOGY, PHYSIOLOGY AND ECOLOGY

Chapter 14

BIOLOGY

14.1. INTRODUCTION

The life cycle of most plant parasitic nematodes is simple and direct. Females lay eggs that hatch into larvae and these larvae are much like adults in appearance and structure. Nematodes usually undergo the first molt before hatching, while the larvae are still within the egg shell. During their growth and development larvae undergo a series of four molts and the periods of growth between molts are called larval stages. The first larval stage is terminated by the first molt, the second larval stage by the second molt, and so on. The fourth molt terminates the fourth larval stage, after which the individual is an adult (Fig. 14.1). There are exceptions to this cycle. Larvae of *Xiphinema index* are reported to emerge from the eggs before the first molt. At the end of each larval stage, the cuticle is shed, including its extensions into the oral opening, excretory pore, vulva and the rectum or the cloaca.

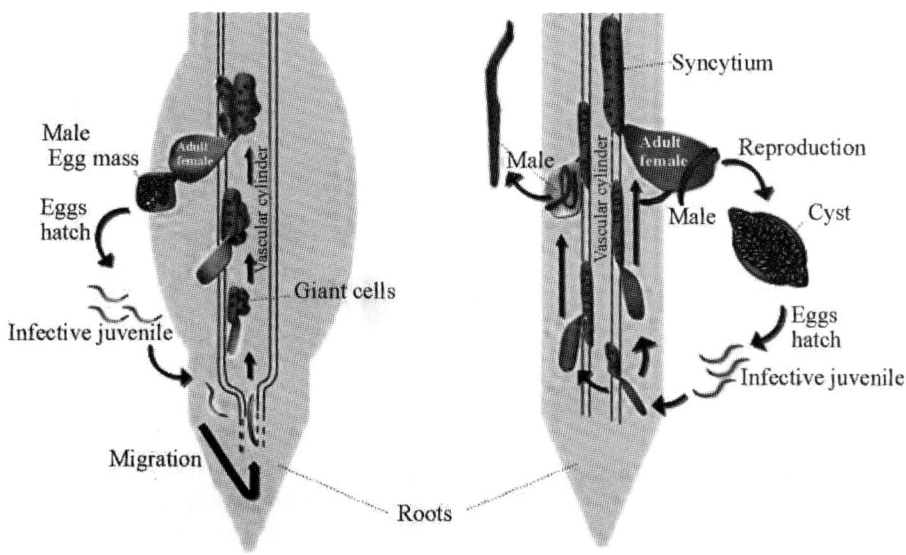

Fig. 14.1. Life cycle of root-knot nematode (*left*) and cyst nematode (*right*) (Abad and Williamson, 2010)

14.2. THE EGG

The majority of nematode eggs are similar in shape (oval) and size irrespective of the size of the adult nematode. There are three layers in a typical nematode egg shell, an inner lipid layer, a middle chitinous layer and an outer vitelline layer (Fig. 14.2) which is about 30 µm thick in *Meloidogyne javanica.* The lipid layer is soluble in various dehydrating agents. It is made up of proteins and lipids and is soluble in fat solvents. This layer is thought to be responsible for the extreme resistance of eggs to adverse environmental conditions. The amount of chitin in the chitinous layer varies in different nematodes. It is synthesized in response to the stimulus of fertilization. The vitelline layer of the egg shell resembles the external cortical layer of the larval cuticle and appears to be bounded by a triple-layered plasma membrane which may be derived from the unit membrane surrounding the oocyte.

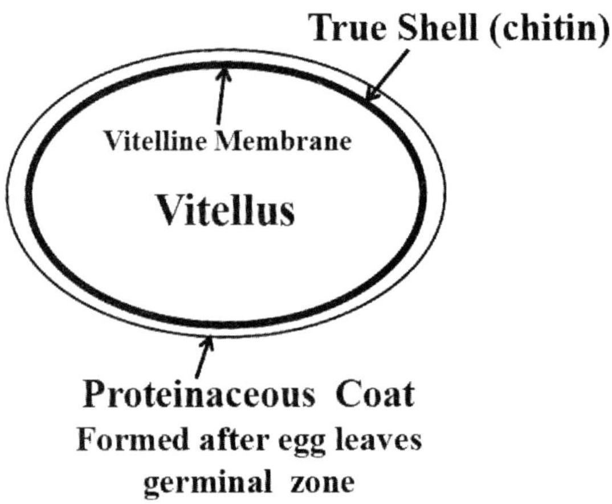

Fig. 14.2. Structure of nematode egg

14.3. EMBRYOLOGY

The first cleavage of the egg gives rise to two equal cells or blastomere which are the first somatic cell (S_1) and the parental germinal cell (P_1) (Fig. 14.3) from which the gonads of the adult can be traced. The second cleavage gives rise, at first, to four cells. This is achieved by the blastomere S_1 dividing longitudinally to form A and B and the blastomere P_1 dividing transversely to form P_2 and S_2. These blastomeres become rearranged to form a rhomboid which is the characteristic shape of the typical nematode four celled stage. The S_1 blastomere is the primary somatic cell and its descendants A and B give rise to most of the nematode's ectodermal cells. The blastomere S_2 also gives rise to somatic tissue and will give rise to endoderm (E), mesoderm (M) and stomoderm (St) tissues. The nematode's gonads will be derived from P_1.

The cells A and B divides to give a, b, a and b, and P_2 divides to give P_3 and S_3. The S_2 cell or E M St divides into E and M St. The dorsal cells derived from A and B continues to divide and eventually give rise to most of the hypodermis, the nervous system and excretory cells. The cell P_3 divides to give P_4 and S_4. Both S_3 and S_4 are ectodermal and give rise to the hypodermis in the posterior region of the nematodes' body.

The endodermal tissue is derived from the descendants of cell E_1 and P_1 divides into P_5 and S_5. The descendants of cell S_5 produce the epithelium which covers the gonads and their ducts and the descendants of P_5, G_1 and G_2 and their descendants proliferate to germ cells only.

The nematode's body wall musculature and its pseudocoelomic cells are derived from the primary mesodermal cell M and its pharynx from the descendants of the cell St. In the early embryonic stages, these primary cells St, M and E lie on the ventral surface of the embryo and are taken within the embryo by a process of gastrulation. This is achieved usually, either by a type of cell overgrowth known as epibody, or more commonly, by a type of invagination in which cell E proliferates a chain of cells that will eventually become the mid-gut, the descendants of cell M forms on either side of these cells, and St proliferates a mass of cells which become invaginated into the interior.

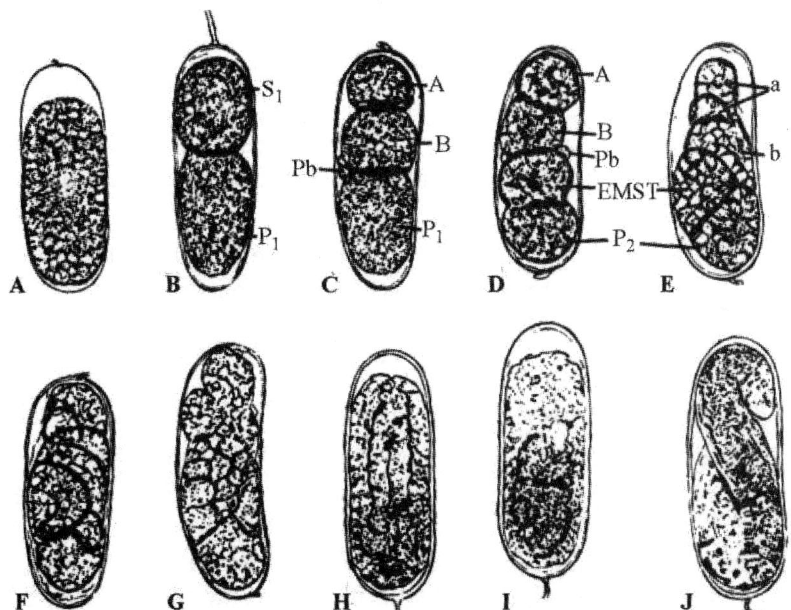

Fig. 14.3. Eggs of nematode in various stages of development. (A) One-celled stage; (B) two-celled stage: (C) three-celled stage showing two polar bodies between cells B and P_1; (D) four-celled stage with a polar body on either side between B and EMST; (E) seven-celled stage; (F-G) Multiple-celled stages; (H) Gastrula; (I) Early 'tadpole' stage; (J) First-stage larva

By the process of gastrulation, the flattened embryo becomes cylindrical and the embryo starts to become worm-shaped. It continues to grow and elongate and eventually develops into larva. The larva molts once before hatching.

14.4. HATCHING

Hatching occurs either in response to a stimulus or stimuli from the host, as in some parasitic forms, or it takes place under normal environment. The eggs of *Globodera rostochiensis* normally hatches in response to stimuli provided by the Solanaceous root exudates, and will only hatch *in vitro* in the presence of a hatching factor isolated from these roots or in aqueous solutions of anhydro tetronic acid, picrolinic acid and sodium metavandate. The eggs of *Heterodera schachtii* will hatch in response to the stimuli of over 50 synthetic hatching agents.

G. rostochiensis cuts a slit in the egg shell by using the tip of its stylet to make a line of close perforations through the shell, which merges to make the slit. The process of hatching in this nematode is largely mechanical and pharyngeal gland secretions are important in this process.

There are many recorded instances of mechanical force being used by the hatching larvae just prior to emergence from the egg. These includes widely separated genera such as *Pratylenchus, Paratylenchus, Nacobbus* and *Meloidogyne.*

The basic mechanism of hatching is similar throughout the Nematoda. Generally, it seems to take place as follows: the stimulus elicits a behavioral response in the larva which brings about increased larval activity, breakdown of the lipid layer, and an increase in plasticity of the egg shell leading to its distortion.

In *Xiphinema*, exudation from the mouth of the nematode appears to bring about changes in the structure of the egg which leads to its rupture.

14.5. MOLTING

Growth in nematodes is associated with molting which normally occurs four times before the adult stage is reached. Growth curves for a number of different species of nematodes have been drawn including *Meloidogyne javanica, Ditylenchus dipsaci* and *Aphelenchus avenae.*

The growth curves are usually sigmoidal in shape. In *Meloidogyne,* growth does not occur between parasitic molts. In most parasitic forms, greatest growth occurs after the last molt and molting tends to occur in the earlier half of the growth curve.

14.5.1. The stimulus

It has been suggested that the neurosecretory cells of nematodes are stimulated to produce secretions which activate glands that produce hormones or enzymes which, in turn, initiate molting. In *Paratylenchus nanus,* root exudations will act as a stimulus to the fourth stage larval molt. In endoparasitic nematodes of plants in which molting does not occur until some growth has occurred within the host, this stimulus may be more complex and may be closely associated with an actual increase in size of the nematode. Thus, the stimulus may come from within the parasite rather than directly from its host and the receptor may function here as a stretch receptor.

14.5.2. The receptor

The receptor in all cases may be cuticular and hypodermal structure of universal occurrence in nematodes, such as the hemizonid. The receptor appears to be associated with neurosecretory activity which leads to the production of an enzyme which initiates molting.

14.5.3. Molting

The nematode cuticle may be shed in one piece as in *Paratylenchus nanus* or the anterior part may be shed separately as a cap. The lining of the amphids, esophagus, excretory duct, phasmids and rectum are shed at each molt.

Prior to molting, the hypodermis increases in thickness due to the accumulation of ribosome-like granules which contains nucleic acid and protein.

Parts of the external cortical layer are first laid down on the surface of the hypodermis at the base of the old cuticle. The new cuticle is formed beneath the new external cortical layer and, in some species the old cuticle is partially resorbed through the new cuticle before it is shed, while in others, it is shed intact.

14.6. SIGNIFICANCE OF LIFE HISTORIES

Life histories of nematodes should be understood when control measures are considered. The wheat gall nematode (*Anguina tritici*) may be effectively controlled by crop rotation with non-host plants, because the emergence of larva from the galls is virtually complete when soil moisture and temperature become favorable. Larvae die when they are outside the gall in the absence of the host.

■■■■

Chapter 15

PHYSIOLOGY

15.1. INTRODUCTION

Investigations on nematode metabolism, biochemistry and nutrition are increasing. The limited research on many aspects of nematode physiology has been partly ascribed to the difficulty in obtaining adequate quantities of specific plant parasitic nematodes. Studies on nematode physiology may give an insight as to why some of the commonly applied control methods are successful.

15.2. CHEMICAL COMPOSITION

Nematodes, like other animals, contain carbohydrates, proteins, lipids, nucleic acid, vitamins, hormones, minerals and numerous other chemicals, but not much is known of the precise kinds or amounts of these substances present in nematodes. Chemical composition of nematodes affects longevity, degree of resistance to temperature extremes, desiccation, atmosphere, osmotic conditions and chemicals.

Glucose, fructose and 15 amino acids were identified from two species of plant nematodes. Plant nematodes generally contain far more lipids than do animal parasites. The cuticle is composed primarily of protein but also contains lipids, polyphenols, enzymes and nucleic acids.

15.3. METABOLISM

Intermediary metabolism in plant nematodes is at present not well understood. Portions of metabolic sequences of sugar breakdown and terminal oxidation have been elucidated, but in no instance have all the steps in any one sequence been demonstrated. Synthesis of amino acids by two species of plant nematodes incubated in ^{14}C-labeled glucose and acetate solutions in the absence of plant tissues has been shown, illustrating that it is not imperative for nematodes to feed on cells in order to take in chemicals.

Nematodes seems able to digest a variety of polymeric plant components outside the body. It has been reported that enzymes that breakdown starch, sucrose, pectin, cellulose, protein and glycosides have been discharged by plant nematodes. Presumably, these enzymes breakdown plant components outside the nematode body before they are ingested by the parasite. Homogenates or extracts of nematodes contained all the above and other digestive enzymes. In addition to being active outside the nematode, such enzymes are very likely active in the further degradation of plant substances in the digestive tract of the nematode. Possession of specific digestive enzymes may be determine if a nematode can parasitize higher plants or fungi or both.

15.3.1. Respiration

Plant nematodes require oxygen. They may move actively in the presence of low oxygen, but such conditions may inhibit development and egg hatch. Egg production and survival of molting nematodes and males were most sensitive to low oxygen tension. Hatching of eggs was less sensitive and movement and survival of pre-adult and mature females were least sensitive to low oxygen.

Oxygen consumption by intact specimens of several species of plant nematodes has been measured by microrespirometry. Rates of oxygen uptake varied according to nematode species, condition of the nematodes and carbon dioxide and osmotic concentrations of the incubation solution.

15.3.2. Temperature

Since nematodes lack a means of controlling body temperature, metabolic and physical activity are dependent on the ambient temperature. In general, the optimum temperature for growth and development of plant nematodes is in the range of 5 to 30°C. Desiccated nematodes can withstand both higher and lower temperatures than fully hydrated ones, probably because enzymes are more resistant to temperature inactivation in the dehydrated than in the hydrated condition. Moisture plays an important role in determining the thermal stresses that nematodes can withstand.

In addition to its direct effects on survival, temperature influences other aspects of nematode biology. Infectivity, sex determination and rate of development of plant nematodes are affected by temperature. Cysts of *Heterodera avenae* are reported to require a period of exposure to low temperatures before eggs will hatch. In soybeans grown at 24°C, the developing *H. glycines* are mostly females, while at 31°C they are mostly males. A few days of exposure of young infected plants to 35°C followed by growth at 24°C was sufficient to cause most of the nematodes to develop as males. Root-knot nematodes produce few viable eggs above 35°C, whereas 25 to 32°C is optimum for egg production. However, *Globodera rostochiensis* develops and produces eggs most rapidly at 18°C. Optimum temperatures for development of the host plant, as well as fluctuating temperatures may also affect the rate of development of nematodes.

15.3.3. Moisture

The natural environment of nematodes is aquatic. The soil solution that covers soil particles and is in soil pores is the medium in which nematodes live and through which they move to contact plant tissues. It is also the medium in which the physiological functions of nematodes such as gas exchange and discharge of excretory products occurs. For nematodes inside plant tissues, ample solution is available to satisfy their requirements.

Water moves freely into and out of the nematode body through the cuticle, although recent findings show that nematodes have at least some control over loss of body water under conditions of water stress. However, when desiccated, eggs and tanned cysts are resistant to water loss. Under conditions of ample moisture, solutes in the nematode body cavity maintain a constant turgor pressure within the body. All nematodes can probably withstand some water stress, but the degree of water stress which is lethal varies among species. Nematodes inhabiting the above-ground parts of plants, as well as certain soil inhabiting nematodes such as the cyst stage are especially resistant to desiccation. Some nematodes will survive freeze drying, while other species are killed when soil containing them is air-dried. When the moisture content of soil approaches the wilting point for plants, activity of nematodes in soil is curtailed.

Most free living and plant nematodes withstand severe and repeated osmotic changes with little or no adverse effects. In hypertonic solutions, nematodes shrink and become inactive because of loss of water. But, when transferred to isotonic or hypotonic solutions, they quickly become turgid and active again. Eggs are not plasmolyzed in hypertonic solutions. Although, solutions of high or fluctuating osmotic concentration may not kill nematodes, movement and egg hatch are inhibited at relatively low solute concentrations (0.1 to 0.2 M). Ions seems to move into and out of the nematode body with changes of osmotic concentration of the ambient solution, but movement of particular ions is somewhat selective. Although water and certain ions pass directly through the cuticle of all nematodes, the excretory system appears to regulate the osmotic concentration of the body fluid as well as functioning in the elimination of certain body wastes.

15.4. DORMANCY AND LONGEVITY

Several species of plant nematodes become dormant during periods of unfavorable environmental conditions and then revive when conditions become favorable again. Nematodes that are parasitic on the above-ground parts of plants survive repeated periods of unfavorable conditions enroute to tissues upon which they will feed. Plant tissues suitable for growth and reproduction of *Anguina tritici* are available only for the short time of flower embryo development, after which weeks or months may elapse before plant tissues are again available. Larvae of *A. tritici* have been revived from galls after 28 years of storage,

Ditylenchus dipsaci after 23 years of storage in infected plant material and *A. agrostis* after four years of storage.

The developmental stage in which nematodes survives periods of dormancy varies with the species. Only second stage larvae of *A. tritici* survives dormancy, while in *Ditylenchus* and *Aphelenchoides* species the resistant stages are the fourth stage pre-adult larvae and adults, respectively. Larvae inside eggs in cysts of *Globodera rostochiensis* and *Heterodera schachtii* remain alive for years when stored dry in the laboratory. Larvae in cysts of certain other cyst forming species are less resistant to drying. Dormant stages of nematodes are particularly resistant to the action of nematicides because of the protective cyst wall in cyst nematodes and the lowered metabolic activity or other physiological factors in both cyst and other nematode species.

In the presence of growing host plants in a suitable environment, most nematodes can be maintained indefinitely. Longevity in the absence of host plants is greatly influenced by the environment. Nematodes of many species can survive in soil for at least a year in the absence of a suitable host, probably on body food reserves. In soil, nematodes survive over a wide temperature range, but they survive longer at lower than higher temperatures. This is caused primarily by slow depletion of body food reserves and low metabolic rate.

Certain plant nematodes such as *Globodera rostochiensis* persist for years in soil in the absence of host plants. However, total exclusion of possible hosts, such as particular weeds, that might maintain populations is not certain. Populations of some cyst nematodes in soil without host plants declined at a rate of 35 to 60 per cent per year, whereas populations of other types of nematodes were reduced by 75 to 95 per cent in the first year. In most experiments, a few nematodes were still being recovered from soil when the tests were terminated. The reduction of populations of plant nematodes in fallow soil thus follows a hyperbolic curve. Yet, even after extended periods of time, a few specimens that are important for perpetuation of the species, may survive to serve as an inoculum to rebuild populations once host plants are again available.

15.5. GROWTH AND SEX DETERMINATION

When food is in ample supply, growth in nematodes is a continuous process except during molts. The cuticle can make limited growth between molts. Since nematodes are determinate in cell number, growth is primarily expressed as an increase in cell size. Growth may be accompanied by changes in body shapes of females, such as those that occur in root-knot, cyst, reniform and several other nematodes; but, in most kinds of nematodes, males and females remain vermiform throughout the life cycle.

Sex determination in nematodes seems to be governed by both genetic and nutritional mechanisms. Results of several studies indicates that when the food supply becomes limited the sex ratio is shifted toward males. Males often predominate in nematode populations developed in the presence of resistant plants, which may be a reflection of nutritional deficiency, possible control of sex by hormones in nematodes has not been investigated.

15.6. NUTRITION AND LABORATORY CULTURING

Several free living and one stylet bearing forms have been propagated *in vitro*. No nematode has been propagated on a completely defined medium, despite intensive efforts towards this goal. All media that support *in vitro* propagation of free living nematodes have failed to support reproduction of plant nematodes.

Although all plant nematodes require living plant tissues for reproduction, there is great variation in the host ranges of different nematodes. For instance, the golden nematode will reproduce only on a very few plants, while the host plants of certain root-knot nematodes number in hundreds. Attempts at *in vitro* culturing of plant nematodes would be most likely to succeed with species that do not induce specialized plant reactions and that have wide host ranges.

Laboratory culturing of plant nematodes on a scale adequate for biochemical studies has been successful with about 12 nematode species. Sterile root cultures, grown on nutrient agar media, were first utilized for culturing nematodes. Plant callus cultures, particularly alfalfa, proved highly useful. This culturing technique is presently the most promising method of propagating large quantities of plant nematodes in the laboratory.

■■■■

Chapter 16

ECOLOGY

16.1. INTRODUCTION

Ecology has been defined as the science of the inter-relationships between plant parasitic nematodes and the environment in which they live. Agricultural land represents a specialized environment, ranging from a dry, barren waste to a moist, bush jungle of plant growth. Plant nematodes are mostly those soil inhabiting species that are capable of withstanding the frequent changes caused by man's agricultural practices. Some nematodes, such as spiral (*Helicotylenchus* sp.), stunt (*Tylenchorhynchus* sp.) and sheath (*Hemicycliophora* sp.) nematodes, can live in a wide variety of habitats. Some, such as the rice root nematodes (*Hirschmanniella oryzae)* in aquatic habitats, are widely distributed but are limited to a particular combination of environmental conditions. Still others, such as sting nematodes (*Belonolaimus* sp.) in the sandy soils of the Southeastern United States, are found only in very special situations.

16.2. VERTICAL DISTRIBUTION

The vertical distribution of nematodes in cultivated soil is generally closely related to the distribution of plant roots and the rhizosphere. The number of plant nematodes is correlated with the distribution of roots of present and previous hosts. Nematodes are mainly concentrated in the top foot of soil and as many as six billion have been estimated in the top inch of an acre of soil. The fallowing of soil for an adequate period will generally reduce the number of nematodes present. The cyst nematodes (*Heterodera* sp.), the most persistent of the plant nematodes, decline in soil at a steady rate of 35 to 60 per cent per year, depending on soil type, moisture and temperature, regardless of the density of the initial population per unit of soil.

16.3. SURVIVAL

Plant nematodes are able to survive despite unfavorable conditions such as cold and dry periods between host crops. They survive unfavorable environments in a dormant condition, which is a quiescent or inactive state that is often associated with a low metabolic rate. The length of the quiescent period is usually limited by the amount of food reserves in the nematode and the environmental conditions. Quiescence may serve to extend a comparatively short life cycle of 20 to 40 days to periods varying from a year for many plant parasites to 20 to 30 years for the stem nematode (*Ditylenchus dipsaci*) and the wheat nematode (*Anguina tritici*).

16.4. POPULATION DYNAMICS

In agricultural soil, the upper population limit for any plant nematode species depends on the nematodes' reproductive potential, the host-plant species and the length of time the nematode remains in an environment favorable for reproduction. Generally, the reproductive potential of the endoparasites and above-ground parasites is greater than that of the ectoparasites. Some nematodes have only one or two generations a year, while others such as root-knot, cyst, lesion and citrus nematodes have several generations during the growing season. The population level is dependent on the nematode's ability to live successfully in soil.

The importance of a nematode as a plant parasite depends largely on whether or not the population limit exceeds the level at which economic damage occurs to a crop plant. This concept of population threshold at which yield loss begins is often used in connection with crop pests to determine the tolerance level of a host crop. For any set of environmental conditions, each host crop has its own tolerance level for a nematode species. Thus, a nematode causes economic damage only if its population density exceeds the tolerance level of the plant grown in the field. For example, the population density of root-knot nematodes is generally much higher than the tolerance level of many host plants, thus accounting for their importance as a nematode pests. Estimates of thresholds vary with seasons and from field to field.

A partial list of available information on threshold densities of certain nematodes causing damage on specific crops is given in Table 16.1.

Table 16.1. Approximate threshold densities of certain nematodes causing damage on specific crops

Nematode	Crop	Threshold density
Belonolaimus longicaudatus	Cotton, groundnut, soybean	25/500 cm^3 soil
Ditylenchus dipsaci	Onion	1-5/500 g soil
Globodera rostochiensis	Potato	31 cysts/100 g soil
Hemicycliophora sp.	Vegetables	50-1300/100 ml soil
Heterodera carotae	Carrot	<10 cysts/100 ml soil, >500 larvae/100 ml soil

Nematode	Crop	Threshold density
H. goettingiana	Peas	11-30 cysts/100 ml soil, 700-2000 larvae/100 ml soil
H. schachtii	Sugar beet	4-10 eggs/g of soil
Meloidogyne arenaria	Tomato, cucumber	100-200/100 ml soil
M. hapla	Strawberry	25/500 cm^3 soil
	Groundnut	100/500 cm^3 soil
M. incognita	Tobacco, tomato	25-50/500 cm^3 soil
	Cucumber	100-200/100 ml soil
M. javanica	Tomato, cucumber	100-200/100 ml soil
Pratylenchus brachyurus	Groundnut	25/500 cm^3 soil
	Tobacco	> 200/500 cm^3 soil
P. penetrans	Daffodil	1-5/500 g soil
	Potato	>500/500 g soil
	Onion	>5/500 g soil
P. vulnus	Roses	>190/100 ml soil
Rotylenchus robustus	Beets, carrots. vegetables	100-1000/100 ml soil
Trichodorus sp.	Beet	20-240/100 ml soil
Tylenchorhynchus dubius	Grasses, grain crops, clover and peas	500-4500/100 ml soil
T. claytoni	Azalea	>100/500 cm^3 soil

16.5. SOIL ENVIRONMENT

All plant nematodes inhabit soil for varying lengths of time during their life cycles. For example, the ectoparasitic nematodes such as *Hemicycliophora arenaria* and *Trichodorus christiei* spend their entire lives in the soil, usually in the rhizosphere of the plant. The more specialized endoparasites such as *Heterodera schachtii* and *Meloidogyne javanica* enter plant tissue and thereby spend less of their lives in the soil and rhizosphere. The above-ground parasites such as *Anguina tritici* and *Ditylenchus dipsaci* are mostly inside the plant tissues and spend very little of their lives in the soil.

The principal factors in the nematode's soil environment are temperature, moisture, texture, aeration and soil solution.

16.5.1. Temperature

Temperature affects nematode activities such as hatching, reproduction, movement, development and survival and also affects the host plant. Most plant nematodes become inactive at a low temperature range of 5 to 15°C, have an optimum range of 15 to 30°C and become inactive at a high temperature range of 30 to 40°C. Temperatures outside these extremes may be lethal.

16.5.2. Moisture

Fluctuating soil moisture due to rainfall or irrigation is one of the chief factors influencing nematode population increases. Although dry conditions may depress nematode activity and resulting populations, eggs of most nematodes survive drying.

Saturated soils are not generally favorable for nematode pests of agricultural crops. However, high populations of some nematodes, such as species of *Dolichodorus, Radopholus* and *Hirschmanniella* are found chiefly in wet locations.

The nematodes are constantly active in soil that contains moisture content between 40 and 60 per cent of field capacity. In dry and wet soils they are quiescent for varying periods. Nematodes need free water films in the soil for hatching and movement.

16.5.3. Soil texture

Texture refers to the relative proportions of different sized particles in soil. A coarse textured soil usually contains a high per cent of sand and has large pores that drain more quickly than the small pores of fine textured soil, which has a high proportion of clay and silt.

Light sandy soils (coarse - textured) are generally more favorable to large populations of nematodes such as cyst, root-knot, lesion and stubby root nematodes. However, the stem, sugar beet cyst and some species of lesion and stunt nematodes are numerous in clay soils. Still others, such as the citrus nematode, occur frequently in both sandy and clay soils.

16.5.4. Soil solutions

The chemical constitution of the soil solution, a major constituent of the soil environment, includes soil salinity, pH, organic matter, fertilizers, insecticides and nematicides. Plant nematodes probably derive few nutrients from the soil solutions. The hatching of eggs and survival of larvae may be influenced by various salts and ions. During dry and wet periods, soil nematodes are subjected to available salt concentrations in the soil solution. However, nematodes can tolerate osmotic pressures up to about 10 atm, at least for short periods. A soil pH ranging between 5.0 and 7.0 has little effect on nematodes. Fertilizers and organic matter may influence nematode populations indirectly by increasing host plant growth. Occasionally, the use of insecticides and nematicides in soil may eliminate some nematode enemies, thus leading to an increase in population of a plant parasite.

16.6. CLIMATE

Rainfall and temperature above soil level are extremely important to the growth and development of both nematodes and plants. These factors are usually responsible for seasonal fluctuations in nematode populations and may even determine the success of a species in becoming established in a new habitat or region. Climatic factors affecting humidity are particularly important to above-ground parasites, which are able to invade seedlings and move upward on plant surfaces covered by water films or droplets. These nematodes may be subjected to severe desiccation and great extremes of temperature owing to the more violent changes in aerial climate compared with soil climate. Perhaps, as an adaptation to this, certain stages of these nematodes, such as second stage larvae in wheat nematodes and fourth stage larvae in stem nematodes, are capable of withstanding long periods of desiccation.

16.7. PLANT ENVIRONMENT

The host plant environment, consisting of either root or stem and leaf tissue, greatly influences the endoparasitic nematodes. The plant tissues that are usually attacked are apical meristems that contain cells with thin walls and offer a chemically rich environment. The epidermis and cell wall offer mechanical barriers to nematode entrance and movement. The plant tissue protects endoparasitic nematodes from the soil environment and is their sole source of food, and the quality and quantity influence nematode growth and reproduction. Thus, host susceptibility, tolerance and resistance to nematodes are dependent on properties of individual plant cells and tissues. The periderm and necrotic areas, which are formed in some plants in response to nematode feeding, may affect nematode growth and reproduction, because the quality and quantity of nutrients are deficient in these areas, or the nematodes may be excluded from suitable plant cells by these areas. In root-knot, cyst and citrus nematodes the host cells are modified to provide specialized feeding sites and their physiological and nutritional dependence on the host become delicately balanced.

16.7.1. The rhizosphere

In addition to serving as a source of food for nematodes, plant roots may also modify the soil environment by lowering the concentration of mineral nutrients, depleting moisture, increasing carbon dioxide, reducing oxygen and contributing a variety of organic substances by exudation and sloughing off of cells. The rhizosphere is a dynamic environment, where the relationships among nematode, host and environment are often of a chemical nature. A root exudate stimulates the hatching of eggs of the cyst nematodes. The exudates significantly increases the hatching of root-knot nematode eggs as compared to a water hatching. They stimulate the metabolism of larvae after hatching and may account for their directional movement towards plant roots.

Root exudates may also inhibit egg hatching or may repel nematodes. Some plants, such as marigold, asparagus and tobacco produce chemicals that repel or even kill some nematodes.

Microorganisms in the rhizosphere may significantly influence nematodes in several crops by antagonism, by competition for food and oxygen, or by excretions that may stimulate or inhibit nematodes.

■■■■

Section V

HOST-PARASITE INTERACTIONS

Chapter 17

FEEDING

17.1. INTRODUCTION

Some nematodes feed on the buds, stems and leaves, but a much larger number feed in or on the roots and other structures that grow below-ground. All plant parasitic nematodes have at the anterior end a stylet or spear, which is a sharp-pointed structure that may be used to puncture plant cells and suck out the protoplasmic content. They feed on individual cells (Fig. 17.1).

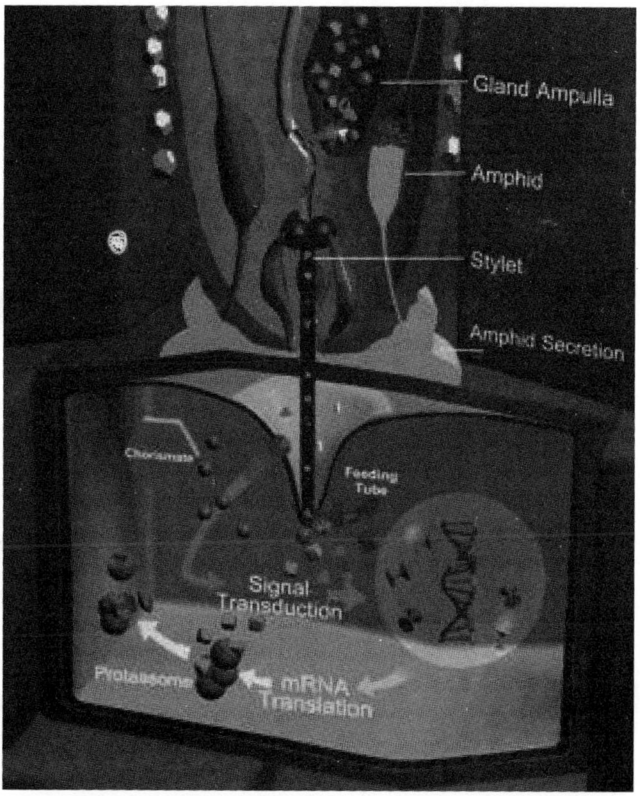

Fig. 17.1. Feeding of a plant parasitic nematode on a plant cell.

17.2. ECOLOGICAL CLASSIFICATION OF PLANT PARASITIC NEMATODES

Plant parasitic nematodes can be divided into three broad categories according to their feeding habits (Fig. 17.2).

- Ectoparasitic nematodes
  - Migratory ectoparasites
  - Sedentary ectoparasites
- Endoparasitic nematodes
  - Migratory endoparasites
  - Sedentary endoparasites
- Semi-endoparastic nematodes
  - Migratory semi-endoparasites
  - Sedentary semi-endoparasites

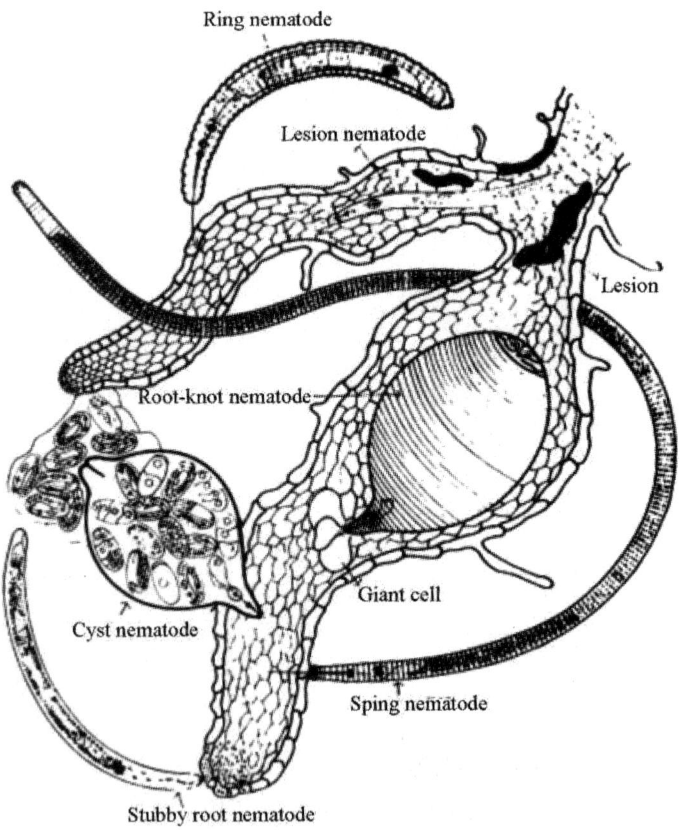

Fig. 17.2. Feeding habits of plant parasitic nematodes.

17.2.1. Ectoparasitic nematodes

These are free living forms that move about freely near or on the surface of the roots intermittently feeding on the epidermis and root hairs near the root tip. This group is in turn subdivided into two:

17.2.1.1. Migratory ectoparasites: These nematodes spend their entire life cycle free in the soil, feeding externally on the roots of host plants (Fig. 17.3). They usually stop feeding and detach themselves when the roots are disturbed. Eggs are deposited in the soil. The nematodes belonging to this group are *Trichodorus* sp., *Paratrichodorus*, *Longidorus* sp., *Xiphinema* sp., *Belonolaimus*, *Dolichodorus*, *Paratylenchus* sp. and *Criconemoides* sp.

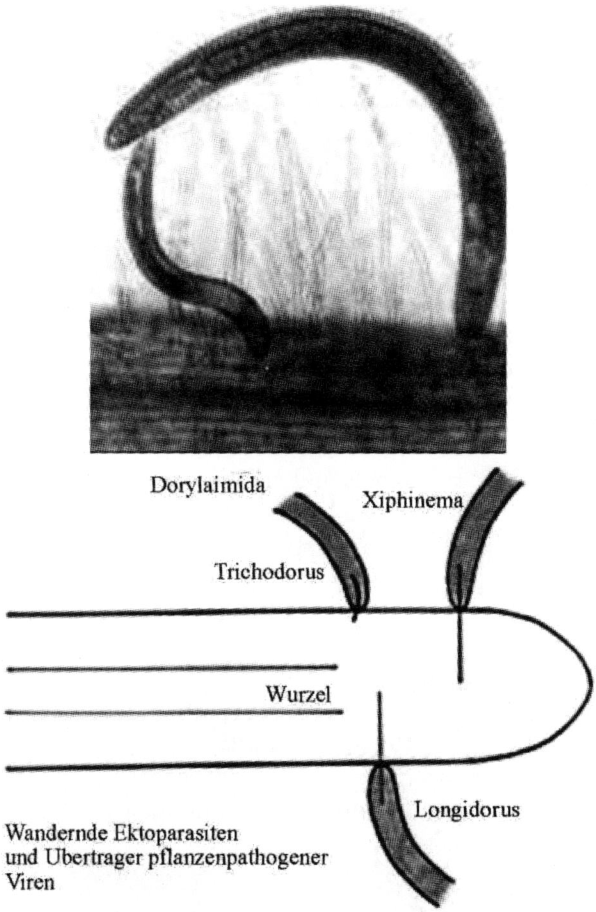

Fig. 17.3. Migratory ectoparasitic nematode

17.2.1.2. Sedentary ectoparasites: They are similar to the above nematodes but differ in that attachment of the adult female to the host roots is more permanent. These have long stylets that penetrate deep into the root. In most cases, these nematodes are essentially sedentary after attachment to the root (Fig. 17.4). The nematodes belonging to this group are the *Hemicycliophora arenaria, Hemicriconemoides* sp. and *Cacopaurus pestis*.

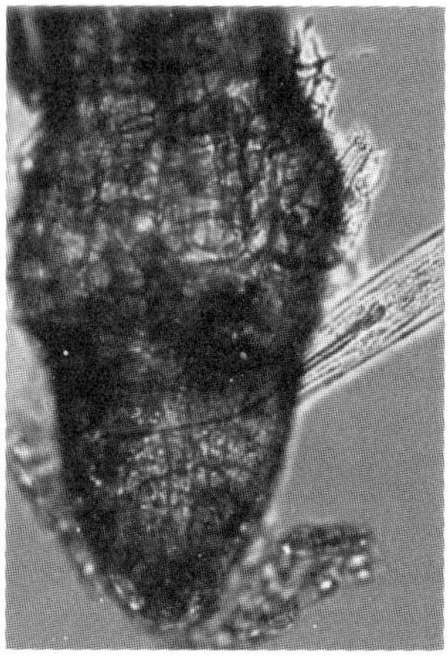

Fig. 17.4 . Sedentary ectoparasitic nematode

17.2.2. Endoparasitic nematodes

These nematodes completely enter the root and lose physical contact with the soil or a large part of their body is in plant tissue. The normal habitat for these nematodes is inside the roots where they move about freely in the tunnels and cavities formed as they feed.

Eggs are deposited in the tunnels and cavities where they hatch, and the life cycle seems to be generally completed. The nematodes in this category seem to be adapted for life within the roots and only move about in the soil while migrating in search of new roots. This category is also subdivided into two groups:

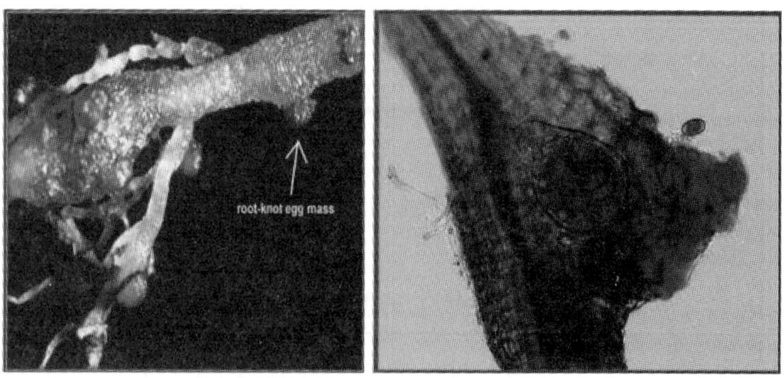

Fig. 17.5. Sedentary endoparasitic nematode (*Meloidogyne* sp.)

17.2.2.1. Sedentary endoparasites: These penetrate rootlets as second stage larvae, become sedentary before molting and remain through the remainder of the life cycle. Eggs are deposited in a matrix as in the case of root-knot nematodes (*Meloidogyne* sp) (Fig. 17.5), or may be retained inside the body of the female, as in cyst nematodes (*Heterodera* sp). The cuticle of the cyst nematodes protects the eggs from adverse conditions.

17.2.2.2. Migratory endoparasites: These are migratory in the cortical parenchyma of host roots. They move through the root tissue, feeding on cells, multiplying and often causing necrosis of root tissues (Fig. 17.6). Overwintering occurs in the roots, or in the soil about the roots, without any known special mechanism for protection from adverse conditions. The nematodes belonging to this group are species of *Pratylenchus, Radopholus* and *Hirschmanniella.*

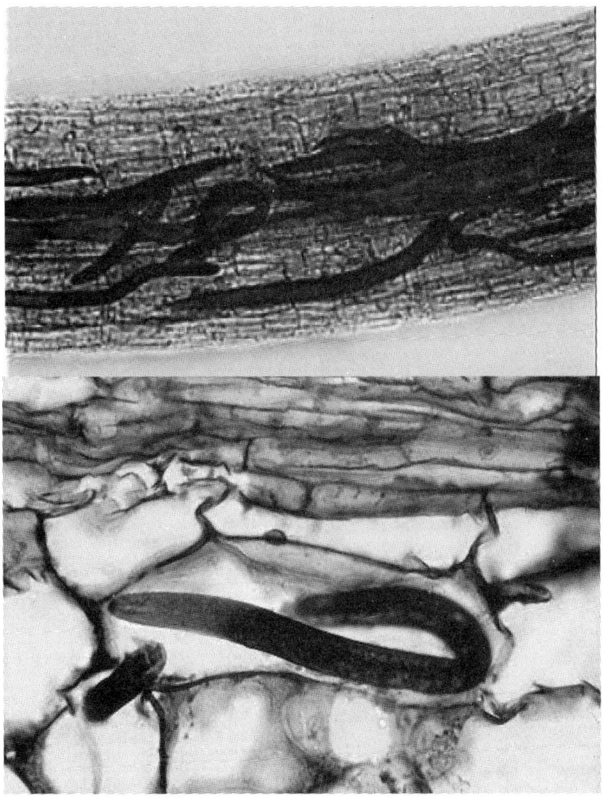

Fig. 17.6. Migratory endoparasitic nematode (*Radopholus* sp.)

17.2.3. Semi-endoparastic nematodes

These nematodes only partially penetrate the root. The head and neck become permanently fixed in the cortex and the posterior portion extends free into the soil. Physical contact with the soil is maintained only by the posterior portion of the body.

17.2.3.1. Sedentary semi-endoparastic nematodes: Once these nematodes penetrate the root, they become immobile. The eggs laid in the soil remain clustered about the posterior of the female until they hatch, and the larvae then move about freely over the root surface or through the soil in search of a feeding site (Fig. 17.7). The nematodes belonging to this group are species of *Tylenchulus, Rotylenchulus* and *Nacobbus.*

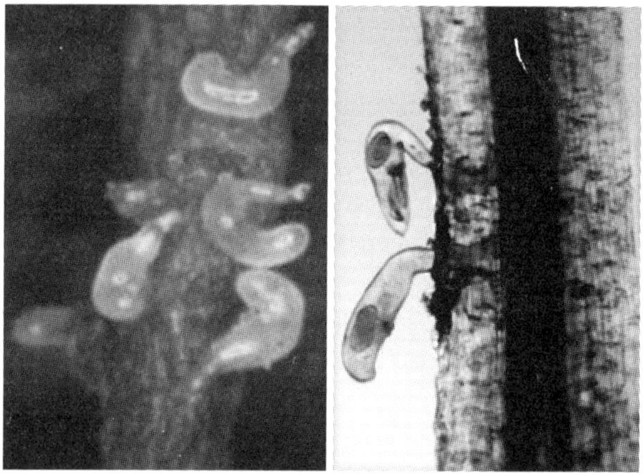

Fig. 17.7. Semiendoparasitic nematode (*Rotylenchulus* sp.)

17.2.3.2. Migratory semi-endoparastic nematodes: They remain vermiform, do not establish a permanent feeding site in plant tissues, rarely the entire body may be inside the plant tissues (Fig. 17.8). The nematodes belonging to this group are species of *Tylenchorhynchus, Helicotylenchus* and *Hoplolaimus*.

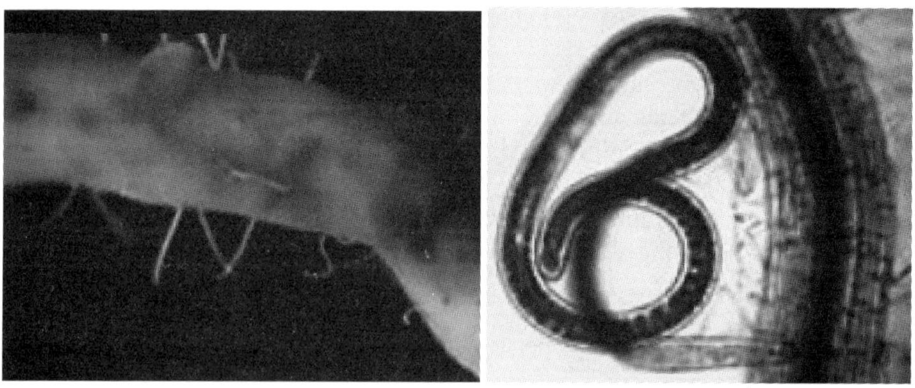

Fig. 17.8. Migratory semi-endoparasitic nematodes

17.3 FEEDING

Behavior leading to feeding or penetration falls into three sequences in most tylenchid species. The nematode moves randomly in the soil, with periods of activity interspersed with quiescent periods, until a host is located. When it detects an attractive chemical stimulus from a plant it moves up the concentration gradient towards the source. The migration is accompanied by searching movements of the head, exploratory probing with the stylet, and the accumulation of secretary material in the ducts of the pharyngeal glands. This is called ' widespread exploration' . When the nematode reaches the surface of the plant, it glides over the surface making intermittent probes with the stylet and occasionally stopping to explore the cell surface more intensively. This is called ' local exploration' and probably represents a search for a suitable spot to attack. The sensory lips are pressed against the surface of the plant cell and rubbed against it with brief pauses when probing with the stylet occurs, during exploration of the cell surface. The nematodes seems to prefer rigid to flexible cell surfaces as most probing occurs here. The mouth and stylet are placed at high angles to the cell surface when a suitable site has been found while the rest of the body usually remains immobile and is often braced against surrounding soil particles or projection on the plant. This allows maximum thrust to occur. The stylet is then repeatedly averted against the cell wall until the cell is punctured; this is carried out more frequently, more rhythmically and more powerfully than exploratory probings. The nematode leaves and continues widespread exploration if the cell contents are not suitable but ingestion or penetration commences if the contents are suitable.

Following penetration of a cell, the contents may be withdrawn immediately through the stylet by means of the pumping action of the pharynx, or secretions from the pharyngeal glands may pass through the stylet into the cell where they bring about extra-corporeal digestion of the cell contents. The wholly or partially digested material is then withdrawn through the stylet by means of the sucking action of the pharynx. *Hexatylus viviparus,* however, feeds passively on the host cells, repeated defecation bringing about a lower pressure in the alimentary tract than in the host cell, with the result that cytoplasm flows from the host cell into the nematode. The time taken for salivation can range from 20 seconds to two hours in different species and the time taken for complete ingestion of the cell contents can range from 30 seconds to one week. *Ditylenchus* have a large median bulb, a well-developed dorsal pharyngeal gland and carry out more extra corporeal digestion. The median bulb is more important than the posterior bulb in ingestion and takes a longer period of time over salivation and ingestion.

Nematodes that enter the tissues of the plant and become endoparasitic first enter through the epidermis lenticels or stomata. When juveniles of *Heterodera* penetrates into the tissues of the host plant the stylet is withdrawn after penetration of a cell and another hole, adjacent to the first, is made in the cell wall. This is repeated several times and as each perforation is made it merges with the previous one until eventually, a slit is made. This is referred to as the ' cutting cycle' . The juvenile nematode then passes through the slit in the wall of the cell and repeats the attack on a cell deeper in the plant. A juvenile of *H. cruciferae* takes about five minutes to cut a slit large enough for it to pass through. When juveniles of *Heterodera* and *Meloidogyne* reaches the vascular bundle within the plant they become sedentary, feeding commences and giant cells are formed.

17.4. METHODS OF FEEDING

Plant nematodes can be grouped into two based on the methods of feeding:

17.4.1. Nematodes that bring about destructive cellular changes

Some nematodes withdraw small amounts of material from host cells and little visible damage results. For example, when *Tylenchorhynchus dubius* feeds upon root hairs, a spherical mass forms at the stylet tip inside the cell. After 30 seconds the bulb pulsates and the spherical mass disappears within one minute. There are no other visible changes in the host cell and it survives the attack.

Many nematodes that feed on the cells of higher plant affect only those cells that they attack. For example, the stylet of *Rotylenchulus* penetrates the cell wall into the cell vacuole and the nematode may remain feeding on this cell for some time. The cell turns brown and dies when the nematode withdraws, but adjacent cells are not affected. Some species of *Trichodorus* and *Hemicycliophora* secrete a feeding tube or plug, which encircles the stylet and attaches the nematode firmly to the feeding site. This enables the nematode to maintain the feeding position for long periods. *Longidorus* possesses a long stylet and this allows the nematode to feed upon the stele while remaining outside the plant. Some nematode can alter the permeability and nature of the contents of cells adjacent to the attacked cell, for example, *Radopholus similis* brings about the disappearance of starch in cells adjacent to attacked cells and *Pratylenchus* affects the permeability of tissues at some distance from itself. *Ditylenchus dipsaci* brings about separation and hypertrophy of cells at some distance from itself.

17.4.2. Nematodes that bring about adaptive cellular changes in the host

Some plant parasitic nematodes bring about non-destructive alterations in the metabolism and structure of host cells and tissues and there are varying degrees of changes brought about by different species.

Juveniles of *Tylenchulus semipenetrans* feeds on superficial cells of roots. The young adult female, unlike the juvenile, penetrates more deeply into the cortex and its head becomes surrounded by six to ten altered cortical cells (nurse cells) upon which it feeds.

These altered cells are normal in size but have a dense cytoplasm, an enlarged nucleus and nucleolus, and no vacuole. After a period of time these nurse cells disintegrate and the nematode moves further along the cortex. A development of this type of feeding is shown by *Nacobbus,* which forms galls on roots. Juveniles invade roots by intracellular migration leaving a tunnel of destroyed cells behind them. The fourth stage juvenile and the female become sessile in the cortex where they induce the formation of a syncytial mass by breaking down the lateral walls of cells adjacent to the head of the nematode. A channel extends from the head of the nematodes into the interior of the syncytial mass and the nematodes feeds on this mass.

Females of *Heterodera* and *Meloidogyne* induces the formation of multinucleate syncytia or giant cells in the cortex or stele of the root (Fig. 17.9). The giant cells produced by the plant in response to attack by *Heterodera* are apparently transfer cells and provide a localized source of protein for the nematode, which increases in weight one hundred times during development. It is suspected that roots containing these giant cells shift their metabolism in the direction of increased protein synthesis and reduced the transport of nutrients to the rest of the plant. This differentiation of unspecialized cells of the root to form transfer cells, which have a specific physiological function in the plant, is thought to be induced by the feeding demands of the nematodes and by the introduction of plant growth regulators into the incipient giant cells. The syncytia induced by *Heterodera* tends to retain remnants of cell walls but many nuclei are present in the syncytium. It is worth noting that certain varieties of potato, which are resistant to *Globodera rostochiensis* do not form giant cells when attacked by this nematode and most juveniles develops into males. Giant cells are not essential for the development of the male. Nematodes in this group all seems to stimulate cells to increase synthesis of products upon which the nematodes feed. Similarly, at low nematode densities on tomato roots most juveniles of *Meloidogyne incognita* developes into females, but at high densities most develop into males. This is brought about by sex reversal as second stage `female' juveniles can develop into males that have two, instead of the normal one, testes.

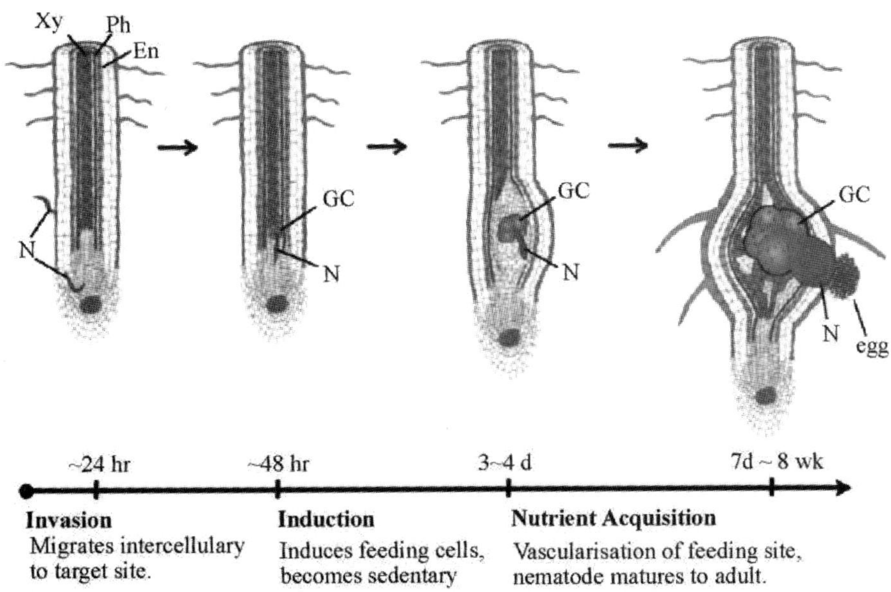

Fig. 17.9. Schematic representation of the root-knot nematode infestation life-cycle.

Root-knot nematode (N) second stage juveniles penetrate a host root and migrate to the target cells within the endodermis (En) where they induce giant cells (GC). These giant cells, which are multinucleate transfer-like cells, become the feeding cells that support the sedentary nematode for the remainder of its life-cycle. The division of surrounding cells during the induction stage probably disrupts vascular continuity, although the exact identity of the proliferating cells and specific nature of this vascular disruption remains to be experimentally verified. Vascularization of the feeding site during nutrient acquisition and restoration of vascular continuity results in an elaborate network of xylem (Xy, pink) and phloem (Ph, green) in the root galls. Once feeding commences, the nematode molts a further three times inside the root and reaches a spherical shape as an adult, after which the females lay approximately 500 eggs on the root surface in a gelatinous egg mass.

Chapter 18

SYMPTOMS

18.1. INTRODUCTION

The recognition of symptoms is important in diagnosis of nematode diseases. However, complete diagnosis cannot be based on symptoms alone. In all cases, the causal organism should be identified.

18.2. SYMPTOMS PRODUCED BY BELOW-GROUND FEEDERS

The nematodes which feed on below-ground plant parts produce symptoms on both above- ground and below-ground plant parts.

The above-ground symptoms produced by below-ground feeders are mostly non-specific such as poor growth, stunting, patchiness (Fig. 18.1) and discolored foliage in a crop.

Fig. 18.1. Patchiness of potato field infected with cyst nematodes

18.2.1. Stunting

Stunting (Fig. 18.2) and slow growth are common in nematode attacks and are often seen as patches in a field. Symptoms such as these are produced, for example, by *Heterodera avenae* in wheat and barley, *Globodera rostochiensis* in potatoes and *Meloidogyne incognita* in peas. The small area of stunted plants may be caused by a non-random distribution of nematodes over the field so that where high nematode populations occurs, disease symptoms are more pronounced. The introduction of nematode infested seeds into uninfested soil might also produce isolated pockets of infestation in a field. Decline and dieback in citrus may be produced by *Tylenchulus semipenetrans.*

Plate 18.2. Stunting of potato plants due to cyst nematode infection. *Left* – Healthy, *Right* - Infected.

18.2.2. Yellowing

This is indicative of nutritional deficiency and with nematode infestations, the cause often lies in root destruction. Such symptoms have been observed in coffee infested with *Pratylenchus coffeae* and black pepper infested with *Radopholus similis* (Plate 18.3). In some infestations of potatoes by *G. rostochiensis,* the foliage appears to be a lighter green. *Aphelenchoides besseyi* produces pale yellow or white tips on the leaves of rice. *Tylenchulus semipenetrans* is reported to induce a fine mottle on the leaves of orange and lemon trees.

Fig. 18.3. Yellowing of black pepper and gladiolus leaves due to burrowing and root-knot nematode infection, respectively

18.2.3. Wilting

Wilting is often the result of root damage which may be caused by nematodes. The plants infested with nematodes tends to wilt in hot weather even though enough moisture is present in the soil. Such symptoms are produced in eggplant infested with the root-knot nematode (Plate 18.4).

Fig. 18.4. Wilting of eggplant infected with root-knot nematodes

The symptoms produced by below-ground feeders on below-ground parts are as follows:

18.2.4. Root galls

Meloidogyne sp. forms characteristic galls on the roots of host plants (Plate 18.5). *Nacobbus batatiformis* and *N. serendipiticus* produces larger galls on beet and tomatoes, respectively. *Ditylenchus radicicola* causes small galls on the roots of oats, barley, rye, wheat and some of the grasses. *Hemicycliophora arenaria* induces galling in lemon roots. *Xiphinema diversicaudatum* causes galls in the roots of roses. *Longidorus* species induces galls in the roots of bur marigold.

18.2.5. Reduced root system

When the nematode feeds on root tips, it often stops growing. A root thus affected usually branches and the growth of these branches also may be stopped by further nematode feeding resulting in stubby root or coarse root.

18.2.5.1. Stubby root: If the branch rootlets manage to attain a moderate length before their growth is stopped, the resulting root system may be composed of numerous short stubby branches often arranged in clusters (Fig. 18.6). This type of symptom is produced by *Trichodorus christiei* on corn.

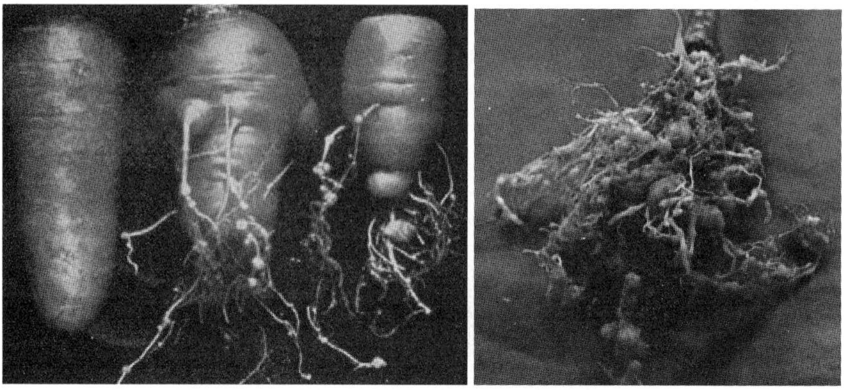

Plate 18.5. Carrot and tomato roots infected by root-knot nematodes.

Fig. 18.6. Abbreviated "stubby-root" symptoms resulting from damage caused by ectoparasitic nematodes.

18.2.5.2. Coarse root: If the growth of lateral roots is stopped just as most of them are breaking through the cortex or while they are very short, an open system made up mostly of the main roots largely devoid of small branch rootlets may result (Fig. 18.7). This type of symptoms is produced by *Belonolaimus longicaudatus* on corn.

Fig. 18.7. Coarse root symptoms on strawberry due to damage by *Belonolaimus longicaudatus*

18.2.6. Root lesions

Necrotic lesions are often visible on the roots of plants infested by *Pratylenchus* spp. *Radopholus similis* causes similar symptoms on citrus. In banana, *R. similis* produces lesions (Fig. 18.8) which girdles the roots, thereby weakening their anchoring power so that at a later stage the trees tends to be uprooted. Lesions are also visible on banana roots attacked by *Helicotylenchus multicinctus*.

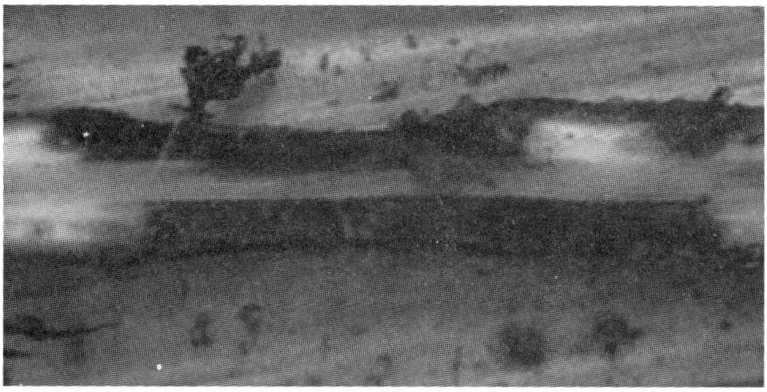

Fig. 18.8. Dark lesions and necrotic roots induced by the burrowing nematode on banana.

18.2.7. Rots of fleshy parts

Nematodes that enter fleshy structures may initiate injury resulting in extensive tissue destruction. While nematodes initiate the damage, the destruction may be continued by other organisms of decay that enter as secondary invaders. The potato rot nematode, *Ditylenchus destructor* and the yam nematode, *Seutellonema bradys* are examples of species that cause this type of injury (Plate 18.9).

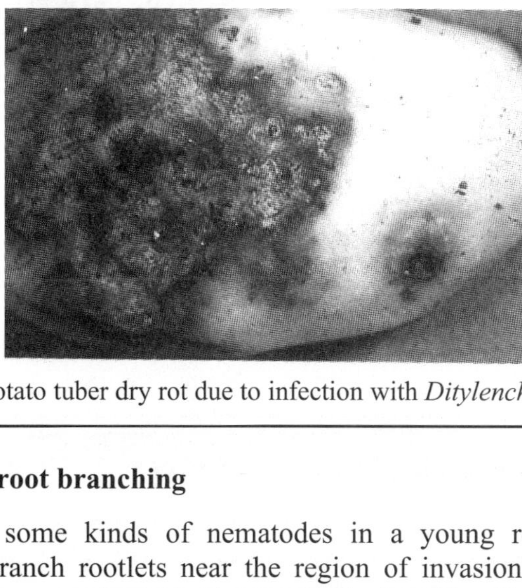

Plate 18.9. Potato tuber dry rot due to infection with *Ditylenchus destructor*.

18.2.8. Excessive root branching

The presence of some kinds of nematodes in a young root stimulates the development of branch rootlets near the region of invasion (Fig. 18.10). The northern root-knot nematode, *Meloidogyne hapla,* has this effect on tomato.

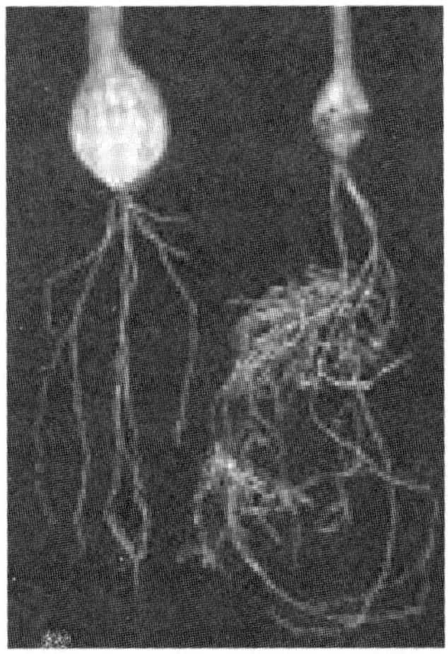

Fig. 18.10. Excessive root branching due to root-knot nematode infection on onion.

18.3. SYMPTOMS PRODUCED BY ABOVE-GROUND FEEDERS

The nematodes that feed on above-ground plant parts produces symptoms only on aerial plant parts.

18.3.1. Dead or devitalized buds

Sometimes a nematode infection may kill a bud or growing point and result in a "blind" plant (Fig. 18.11). This happens occasionally with strawberry plants infected with *Aphelenchoides fragariae.*

Fig. 18.11. Death of the strawberry crown bud due to infection with *Aphelenchoides fragariae.*

18.3.2. Crinkled and distorted stems and foliage

When an infected growing point is not killed but continues to function, the stems, foliage and other structures that develop from it are likely to be crinkled, twisted and distorted (Fig. 18.12). The wheat seed gall nematode, *Anguina tritici* which during its larval stages lives in the growing point of the wheat plant may cause this type of injury.

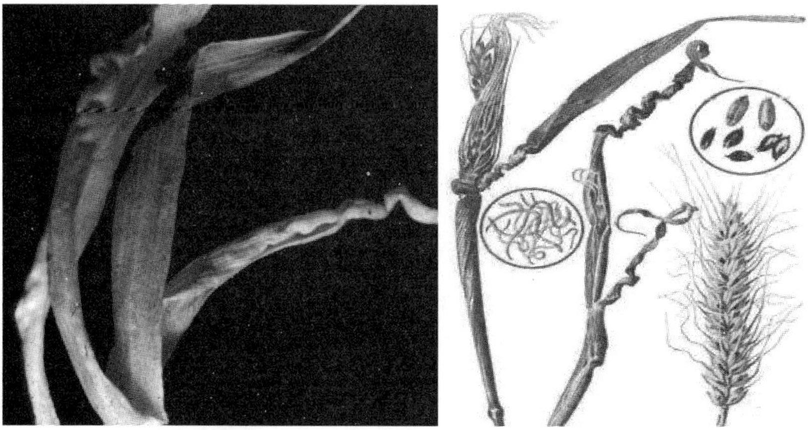

Fig. 18.12. Crinkled and distorted stems and foliage of wheat infected with *Anguina tritici.*

18.3.3. Seed galls

These are caused by species of the genus *Anguina* on grasses and cereals. When the flower heads begin to form, almost fully grown nematode larvae enter the flower primordia where they quickly mature and reproduce. Instead of developing into a normal seed, an infected flower primordium develops into a gall containing a large number of larval nematodes capable of surviving long periods of desiccation (Plate 18.13).

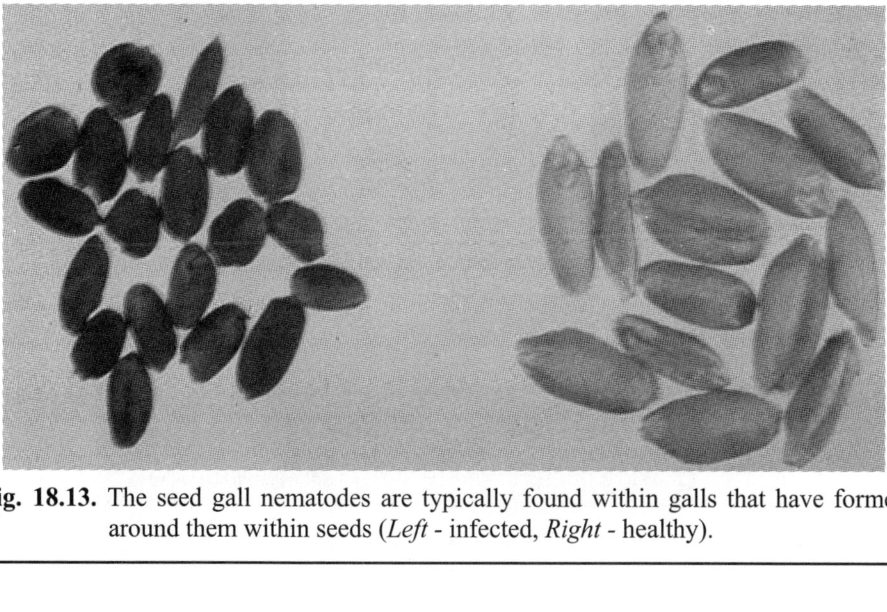

Fig. 18.13. The seed gall nematodes are typically found within galls that have formed around them within seeds (*Left* - infected, *Right* - healthy).

18.3.4. Necrosis and discoloration

Some nematodes live and feed within the tissues of stems and leaves and cause varying degrees of necrosis and discoloration. The "red-ring" disease of coconut palms, caused by *Bursaphelenchus cocophilus*, is an example of this type of injury occurring in the trunk of a tree (Plate 18.14).

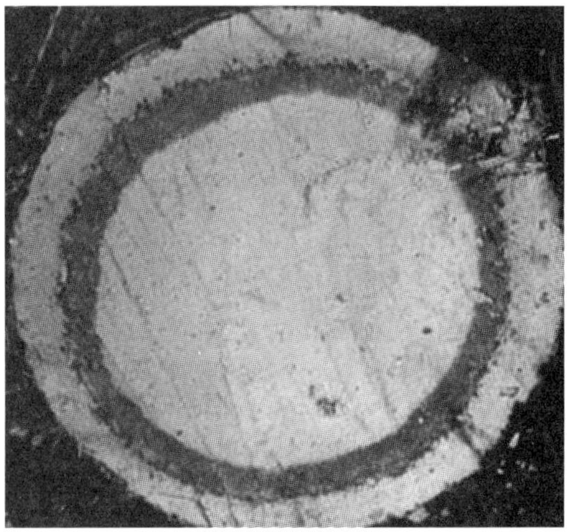

Plate 18.14. Cross section of coconut stem showing `red ring' symptoms caused by *Bursaphelenchus cocophilus.*

18.3.5. Leaf lesions

These are caused by nematodes that penetrate the foliage of broad leaved plants, often through the stomata, to feed on and destroy the leaf parenchyma. The chrysanthemum foliar nematode, *Aphelenchoides ritzemabosi* is an example of species that cause leaf lesions (Plate 18.15).

Fig. 18.15. Chrysanthemum leaves infected by *Aphelenchoides ritzemabosi.*

18.3.6. Leaf galls

Anguina balsamophila and A. millefolii produces galls on the leaves of their host plants (Fig. 18.16). These are comparable to seed galls and they serve as structures in which the parasites mature and reproduce.

Fig. 18.16. Leaf galls produced by *Anguina balsamophila.*

Chapter 19

PHYSIOLOGICAL RACES

19.1. INTRODUCTION

Races of nematodes are of obvious importance to the plant breeder who is trying to produce crops resistant to nematode attack and to the nematologist who relies to some extent on the pathological reactions of the plant to identify the nematode. Successful nematode control by crop rotation also depends on knowing which crops, a particular nematode will attack. A consideration of races in host parasite relationships is therefore important, especially in management problems.

Biological or physiological races are generally host races, showing none or only small morphological differences but definite difference in biology such as food preferences. Races often comprises several biotypes, each of which is capable of a change in virulence by genetic mutation or recombination to give rise to new biotypes and new races. New biological races can also arise by selection among different biotypes within a race.

19.2. PHYSIOLOGICAL RACES

There are numerous examples of plant parasitic nematodes where races are known to occur.

19.2.1. *Ditylenchus dipsaci*

More than 30 races, differing in their host ranges and named after their preferred hosts or those plants on which they were first detected, have been described. Many races are polyphagous, attacking plants from many different families, while others appear to have a limited host range.

19.2.2. *Globodera rostochiensis*

At present five pathotypes are differentiated by their ability to overcome resistance in potato varieties bred from *Solanum andigenum, S. kurtzianum, S. verni* and *S. tuberosum* (Table 19.1).

Table 19.1. Pathotypes of *Globodera rostochiensis*

Differential host	Nematode pathotypes*				
	1	**2**	**3**	**4**	**5**
Solanum tuberosum	+	+	+	+	+
ex *andigena* CPC 1673	–	+	+	–	+
ex *kurtzianum* 60.21.19	–	–	+	+	+
ex *vernei* 58.1642/4	–	–	–	+	+
ex *vernei* 62.33/3	–	–	–	–	+
ex *vernei* 65.346/19	–	–	–	–	–

* (+) indicates ability to produce females, (–) indicates inability to produce females

19.2.3. *Globodera pallida*

At present, four pathotypes in *G. pallida* are differentiated using the following differentials (Table 19.2):

Table 19.2. Pathotypes of *Globodera pallida*

Differential host	Nematode pathotypes*			
	1	**2**	**3**	**4**
Solanum tuberosum	+	+	+	+
ex *multidissectum* P 55/7	–	+	+	+
ex *vernei* 62.33/3	–	–	+	+
ex *andigena* x *multidissectum* D 47/1	–	–	–	–
ex *andigena* x *multidissectum* D 47/2	–	–	–	–

* (+) indicates ability to produce females, (–) indicates inability to produce females

19.2.4. *Heterodera avenae*

Two races have been distinguished in Denmark, Sweden and Ireland; three in England and four in the Netherlands and Germany (Gair, 1965).

At least 4 races (or biotypes) of *Heterodera avenae* are known: barley varieties may be resistant to one and not the others whereas oats and wheat appears to be resistant to all or none. Drost barley is resistant to race No. 1, LP191 is resistant to Nos. 1 and 2 and Morocco is resistant to others.

Studies conducted at Jaipur (Mathur *et al.,* 1974) indicated the presence of at least 5 biotypes in Rajasthan, characterized by the following host preferences (Table 19.3):

Table 19.3. Biotypes of *Heterodera avenae* (Mathur *et al.,* 1974)

Biotypes	Formation of cysts on					
	Barley			Oat	*Lolium perenne*	*Dactylis glomerata*
	191	Herta	Drost	Sun II		
1	–	–	–	–	+	–
2	+	+	–	–	–	–
3	–	+	–	–	–	–
4	–	+	–	–	–	+
5	–	+	+	–	–	–

19.2.5. *Heterodera glycines*

Studies by Epps and Golden (1967) indicate that there are two, possibly three, biological races in the United States.

Four races of the soybean cyst nematode, *Heterodera glycines* have been recognized on the basis of the ability to reproduce on three resistant soybeans, *Glycine max* cultivars or breeding lines (Golden *et al.*, 1970). All four races reproduce freely on the susceptible cultivar Lee. Race-3 cannot reproduce on any of the resistant soybeans; race-I reproduces only on P.I. 88788 resistant soybeans; race-2 on P.I. 88788 and Pickett, and race-4 on all three resistant soybeans, P.I. 88788, Pickett, and P.I. 90763.

19.2.6. *Radopholus similis*

Two races, which are found in Florida, are separated by their ability to parasitize citrus and banana, the "banana race" attacking banana but not citrus and ``citrus race" attacking both. There is an evidence of a third race in Florida (Du-Charme and Birchfield, 1956). Now the citrus race has been included as separate species *Radopholus citrophilus* (Huettel *et al.*, 1984).

Venkitesan (1976) reported four host types of *R. similis* based on differential hosts. Population isolated from pepper did not infect *Citrus sinensis, C. reticulata, C. aurantifolia, Coffea arabica,* coconut and areca nut; but infected banana only. The banana isolates infected pepper, coconut and areca nut. Furthermore, the coconut isolates infected pepper and banana but not areca nut. The arecanut isolate inturn infected pepper and banana but not coconut.

19.2.7. *Belonolaimus longicaudatus*

At least 2 and possibly 3 physiological strains differing in their ability to attack citrus and groundnut have been discovered in Florida (Abu-Gharbieh and Perry, 1970):

- The Fuller's Crossing population reproduced well on and caused injury to rough lemon and tomato but not strawberry;
- The Gainesville population reproduced well on and caused injury to peanut, strawberry and tomato but not on rough lemon;
- The Sanford population reproduced well on and caused injury to peanut, reproduced well on tomato, but neither reproduced upon nor caused injury to strawberry or rough lemon.

19.2.8. *Rotylenchulus reniformis*

At least 2 races, A and B, have been detected in India, the former multiplying on cowpea, castor and cotton and the latter reproducing only on cowpea (Dasgupta and Seshadri, 1971).

19.2.9. *Meloidogyne incognita*

Taylor and Sasser (1978) reported four races of *M. incognita.* Race 1 did not reproduce on root-knot resistant tobacco (cv. NC 95) or cotton. Race 2 reproduced on tobacco but not on cotton. Race 3 reproduced on cotton but not on tobacco. Race 4 reproduced on both tobacco and cotton (Table 19.4).

Table 19.4. Identification of the races of *Meloidogyne incognita* using the differential host test (Sasser and Carter, 1985; Hartman and Sasser 1985).

Nematode races	Cotton cv. Deltapine	Tobacco cv. NC 95	Pepper cv. California Wonder	Watermelon cv. Charleston Gray	Peanut cv. Florunner	Tomato cv. Rutgers
Race 1	–	–	+	+	–	+
Race 2	–	+	+	+	–	+
Race 3	+	–	+	+	–	+
Race 4	+	+	+	+	–	+

19.2.10. *Meloidogyne arenaria*

Taylor and Sasser (1978) reported 2 races of *M arenaria.* Race 1 reproduced on peanut while Race 2 did not (Table 19.5).

19.2.11. *Tylenchulus semipenetrans*

Four biotypes were differentiated on the basis of infectivity on sweet orange, citrange, *Poncirus trifoliata*, grapevine and olive (Baines *et al.,* 1974). A "grass strain", which parasitized *Andropogon rhizomatus* but failed to infect citrus, was reported by Stokes (1969).

Table 19.5. Identification of the races of *Meloidogyne arenaria* using the differential host test (Taylor and Sasser, 1978).

Nematode races	Cotton cv. Deltapine	Tobacco cv. NC 95	Pepper cv. California Wonder	Watermelon cv. Charleston Gray	Peanut cv. Florunner	Tomato cv. Rutgers
Race 1	–	–	+	+	–	+
Race 2	–	+	+	+	–	+

The citrus biotype infects many genera in the Rutaceae family, including *Citrus* sp., Troyer and Carrizo citrange (*Citrus sinensis* x *Poncirus trifoliata*), as well as olive (*Olea europaea*), grape (*Vitis vinifera*) and persimmon (*Diospyros* sp.), but reproduces poorly on *Poncirus trifoliata* and some hybrids of this genus. The host range of the Mediterranean biotype is very similar to the citrus biotype, with the exception of olive. The Poncirus biotype reproduces on *Citrus* spp., *P. trifoliata* and hybrids of *P. trifoliata*, as well as grape, but not olive. This biotype has been identified in California, Japan and Israel. The citrus biotype occurs in California and Italy. The Mediterranean biotype has been identified in Mediterranean countries and South Africa. The grass race is presently known only from Florida. It is only known true host is *Schizachyrium rhizomatus*.

It is clear from this brief review that biological races in plant parasitic nematode species are by no means a rarity; in fact it is possible that they may occur in most species.

■■■■

Chapter 20

HISTOPATHOLOGY

20.1. INTRODUCTION

Lesion formation, necrosis, cell hypertrophy, hyperlasia, formation of tyloses and abnormal cell division have been observed in nematode infested plants. Brief descriptions of the histopathology of plants associated with infestations of a few selected genera will help to illustrate these phenomena.

20.2. HISTOPATHOLOGY

20.2.1. *Meloidogyne*

After invasion, the second stage larvae migrate intercellularly and intracellularly to the site of feeding (phloem) and subsequent nematode growth. The first reaction of the plant to nematode feeding is giant cell formation (Fig. 20.1). Each larva is associated with about 4 to 9 giant cells which are multinucleate. Root galls are produced due to hypertrophy and hyperplasia of cells.

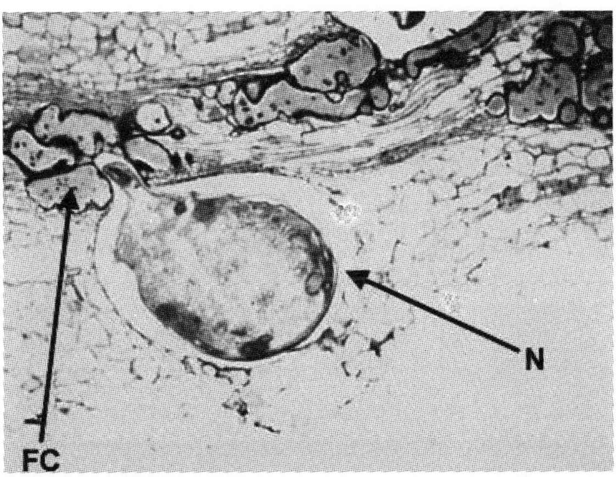

Fig. 20.1. Cross section of a root showing the feeding of a root-knot nematode on giant cells. The arrow labelled as N is pointing to the adult nematode and the arrow labelled as FC is pointing to the multinucleate feeding cells or giant cells (Courtesy H. A. Johnson)

20.2.2. *Heterodera*

The giant cells occur near the head of a larva but root galls are not formed (Fig. 20.2).

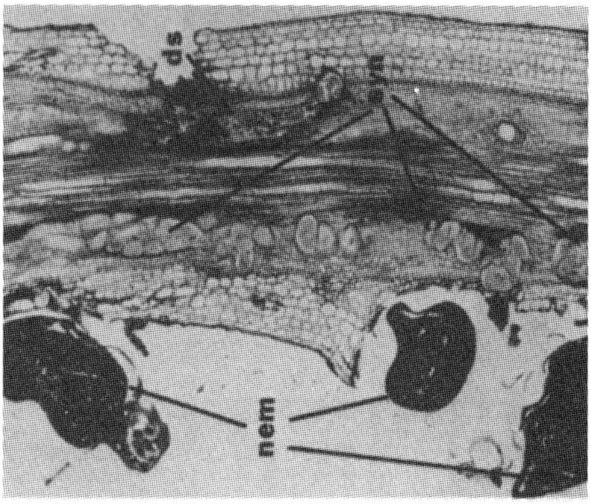

Fig. 20.2. Syncytium induced by the potato cyst nematode *Globodera rostochiensis* in the roots of potato.

20.2.3. *Ditylecnhus*

The larvae inhabits the intercellular spaces of parenchymatous and cortical tissues of leaves and stems where they cause swelling by hypertrophy and hyperplasia of the cells. The middle lamella of cell walls is dissolved and the cells are separated with subsequent formation of large air spaces. When most of the cytoplasm is withdrawn, the cell walls collapses and a cavity forms in the tissue around each nematode.

20.2.4. *Aphelenchoides*

The first reaction of the leaf after the nematode invasion is the production of patches of an opaque brown substance within the cells. Eventually browning extends to the epidermis, few chloroplasts remains and cells in the mesophyll are ruptured, producing large cavities.

20.2.5. *Anguina*

These nematodes produce well defined galls in leaves, stems and inflorescences. The first signs of swelling in the tissues occurs in the mesophyll and in the 2 epidermal layers. The mesophyll cells show marked hypertrophy, their cytoplasm is granular and the plastids are grouped around the periphery of the cell. The epidermal cells immediately above the gall undergo some multiplication and enlargement forming a layer 2 to 3 cells deep, whereas those of the lower epidermis become enlarged and stretched.

20.2.6. *Radopholus*

As the nematodes move through the roots, tunnels and cavities are formed in the cortex and stele. The nematodes pass through the endodermis to reach the phloem cambium ring, where they accumulate in large numbers. Wound gum accumulates next to the stele in the invaded area giving an amber color to the older parts of the lesion. Some cells near the nematode show hypertrophy and hyperplasia (Fig. 20.3).

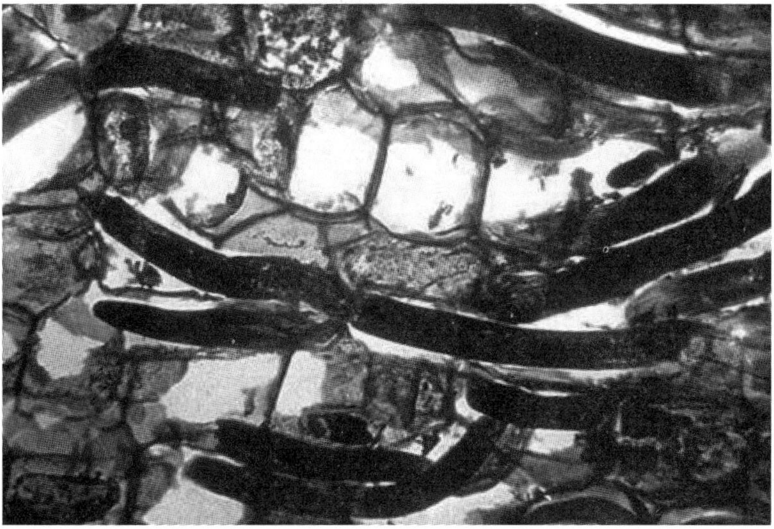

Fig. 20.3. Cavity created in fibrous root cortical tissue by *Radopholus similis.*

20.2.7. *Rotylenchulus*

The permanent feeding site of adult female reniform nematodes is usually initiated in an endodermal cell, although cortical cells near to the endodermis and pericycle cells may be modified first. The initial cell (prosyncyte or feeding cell), usually located near a protoxylem pole expands to 4-5 times its normal size. The nucleus enlarges, cytoplasmic organelles increase and lysis of wall occurs centripetally to the pericycle, so that a syncytium develops (Fig. 20.4). This is extended by incorporation of 5-6 pericycle cells in all directions around the initial cell and these also expand. Beyond the syncytial cells (or feeding zone other epicycle cells are stimulated to fill with cytoplasm, but these remain discrete and uninucleate. The modified cells, 100-200 in all, form a concave diamond shaped region centered on the initial cell. Wall fragments are rounded and walls thickened by polysaccharide deposition. However, wall ingrowths are absent.

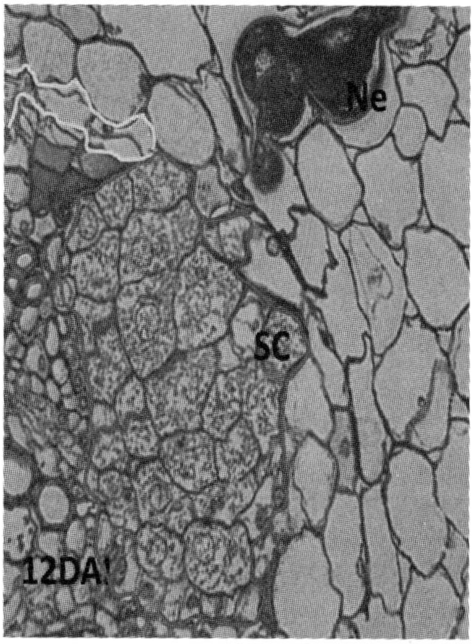

Fig. 20.4. Cross section of a root showing the feeding of a reniform nematode (Ne) on syncytium cells (SC) (circled in red line).

20.2.8. *Tylenchulus*

The nurse cells associated with adult citrus nematodes are located in the root cortex. About 6 to 10 modified cells surround the head of the nematode. The cells are discrete, contain an enlarged nucleus and are filled with cytoplasm. The nematode appears able to move its head, located in an empty cell in the middle of the feeding site, to each of the nurse cells (Fig. 20.5).

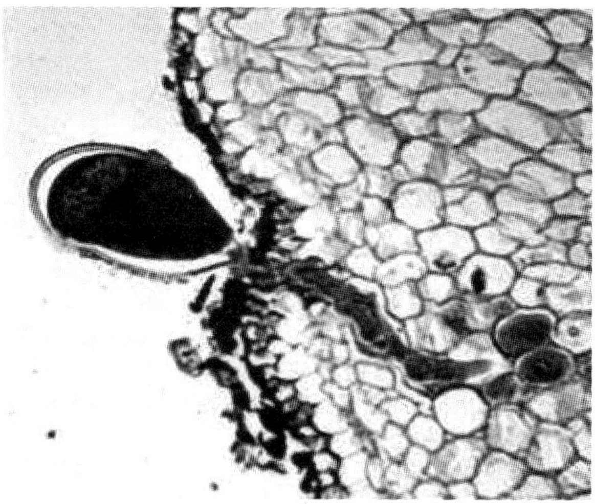

Fig. 20.5. Histopathology of *Tylenchulus semipenetrans*.

20.2.9. *Nacobbus*

It induces the formation of syncytia within galls. The syncytia develop as a spindle or crescent shaped group of interconnected cells 2-3 mm long. The stimulation of cell division that leads to the galling appears to take place before wall breakdown and cell fusion. On incorporation of a new cell into the syncytium, the vacuole is replaced by active cytoplasm, whorls of endoplasmic reticulum and numerous organelles and nuclei and nucleoli enlarge (Fig. 20.6). A notable feature is the presence of large starch grains within plastids, both in syncytium and in gall cells which suggest that sucrose offloading from the phloem is enhanced. Within syncytia, wall fragments are evenly thickened, but wall ingrowths do not occur.

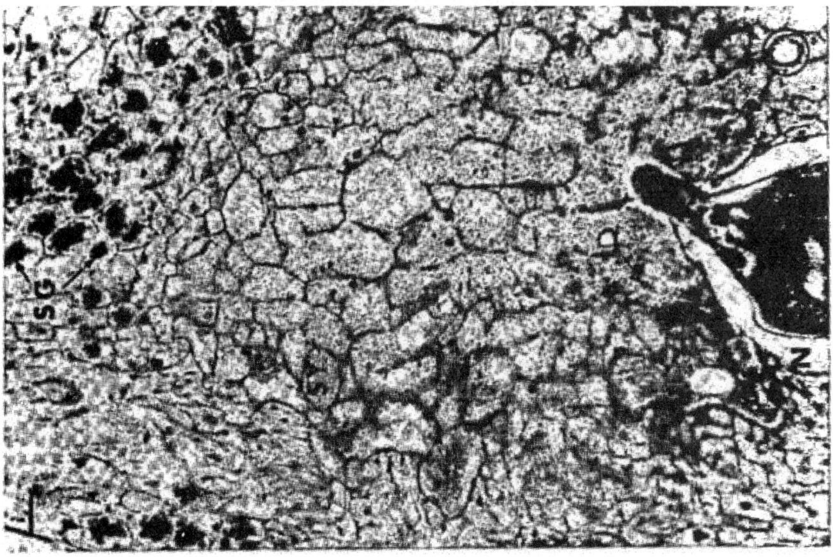

Fig. 20.6. Histopathological changes induced by *N. aberrans* in the susceptible potato cultivar. The syncytium (SY) 30 days after inoculation with accumulation of starch grains (SG) around it.

20.2.10. *Meloidodera*

It appears that the cystoid nematode induces possibly only one giant cell located either in the root cortex or in undifferentiated regions of meristematic tissue associated with mycorrhizae.

20.2.11. *Xiphinema*

The feeding structures induced by *X. index* have some structural parallels with those induced by root-knot nematodes and, in both cases, are produced as a result of repeated rounds of nuclear division without cytokinesis (Fig. 20.7) (Griffiths *et al.*, 1982)

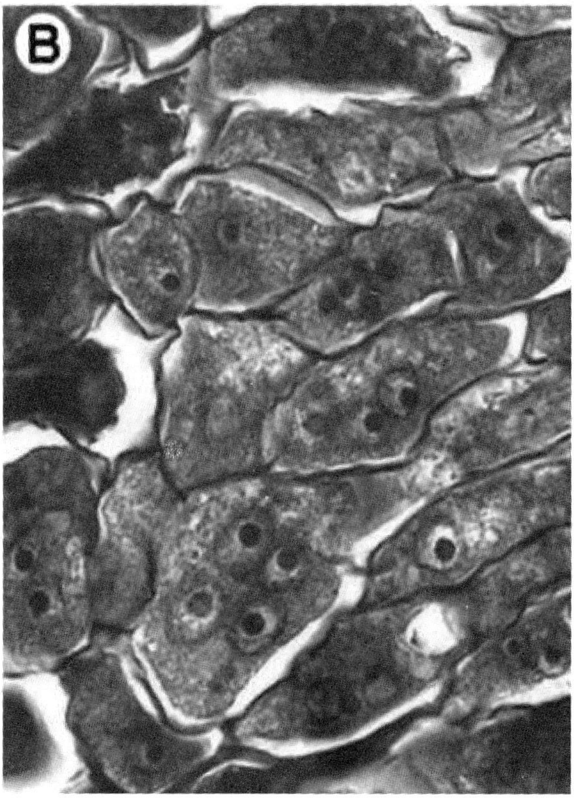

Fig. 20.7. Multinucleate cells induced by *Xiphinema index* in grapevine rootstock Paulsen 1103 (*Vitis berlandieri* x *V. rupestris*).

SECTION - VI
INTERACTION WITH OTHER MICRO-ORGANISMS

Chapter 21

INTERACTION WITH FUNGI

21.1. INTRODUCTION

Many root diseases are caused by nematodes in combination with other soil-borne pathogenic fungi. In order to control disease complexes, the causal agents must be identified and their relative importance determined. Because of the very considerable damage caused by disease complexes, significant breakthroughs in research could be very rewarding in increasing crop yields.

21.2. WILT DISEASE INTERACTIONS

21.2.1. Root-knot and wilt disease interactions

Many reports of nematode fungus interrelationships have involved a nematode pathogen along with one of the wilt inducing fungi. Atkinson (1892) first observed that infection by root-knot nematodes (*Meloidogyne* spp.) increased the severity of *Fusarium* wilt in cotton. Varieties of cotton resistant to *Fusarium* wilt, succumbed to wilt when the root-knot nematode was present in the soil.

Bazan de Segura and Aguilar (1955) have indicated that root-knot nematodes increases damage by the *Verticillium* wilt fungus on cotton.

Jenkins and Coursen (1957) induced wilting in *Fusarium* wilt-resistant tomato variety Chesapeake only when root-knot nematodes were present along with fungal inoculum. Furthermore, when *Meloidogyne hapla* was combined with the fungus, only 60% of the plants wilted, whereas *M. incognita acrita* promoted wilt in 100% of the plants.

Minton and Minton (1963) sectioned roots of cotton seedlings inoculated with *Fusarium oxysporum* f. sp. *vasinfectum* and *M incognita acrita* and found that the fungus grew well in sloughing epidermal cells, decaying cortical cells, xylem and nematode induced giant cells. Decay of nematode infected roots exposing the xylem was thought to be the cause of increased wilt incidence.

Melendez and Powell (1967) reported the root-knot and *Fusarium* wilt interaction in tobacco. Giant cells and adjacent xylem elements in both *Fusarium* - susceptible and - resistant plants were extensively colonized by the fungus.

Occasionally, the female nematode and gelatinous matrix of the egg mass were invaded by the fungus.

In some hosts, the nematodes apparently are capable of maximum predisposition only after 3 to 4 weeks. This time interval is sufficient to permit the formation of galls, along with the accompanying morphological and physiological changes. In fact, this relationship appears so in tobacco that the appearance of *Fusarium* wilt symptoms is more closely related to the time of nematode inoculation than to time of fungus inoculation (Fig. 21.1). Goode and McGuire (1967) found that the root-knot nematode infection enables certain races of *F. oxysporum* f. sp. *lycopersici* to attack tomato varieties ordinarily resistant to those races and suggested that the fungus mutates within the host.

Fig. 21.1. The simultaneous occurrence of both root-knot nematode (*Meloidogyne* sp.) and *Fusarium* wilt disease (*Fusarium oxysporum*) causing enhanced disease development and tomato yield loss.

21.2.2. Cyst nematode and wilt disease interactions

Certain varieties of soybean, inoculated with the *Fusarium* fungus and *Heterodera glycines*, wilt to a greater extent than those predisposed by *M. incognita.*

21.2.3. Burrowing nematode and wilt disease interactions

Incidence and losses due to Panama wilt caused by *F. oxysporum* f. sp. *cubense* is enhanced in association with the burrowing nematode, *Radopholus similis* under high nematode population (Fig. 21.2).

21.2.4. Sting nematode and wilt disease interactions

Nematodes other than sedentary endoparasites are also capable of forming disease complexes with *Fusarium* fungi. The sting nematode, *Belonolaimus gracilis* is as efficient as root-knot nematodes in promoting wilt in cotton.

Fig. 21.2. Burrowing nematode and Panama wilt disease complex in banana.

21.2.5. Lance nematode and wilt disease interactions

In Peas, early yellowing and root rot disease is dependent upon the presence of both *Hoplolaimus uniformis* and *F. oxysporum* f. sp. *pisi* race 3.

21.3. ROOT ROT DISEASE INTERACTIONS

Nematodes are important and, at times vital to development of root rots caused by fungi. Associations of this type may have greater overall importance than complexes involving nematodes with wilt inducing fungi.

21.3.1. Lesion nematodes and root rot disease interactions

Root-lesion nematodes interact with the black shank fungus in tobacco varieties susceptible to this fungus disease. Significantly, black shank symptoms are more severe if both pathogens are added to plants simultaneously or one week apart than if the nematode precedes the fungus by four weeks.

Migratory endoparasitic nematodes are involved in several root rot disease complexes. Root rot of winter wheat in Canada, in which *Pratylenchus minyus* combines with *Rhizoctonia solani,* significantly reduced wheat growth.

Pratylenchus penetrans and *Trichoderma viride* caused more reduction in root and shoot growth in lucerne than either organism alone.

21.3.2. Sting nematode and root rot disease interactions

Migratory ectoparastic nematodes may enhance the severity of certain fungus root rots. Chrysanthemum roots inoculated with *Belonolaimus longicaudatus* and

Pythium aphanidermatum develop symptoms of root rot earlier and more extensively than those inoculated with the fungus alone.

21.3.3. Citrus nematode and root rot disease interactions

Tylenchulus semipenetrans increases root decay by *Fusarium* sp. in lemon.

21.3.4. Root-knot nematodes and root rot disease interactions

Research on tobacco has presented evidences of the importance of root-knot nematodes in root decay interactions and has shown that the status of normally weak fungal pathogens may be elevated to major significance by nematode infection. *Pythium* and *Rhizoctonia* are not regarded as important pathogens after tobacco plants have progressed beyond juvenile stages. However, if either of these fungi is added to plants which have been previously infected by *M. incognita* (3 to 4 weeks), they are capable of invading the roots and causing extensive decay. On the other hand, little damage results if the fungus and nematode are applied simultaneously and virtually no decay occurs resulting from the fungus alone (Powell and Batten, 1967).

21.3.5. Cyst nematode and root rot disease interactions

Cyst forming nematodes are capable of promoting increases in certain root rots caused by fungi. *Globodera rostochiensis* has been reported to interact with both *Rhizoctonia solani* and *Colletotrichum atramentarium* in tomato.

21.4. SEEDLING DISEASE INTERACTIONS

Fungi and nematodes combined to increase damage from disorders which may be classified as seedling diseases.

21.4.1. Root-knot nematodes and seedling disease interactions

Taylor and Wylie (1959) found that *M. javanica* and *M. hapla* in combination with *Rhizoctonia solani* decreased the emergence of soybean seedlings. There was only 2% emergence with the fungus and *M. hapla*, 17% with the fungus and *M. javanica* and 55% when the fungus occurred alone.

Reynolds and Hanson (1957) showed that damping off of cotton seedlings due to *R. solani* was more severe in plants infected with *M. incognita acrita*. Damping-off in nematode infested soil was significantly reduced by application of nematicides.

Cotton seedlings grown in soil infested with *M. arenaria, M. hapla* or *M. incognita* are susceptible to *R. solani* longer than those grown in the absence of nematodes. Plants are susceptible longer to *Pythium debaryanum* only when either *M. incognita* or *M. hapla* is present.

21.4.2. Citrus nematode and seedling disease interactions

Under greenhouse conditions, the citrus nematode apparently interacts with *F. solani* to reduce citrus seedling growth, and the effects of the pathogen combination is noticeably greater than that of either pathogen alone. This may be important in subsequent root rot in citrus.

21.4.3. Reniform/lance nematodes and seedling disease interactions

Cotton seedlings grown in soil infested with *Rotylenchulus reniformis* or *Hoplolaimus tylenchiformis* are susceptible to *R. solani* longer than those grown in the absence of nematodes.

21.4.4. Stunt nematode and seedling disease interactions

Khan *et al.* (1971a) showed that *Tylenchorhynchus brassicae* alone did not affect the percentage of emergence of cauliflower seedlings. *Rhizoctonia solani* on the other hand was highly destructive by itself, but the adverse effect on the emergence of seedlings was far greater when the nematode infestation occurred along with fungal infection.

21.5. MANAGEMENT OF NEMATODE - FUNGAL DISEASE COMPLEXES

21.5.1. Breeding nematode - fungal resistant varieties

The breakdown of disease resistance in the presence of nematodes must be taken into account by plant breeders in developing disease resistant varieties of crop plants. It is obviously necessary to attempt to incorporate resistance to nematodes with resistance to disease in the same variety. Considerable progress has been achieved by Powell and Nusbaum (1958) in evolving root-knot and black shank resistant varieties of tobacco.

21.5.2. Soil fumigation with chemicals having both nematicidal and fungicidal value

Some of the new, highly active nematicides such as aldicarb and formulations of materials having both nematicidal and fungicidal value may find a useful place in controlling disease complexes. Vorlex, Vorlex 201, Terr-O-cide and experimental compound No. 2680 show unusual promise.

Metham, dazomet, D-D + MENCS mixtures are also the most promising compounds for multiple pest control. In several experiments, these compounds controlled several different species of nematodes and soil fungi.

21.6. NEMATODES AND ARBUSCULAR MYCORRHIZAL FUNGI INTERACTIONS

Arbuscular mycorrhizal fungi have an antagonistic effect on the population of plant parasitic nematodes. Several investigators have obtained a significant reduction in number of *Meloidogyne incognita* juveniles that developed into adults on plants inoculated with *Glomus mosseae*. *G. fasciculatum* was found to reduce *Globodera rostochiensis* levels on potato and increase the resistance of tomato plants to *Rotylenchulus reniformis*. The reproduction of *Heterodera solanacearum* on tobacco plants was suppressed by *Gigaspora gigantea*. *G. marginata* also significantly reduced the numbers of *Pratylenchus brachyurus* on cotton. When rough lemon seedlings infected with *Tylenchulus semipenetrans* were transplanted into soil infested with *Glomus mosseae,* the mycorrhizal fungus infection increased seedling growth compared to non-mycorrhizal seedlings.

Conversely nematodes have been shown to have adverse effect on mycorrhizal fungi. *Aphelenchus avenae* can prevent the formation of mycorrhizal associations. The roots may be damaged sufficiently to block the symbiotic plant fungal association when roots are subjected to attack by plant nematodes (*Pratylenchus, Meloidogyne,* and *Xiphinema*). A study of mycorrhizal fungi and root-knot nematodes on soybeans showed that the nematodes depressed the reproduction of mycorrhizal fungi and reduced soybean yields. Nematodes reproduced more rapidly in the presence of the fungi than without them.

The results seem to indicate that the fungal host plant combination is critical, with each fungal isolate producing different host reactions to a particular plant parasitic nematode (Table 21.1).

Table 21.1. Nematode fungal interactions

Part of plant diseased	Fungi	Nematodes	Host	Role of nematode
Aerial meristem	*Dilophospora alopecuri*	*Anguina tritici*	Wheat	Essential
Hypocotyl (damping off)	*Rhizoctonia solani*	*Meloidogyne javanica*	Soybean	Assists
		M. incognita acrita	Cotton	"
	Pythium debaryanum	*M. incognita acrita*	Cotton, Tobacco	"
	Botrytis allii	*Ditylenchus dipsaci*	Onion	"
Leaves	*Phytophthora infestans*	*D. destructor*	Potato	"
	Alternaria tenuis	*Meloidogyne incognita*	Tobacco	"

Part of plant diseased	Fungi	Nematodes	Host	Role of nematode
Root (Cortical rot)	*Rhizoctonia solani*	*Globodera rostochiensis*	Potato, Tomato	"
		Radopholus similis	Banana	"
		Pratylenchus minyus	Wheat	"
	Cylindrocarpon radicicola	*P. penetrans*	Tulip, Narcissus	"
	Aphanomyces euteiches	*Tylenchorhynchus martini*	Pea	"
Root (Vascular wilt)	*Fusarium oxysporum* f. sp. *vasinfectum*	*Meloidogyne hapla*	Lucerne	"
		M. incognita acrita	Cotton	"
		Rotylenchulus reniformis	"	"
		Belonolaimus gracilis	"	"
		B. longicaudatus	"	"
		Radopholus similis	Banana	Essential
	F. oxysporum f. sp. *cubense*	*Meloidogyne incognita*	Tobacco	Assists
	F. oxysporum f. sp. *nicotianae*	*Tylenchorhynchus claytoni*	Tobacco	"
	F. oxysporum f. sp. *pisi* Race 1	*Pratylenchus penetrans*	Pea	"
	F. oxysporum f. sp. *pisi* Race 3	*Hoplolaimus uniformis*	Pea	"
	F. oxysporum f. sp. *lycopersici*	*Meloidogyne* spp.	Tomato	"
	F. oxysporum f. sp. *tracheiphilum*	*M. javanica*	Cowpea	"
	F. oxysporum f. sp. *dianthi*	*Meloidogyne* spp.	Carnation	"
	Fusarium sp.	*Heterodera glycines*	Soybean	"
	Phytophthora parasitica var. *nicotianae*	*M. incognita acrita*	Tobacco	"

Part of plant diseased	Fungi	Nematodes	Host	Role of nematode
	Verticillium albo-atrum	*M. incognita acrita*	Cotton	"
		Pratylenchus penetrans	Brinjal	"
	V. dahliae	*Globodera rostochiensis*	Potato	"

■■■■

Chapter 22

INTERACTION WITH BACTERIA

22.1. INTRODUCTION

There are relatively few described complexes involving nematodes and bacteria when compared to the number including nematodes and fungi. There are valid cases, however, of interaction between bacteria and nematodes which have real economic consequences.

22.2. WILT DISEASE INTERACTIONS

22.2.1. Root-knot nematodes and wilt disease interactions

Some of the most striking complexes between nematodes and bacteria include those bacteria which induce wilting in the host as one of the primary symptoms. It has been known for a long time that root-knot nematodes, present along with *Ralstonia solanacearum*, greatly increase the incidence of bacterial wilt of several host plants. In tomato, both the root-knot nematode, *M hapla*, and the spiral nematode, *Helicotylenchus nannus,* contribute to an increase in wilt development. *H. nannus* is not quite as effective as *M. hapla* in promoting wilt but still exercises a significant effect.

Fig. 22.1. Root-knot nematode and bacterial wilt disease complex in banana

M. incognita greatly influences bacterial wilt development in tobacco. Plants exposed to nematodes 3 or 4 weeks before exposure to *R. solanacearum* developed more severe wilt symptoms than when plants were exposed to only bacteria or to both pathogens simultaneously. Giant cells in jointly infected roots harbored bacteria like inclusions and degenerated rapidly. Root decay soon followed.

Association of *M. incognita* and *R. solanacearum* has been reported causing severe damage on banana crop (Fig. 22.1).

Eggplant is prone to many soil-borne diseases among which the bacterial wilt (*R. solanacearum*) in combination with root-knot nematode (*M. incognita*) takes heavy toll every year all over the world (Fig. 22.2).

Fig. 22.2. Root-knot nematode and bacterial wilt disease complex in brinjal.

22.2.2. Stem nematode and wilt disease interactions

Bacterial wilt of lucerne, caused by *Corynebacterium insidiosum,* is increased by the stem nematode *Ditylenchus dipsaci.* The basis for this interaction may lie in transmission of the bacteria by *D. dipsaci.* Even varieties which have high resistance to wilt become diseased in the presence of *D. dipsaci.* On the other hand, varieties with nematode resistance remain relatively free from wilt when exposed to both pathogens.

22.3. ROOT DISEASE INTERACTIONS

A number of bacterial pathogens normally inhabit soil and may become pathogenic on roots of susceptible plants. Symptoms of such infections may manifest as galling or as root decay.

22.3.1. Root-knot nematodes and root disease interactions

Crown gall of peach, caused by *Agrobacterium tumefaciens,* is increased by high populations of *M. javanica.*

Some bacteria, usually regarded as secondary invaders, appear to be implicated in necrosis of tomato roots infected by *M. incognita.* Reduction in foliage growth occurs only when plants are inoculated with nematodes and maintained septically. Galls on such plants become necrotic and, upon isolation, yield many bacterial colonies.

22.3.2. Lesion nematode and root disease interactions

A new strain of *Agrobacterium rhizogenes* frequently associated with *Pratylenchus vulnus* causes serious losses in southern California.

22.4. FOLIAGE DISEASE INTERACTIONS

The "tundu" or "yellow slime" disease of wheat is a complex requiring both the nematode, *Anguina tritici* and the bacterium, *Rathayibacter tritici* for expression of the complete disease syndrome. The cauliflower disease in strawberry involving *Aphelenchoides ritzemabosi* and *Rhodococcus facians* provides another striking example of a nematode bacterium complex.

22.5. NEMATODE RHIZOBIA INTERACTIONS

The role of nematodes in the symbiosis between rhizobia and legumes is not well known. In the field, soybeans infected with *Heterodera glycines* usually have few bacterial nodules. On the other hand, neither *H. trifolii* nor *Meloidogyne javanica* have much effect on nodulation in white clover. Both endoparasites reproduce well in nodules. However, nodules containing *M. javanica* deteriorate more rapidly than those free of nematodes. Thus the concurrent nematode infection reduces the total benefit from the nitrogen fixing bacteria.

The nematode bacterial interactions have been summarized in Table 22.1.

Table 22.1. Interaction of nematodes with bacterial diseases

Part of plant diseased	Bacteria	Nematodes	Host	Role of nematode
Aerial meristem	*Rathayibacter tritici* (*Corynebacterium*) *tritici*	*Anguina tritici*	Wheat	Essential
	Rhodococcus (*Corynebacterium*) *facians*	*Aphelenchoides ritzemabosi*	Strawberry	"
Leaves	*Xanthomonas begoniae*	*A. fragariae*	Begonia	Assists
Crown buds (Soft rot)	*Erwinia rhaponticum*	*Ditylenchus dipsaci*	Rhubarb	"

Part of plant diseased	Bacteria	Nematodes	Host	Role of nematode
Crown buds (Vascular wilt)	*Corynebacterium insidiosum*	*D. dipsaci*	Lucerne	Essential & assists
Root (galls)	*Agrobacterium tumefacians*	*Meloidogyne hapla*	Raspberry	Assists
Root (vascular wilt)	*Ralstonia solanacearum*	*M. incognita acrita*	Tobacco	"
		M. hapla	Tomato	"
		Helicotylenchus nannus	"	"
	Burkholderia caryophylli	*Meloidogyne* spp.	Carnation	"
		Helicotylenchus nannus	Tomato	"

■■■■

Chapter 23

INTERACTION WITH VIRUSES

23.1. INTRODUCTION

Although nematodes had long been suspected as possible vectors of plant viruses, the first experimental evidence of such an association was not evident until 1958 when Hewitt, Raski and Goheen showed that *Xiphinema index* transmitted fan leaf virus of grapevine. At present 30 nematode species are known to transmit 15 serologically distinct viruses (MacFarlane *et al*., 2016). Only species of the dorylaimoid genera *Xiphinema, Longidorus, Paralongidorus, Paratrichodorus* and *Trichodorus* have thus far been implicated as vectors.

Nematode transmitted viruses can be divided into two groups on the basis of particle shape. Cadman (1963) named those with polyhedral particles, NEPO viruses, i.e. Nematode transmitted viruses with polyhedral particles. Viruses in this group are frequently referred to as ring spot viruses and are transmitted by *Xiphinema, Longidorus* and *Paralongidorus* species. The second group includes the viruses with rod-shaped or tubular particles for which Harrison (1964) derived the name NETU viruses, i.e. Nematode transmitted viruses with tubular particles. They are also called TOBRA viruses. The two viruses in this group, tobacco rattle virus and pea early browning virus, are transmitted by *Trichodorus* and *Paratrichodorus* species.

23.2. NEPO VIRUSES

These are transmitted by species of *Xiphinema, Longidorus* and *Paralongidorus* and include arabis mosaic and strawberry latent ring spot viruses (Table 23.1). Their chemico-physical properties are similar, and all possess polyhedral particles about 27-30 nm diameter, but they share no antigenic determinants and hence are readily identifiable by serological methods.

Diseases caused by NEPO viruses typically show the "shock" and "recovery" phases that are characteristic of tobacco ring spot disease in tobacco, i.e. the first parts of the plants to become systematically infected show severe symptoms but later growth, although infected, appears healthy.

Table 23.1. NEPO Viruses and their nematode (Longidoridae) vectors

Virus	Acronym	Nematode
Arabis mosaic virus	ArMV	*Xiphinema diversicaudatum*
Artichoke Italian latent virus	AILV	*Longidorus apulus, L. fasciatus*
Cherry rasp leaf virus	CRLV	*X. diversicaudatum, X. californicum, X. rivesi*
Cherry rosette virus	CRV	*L. arthensis*
Grapevine fan leaf virus (Fig. 23.1)	GFLV	*X. index, X. italiae*
Mulberry ring spot virus	MRSV	*L. martini*
Peach rosette mosaic virus	PRMV	*L. diadecturus, X. diversicaudatum, X. rivesi*
Raspberry ring spot virus	RRSV	*L. elongates, L. macrosoma, Paralongidorus maximus*
Strawberry latent ring spot virus	SLRSV	*X. diversicaudatum*
Tomato black ring virus	TBRV	*L. attenuates, L. elongatus*
Tomato ring spot virus	ToRSV	*X. diversicaudatum, X. bricolensis, X. californicum, X. intermedium, X. rivesi, X. tarjanense*
Tobacco ring spot virus	TRSV	*X. americanum, X. californicum, X. intermedium, X. rivesi, X. tarjanense*

Fig. 23.1. Grapevine infected with fan leaf virus disease.

Transmission of virus through the seeds of infected plants is characteristic for the NEPO group of viruses. Tomato black ring and raspberry ring spot viruses transmitted by *L. elongatus* are commonly present in seeds of various weed species in soil from outbreak areas, but it is rarely so with other NEPO viruses transmitted by *Xiphinema* spp. Infected weed seeds appear to form an important reservoir of infection for raspberry ring spot and tomato black ring viruses and serve as a means whereby these viruses perpetuate in soils. These viruses do not persist for more than about 8 weeks in their vector, *L. elongatus*, whereas the viruses transmitted by *Xiphinema* are rarely found in weed seeds from infested soils but persist in their vectors for 8 months or more.

In British Columbia, the tomato ring spot virus (Fig. 23.2) and vector *Xiphinema americanum* occurring only in clay loam soils cause serious problem. Raspberries plantings so infected begin to decline about three years after planting.

Fig. 23.2. Tomato ring spot virus in Raspberry.

23.3. NETU OR TOBRA VIRUSES

The only two viruses in this group include tobacco rattle virus and pea early browning virus, which are transmitted by *Partrichodorus* and *Trichodorus* species (Table 23.2).

Table 23.2. NETU/TOBRA viruses and their nematode (Trichodoridae) vectors

Virus	Acronym	Vector
Pea early browning virus	PEBV	*Paratrichodorus anemones, P. pachydermus, P. teres, Trichodorus primitivus, T. viruliferus*
Tobacco rattle virus	TRV	*P. allius, P. anemones, P. hispanus, P. minor, P. nannus, P. pachydermus, P. teres, P. tunisiensis, T. cylindricus, T. primitivus, T. similis, T. viruliferus*
Pepper ring spot virus	Pep RSV	*P. minor*

Tobacco rattle virus infects a large number of crops and weed hosts in Europe and in North and South America. Preparations of the virus contain tubular particles of two main lengths, about 75 and 185 nm, but only the long particles are infective. Although the shorter particles on their own are not infective they nevertheless appear to be essential for the production of stable virus.

Corky ring spot disease of potato is caused by tobacco rattle virus (Tobravirus), which is vectored by stubby-root nematodes (*Paratrichodorus* spp. and *Trichodorus* spp.) (Fig. 23.3). The condition is sometimes referred to as spraing, a Scottish word meaning a bright streak or stripe.

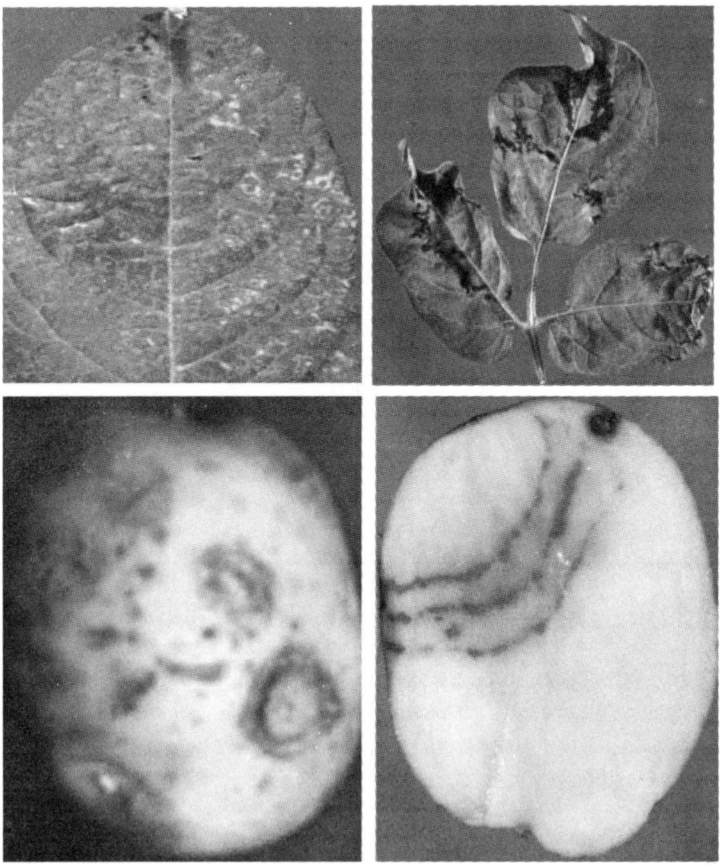

Fig. 23.3. Left – Yellow ring spots induced by TRV on potato leaves and tubers. Note leaf twisting with necrosis induced by TRV.

Pea early browning virus occurs naturally in pea and alfalfa crops in the Netherlands and in Britain. The virus also has straight tubular particles of two main lengths, about 105 and 210 nm, and like tobacco rattle virus it does not readily infect plants systemically.

23.4. RELATIONSHIP BETWEEN VIRUSES AND VECTOR NEMATODES

23.4.1. Transmission

The adult and juvenile stages of most nematode vector species transmit virus and probably with equal efficiency.

23.4.2. Acquisition and inoculation

Teliz *et al.* (1966) propose the terms "acquisition access threshold" to express the minimum time required for a nematode to have access to a virus source in order to become viruliferous and "inoculation access threshold" for the minimum time that a viruliferous nematode must have access to a healthy plant for transmission to occur.

Acquisition access thresholds of an hour have been established for *X. americanum* and tomato ring spot virus (Teliz *et al.*, 1966) and for *P. allius* and tobacco rattle virus and 15 minutes for *X. index* and grapevine fan leaf virus. With several vector-virus combinations, it has been demonstrated that an increase in the acquisition access period results in a proportionate increase in the frequency of transmission obtained.

Inoculation access threshold periods of less than 24 hours have been established for several nematode vectors, and as with acquisition of virus, frequency of transmission increases with time.

23.4.3. Persistence

Grapevine fan leaf virus can persist in its vectors, *X index*, for up to 8 months (Taylor and Raski, 1964). *X americanum* was shown to transmit tobacco ring spot virus after a starvation period of 49 weeks and *X. diversicaudatum* to transmit strawberry latent ring spot and arabis mosaic viruses after 84 and 112 days, respectively. When virus-carrying *Xiphinema* are allowed to feed on virus immune hosts, persistence of virus in the vectors may also be for relatively long periods: *Xiphinema index* retained grapevine fan leaf virus for 12 weeks when allowed to feed on virus immune fig.

Longidorus vectors may not retain infectivity for as long as *Xiphinema.* Raspberry ring spot and tomato black ring viruses persist in *L. elongatus* for a maximum period of about 8 weeks (Fig. 23.4).

Sol (1963) found that the tobacco rattle virus was retained by *P. pachydermus* for a least 36 days and van Hoof (1964) reported transmission of the virus with *P. pachydermus* in soil stored for 10 months at temperatures fluctuating between 3° and 20°C. *P. allius* remained infective with the tobacco rattle virus for 20 weeks when maintained without a host, but for 27 weeks when feeding on virus immune sweet pea.

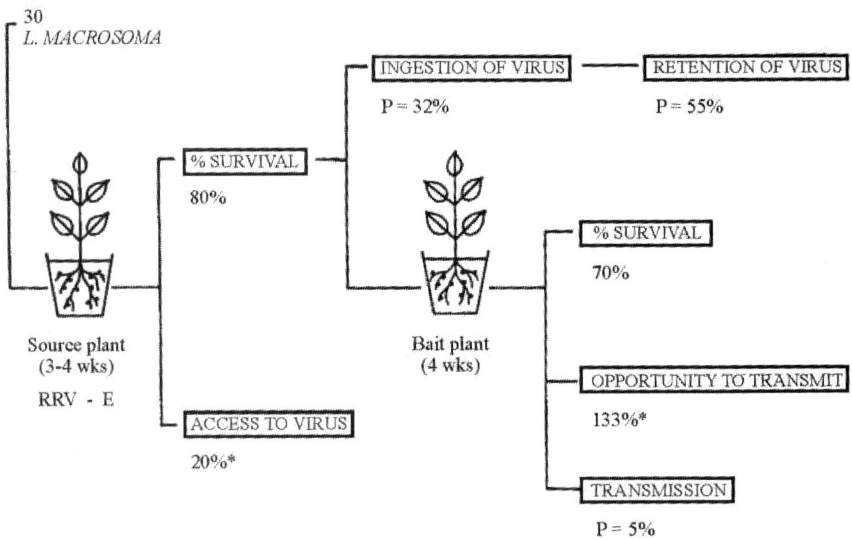

Fig. 23.4. Transmission of raspberry ring spot virus English strain RRV-E by *Longidorus macrosoma*

Despite the long retention of viruses in their nematode vectors, the few experimental data available indicate that nematode transmitted viruses do not persist through the egg or through the molt.

23.5. MECHANISM OF TRANSMISSION

Apparent differences in the retention of virus by *X. index* and *L. elongatus* led Taylor and Raski (1964) to suggest that there may be at least two different mechanisms of transmission among the NEPO viruses, a close biological association existing between virus and vector in *Xiphinema* species, where as in *Longidorus* the virus is mechanically retained. Taylor and Robertson (1969) showed by means of electron microscopy of their sections that the inner surface of the guiding sheath of *L. elongatus* is probably the main site of accumulation and retention of raspberry ring spot and tomato back ring viruses. Particles of tobacco rattle virus have also been found in close association with the cuticle lining the lumina of the pharynx and the esophagus (Fig. 23.5) of *P. pachydermus,* again suggesting a mechanical but specific mechanism of retention of the virus. Virus particles apparently specifically absorbed onto the cuticle lining the stylet extension and the esophagus in both *X. index* and *X. diversicaudatum*, and the transmissible virus is retained at these loci and not within the tissues (Fig 23.6).

Members of Tylenchida have been suggested and tested as vectors but with negative results. If our present understanding of the mechanisms of transmission is correct, these would seem unlikely candidates. It appears that secretions from the dorsal esophageal gland flowing over the virus particles adsorbed to the esophageal lumen are necessary to release the particles and introduce them into the plant cell. While the opening of this duct in Dorylaimida is into the lumen of the basal bulb, in Tylenchida it empties into the lumen just behind the stylet knobs. Unless virus particles were absorbed within the stylet lumen, there would be no opportunity for injection into the plant cell during salivation.

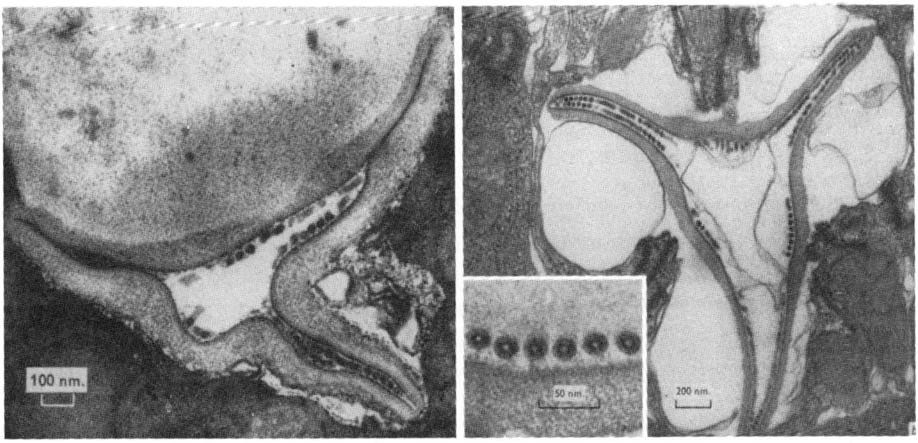

Fig. 23.5. Cross section of the pharynx of *Paratrichodorus* showing tobacco rattles virus particles in close association with the cuticle lining the pharyngeal lumen.

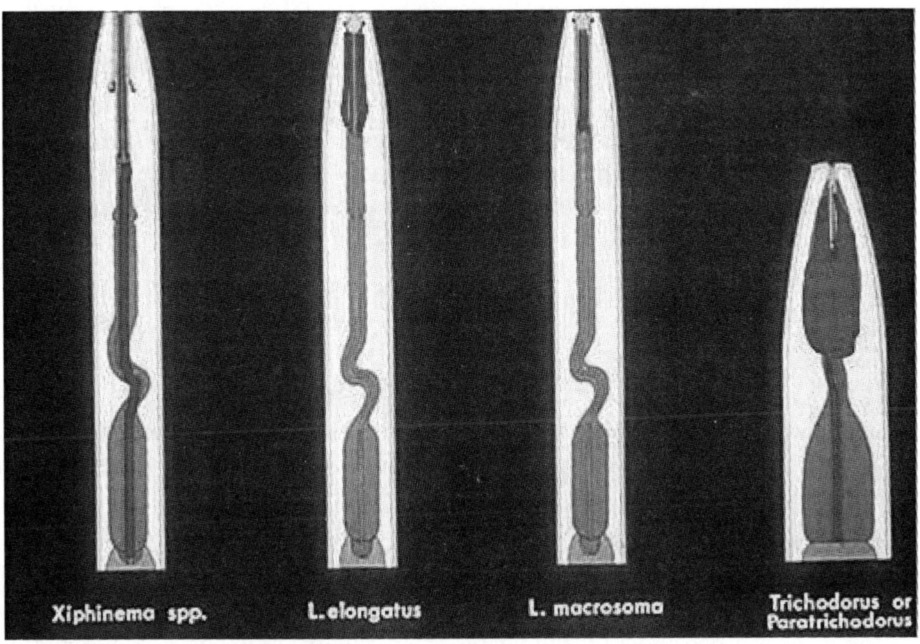

Fig. 23.6. Viral particles (in red) in the pharynx-stylet tubes of plant virus vector nematodes

Transmission involves the adsorption and later detachment of virus particles at the site of retention within the nematode vector. Disassociation probably occurs when secretions from the esophageal glands are passed anteriorly through the lumen of the esophagus and feeding apparatus during the initial phases of feeding. This is considered to modify the pH within the lumen, thus altering the surface charge of the virus particles and bringing about disassociation from the site of retention.

23.6. ECOLOGY AND MANAGEMENT

23.6.1. Use of virus-free planting material

The distribution and use of infected planting material is probably the most important way by which nematode transmitted viruses are disseminated. Grapevine fan leaf virus has almost certainly been distributed in this way from Europe to new areas such as North America. Ornamental bulbs are an effective way of disseminating tobacco rattle virus and probably also other nematode transmitted viruses that appear to commonly infect commercial Narcissus stocks.

Measures may be taken to prevent virus and vectors reaching new sites, by using only virus- free planting material.

23.6.2. Weed management

Because several virus-transmitting nematodes are polyphagous, weed control is essential in any type of management practice. Several weeds, including kochia, prickly lettuce, henbit, nightshade, common chickweed, and annual sow thistle, were shown to be hosts of *P. allius* and Tobacco rattle virus (Boydston *et al.*, 2008). Many broad leaf weeds are hosts of *X. americanum* (Powell *et al.,* 1982), and dispersal of Tomato ring spot virus is aided by the fact that the virus is transmitted by the pollen and seeds of many of its weed hosts (Brunt *et al.*, 1996).

Weed seeds may provide an effective means of disseminating virus and enable them to become established in areas where vectors are present. The infected weed seeds are the most important over wintering sources of raspberry ring spot and tomato black ring viruses in a disease outbreak.

Infected weed seeds appear to play no comparable role in the survival and dissemination of *Xiphinema* transmitted viruses. Nevertheless, soils remain infested for long periods.

In Scotland, weed control resulted in some reduction of infection in a strawberry crop grown on raspberry ring spot virus infected soil. Weed control may, however, be more effective in preventing the spread of virus infection when the crop itself is a poor host for the vector e.g. raspberry crop and *L. elongatus.*

23.6.3. Crop rotation

Rotation to tall fescue (*Festuca arundinacea*) prevented the reoccurrence of ToRSV, although population densities of *X. americanum* remained high (Pinkerton and Martin, 2006). In the same study, rotation with rapeseed (*Brassica napus*) was as effective as fumigation with methyl bromide in preventing reinfection of raspberry by ToRSV for 3 years. Two years of continuous corn, grain sorghum, or wheat followed by a soybean cultivar resistant to Soybean severe stunt virus reduced *X. americanum* population densities and the severity of disease (Evans *et al.*, 2007).

23.6.4. Resistant varieties

TRV resistance has been available for many years and is not uncommon in European-type cultivars. Some cultivars are immune to infection, and this form of resistance is considered to exist in cultivars such as Climax, Record, Saturna, Bintje, and Arran Pilot.

With long term crops such as grapevines and raspberry, the planting of disease escaping or virus immune cultivars is an attractive solution to the problem of nematode transmitted viruses.

23.6.5. Chemical control

Nematicides have so far proved the most satisfactory way of dealing with nematode vectors. Good control of nematodes has been obtained with several granular formulations of nematicides.

■■■■

Section VII
MANAGEMENT METHODS

Chapter 24

REGULATORY METHODS

24.1. PLANT QUARANTINES

Regulatory control of plant nematodes is simply the legal enforcement of measures planned to prevent them from spreading or, having spread, from multiplying sufficiently to become intolerably troublesome. The principle involved in enacting quarantines is exclusion of the nematode from entering into an area which is not infested.

Quarantines are traditionally used to restrict the movement of infected plant materials and contaminated soil into a state or country. Many countries maintain elaborate organizations to intercept plant shipments containing nematodes and other pests. Diseased or contaminated materials may be treated to kill nematodes, or entry may be barred. Quarantines are also maintained to prevent the movement of infected plants and soil out of areas known to have nematodes. Although quarantines are useful especially when a disease is first discovered, they are very expensive and usually cannot prevent the spread of a nematode induced disease.

24.1.1. Quarantine restrictions in India

The Destructive Insects and Pests Act, 1914 and the rules framed there under "Rules for regulating the import of plants etc. into India" restricts the movement of plants and plant material to prevent the introduction into and the transport from one state to another in India of any insect, fungus or other pest which is or may be destructive to crops. The rules permit the Plant Protection Advisor to the Government of India or any officer authorized by him to undertake inspection and treatments. Specific regulations for nematodes have been made against the cyst nematodes of potato (*Globodera rostochiensis, G. pallida*) and the red ring nematode (*Bursaphelenchus cocophilus*) of coconut. Domestic quarantine regulations have also been imposed to restrict the movement of potato both for seed and table purposes in order to prevent the spread of potato cyst nematode from Tamil Nadu to other states and Union Territories in India. In spite of these restrictions, the potato cyst nematode has already spread to Himachal Pradesh.

24.1.2. Quarantine restrictions in USA

In the United States, the soybean cyst nematode (*Heterodera glycines*) has been subjected to Federal Quarantine during 1954. There is a State Quarantine against the golden nematode of potato enacted in the year 1941 in New York State (Long Island). The burrowing nematode of citrus (*Radopholus similis*) was brought under state Quarantine Act in Florida during the year 1953.

24.1.3. Prevention of spread

Nematodes do not move more than a few centimeters a year of their own accord and most spread is passive. They may be spread in soil or in the roots of a host plant (e.g. *Heterodera, Meloidogyne*, and *Pratylenchus*) or in the aerial parts of plants including the seed (*Aphelenchoides, Ditylenchus dipsaci, Anguina*).

24.1.3.1. Spread with soil: Any means whereby soil is moved will result in soil-inhabiting nematodes being spread and some species, e.g. *Heterodera* spp. will remain viable even in the soil adhering to farm implements, labors' feet, seed potatoes and transplants grown in infested soil. Hence farm implements should be thoroughly cleaned before taking them to other fields. Laborers should wash their feet after working in an infested field. Seed potatoes should be freed of adhering soil before transporting them elsewhere. Seedlings should not be raised in infested fields for transplanting elsewhere.

24.1.3.2. Spread with seed: Several nematodes are known to be seed-borne, e.g., *Ditylenchus dipsaci* on onion and lucerne, *Aphelenchoides besseyi* on rice and *Anguina tritici* on wheat. Efficient seed cleaning methods can considerably reduce the amount of nematode infested debris with seed. With *Anguina tritici,* modern seed cleaning methods eliminate the galls from the seed. However, *D. dipsaci* and *A. besseyi* are probably not so readily eliminated from seed as *Anguina.* Ideally there are three steps which should be taken to ensure the production of clean seed.

- Only nematode free seed should be used for growing seed crops.
- This seed should be sown on land where the nematode is not known to exist.
- The crop should be inspected frequently for the presence of the nematode.

Seed-borne nematodes can also be managed by fumigation with methyl bromide. Hot water treatment of bulbs and other planting stock is another method that can be effectively used to reduce the spread of nematodes through seed and planting material.

24.1.4. Prevention of multiplication

Nematode multiplication can be prevented by removing the nematode-infested farm from agricultural use. This land can be used for construction of buildings. Nematode multiplication can also be prevented by prohibition of the cultivation of host crops and by directly controlling the nematode by chemical means. Crop rotation can also prevent the nematode multiplication.

24.2. SEED CERTIFICATION

At a more practical level, certified plant material, nematode-free planting stock and clean farm equipment should be standard IPM practices. This issue is particularly important for vegetatively propagated crops such as banana, potato, and crops for which seedlings are transplanted. *In vitro* propagation of banana has proved to be an effective means of providing clean planting stock and should reduce the need for nematicides (Sarah, 1989).

In addition, a number of nematodes are disseminated with crop seeds, including *Anguina tritici* on wheat and *Ditylenchus dipsaci* on alfalfa and a number of other crops.

Certified alfalfa seed should be used to minimize the chances of contaminating previously uninfested land with *Ditylenchus dipsaci.* Until the importance of seed as a source of transmission of stem and bulb nematode was understood, as many as 17,000 nematodes were found per pound.

The production of vegetative plant-propagation material that is certified to be nematode-free is accomplished by growing clean plants in clean soil. Strict sanitation must be practiced, and both the plant material and rooting medium must be periodically checked for the presence of pathogenic nematodes. Vegetative seed of banana (suckers) free of the burrowing nematode, potato seed pieces free of the cyst nematode, garlic cloves free of the stem nematode and wheat seeds free of seed galls can be produced commercially.

Despite failures and shortcomings, regulatory action against some plant parasitic nematodes has undoubtedly reduced the rate of their spread if not actually confined them.

Chapter 25

PHYSICAL METHODS

25.1. INTRODUCTION

Plant parasitic nematodes can easily be killed in the laboratory by the application of heat, irradiation, osmotic pressure, etc. It is more difficult, however, to employ such methods to large quantities of soil and plant material. Physical treatments may involve hazards for plants or operators but, except possibly with radiation treatments, there are unlikely to be residual effects.

25.2. HEAT

Controlling nematodes by killing them with heat is one of the oldest methods and is probably the most successful physical control measure developed so far. It is used for killing the nematodes in the soil and within the plant tissues.

25.2.1. Heat treatment of soil

Sterilization of soil by steam is a long established practice in greenhouse beds, seed beds and other small areas. Insects, weed seeds, nematodes, bacteria and fungi are killed by steam sterilization. The most important nematode pests controlled by greenhouse steaming are golden nematode of potato attacking tomatoes and root-knot nematodes on tomatoes, cucumbers, lettuces and some ornamentals.

Fig. 25.1. Steaming of soil

Steam is generally introduced through perforated pipes buried in the soil and heating takes place by condensation on the soil particles (Fig. 25.1). It is essential that the soil surface should be covered during steaming. Usually plastic sheets are employed and the formation of a bubble of steam under them indicates that treatment is complete.

Small quantities of soil can be sterilized in an autoclave at 30 lb. pressure per square inch for 30 minutes.

25.2.2. Hot water treatment of planting material

It is widely used for the killing of nematodes within tissues before planting and has proved useful with nematode infested bulbs, corns, tubers, seeds and roots of plants such as chrysanthemums, strawberries, bananas and citrus (Fig. 25.2). First, the time temperature mortality curves of the nematode have to be determined. This entails immersing the nematodes into thermostatically controlled water baths at different temperatures for different lengths of time. Information must also be obtained on the plants reaction to temperature because if the treatment is to be successful the plants must be able to tolerate temperatures lethal to nematodes (Table 25.1).

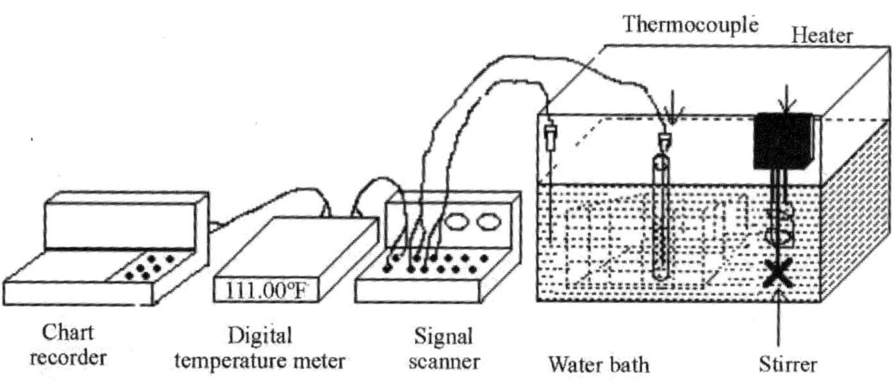

Fig. 25.2. Hot water treatment of planting material for nematode management

Table 25.1. Time and temperature recommendations for hot water treatment for denematizing planting stock

Nematode	Planting stock	Time min	Temp °C
Tylenchulus semipenetrans	Citrus rooted cuttings	10	46.7
		25	45.0
Radopholus similis	Banana suckers	10	50.0
	Citrus rooted cuttings	10	50.0
Aphelenchoides besseyi	Rice seeds	10	52.0

Nematode	Planting stock	Time min	Temp °C
A. ritzemabosi	Chrysanthemum stools	15	47.8
		30	43.0
A. fragariae	Easter lily bulbs	60	44.0
	Strawberry runners	10	46.0
Pratylenchus penetrans	Dormant strawberry plants	7.5	51.0
		17.5	49.4
Ditylechus dipsaci	Narcissus bulbs	240	43.0
	Onion	120	43.5
D. destructor	Irish bulbs	180	43.0
Anguina tritici	Wheat seeds	10	54.0
		30	50.0
Meloidogyne spp.	Cherry rootstocks	5-10	50.0-51.1
	Sweet potatoes	65	46.7
	Peach rootstocks	5-10	50.0-51.1
	Tuberose tubers	60	49.0
	Grapes rooted cuttings	10	50.0
		30	47.8
	Begonia tubers	30	48.0
		60	45.0
	Caladium tubers	30	50.0
	Yam tubers	30	51.0
	Ginger rhizomes	10	55.0
	Strawberry roots	5	52.8
	Rose roots	60	45.5
	Potato tubers	120	46.0-47.5

Since most animal enzymes are inactivated after short exposure to temperatures near 50°C, death can probably be explained on this basis.

25.3. SOIL SOLARIZATION

Another method of killing pests with heat is by soil solarization. This technique consists of covering moist soil with a plastic film during periods of intense sunshine and heat (Fig. 25.3), thereby capturing radiant heat energy from the sun, causing physical, chemical and biological changes in the soil. The soil temperature is increased to levels (by 5-15°C) lethal to many soil-borne fungi, bacteria, nematodes and weed seeds. The short wave incident solar radiation penetrate the polyethylene sheet but the long wave radiation is prevented from soil, thus trapping the heat (Fig. 25.4) resulting in thermal inactivation and production of heat shock proteins and irreversible heat injury. Solarization also improves plant nutrition by increasing the availability of N and other essential elements. This practice can be profitably used on a small scale and is especially useful for nursery beds.

Soil solarization is one of the effective ways of suppressing root-knot nematode populations (*Meloidogyne javanica, M. indica*) on citrus and can be employed mostly during hot weather days. The nursery area should be plowed well and levelled after breaking the clods. The prepared field may be covered with polythene sheet of 100 gauze thickness for about 7-10 days before sowing.

Fig. 25.3. Plastic sheets laid by machines in strips for soil solarization.

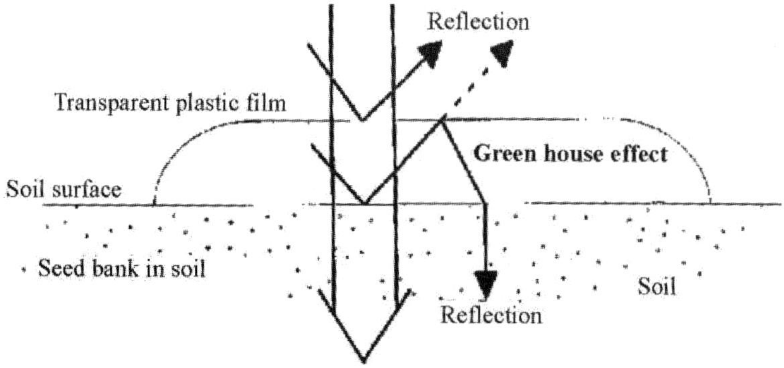

Fig. 25.4. Concept of solarization

A 37-100% reduction in population of *Macroposthonia xenoplax* and other nematodes on peach has been achieved by soil solarization in South Africa (Barberecheck and Broembsen, 1986).

Soil solarization using tarping with Linear Low Density Poly Ethylene (LLDPE) 100 gauze film during summer for 15 days (in between April 15th to June 15th) in tomato nursery at Anand gave 66 and 92% reduction of root-knot nematodes and weeds, respectively. This practice produced 615% more seedlings over control giving net profit of Rs. 7,661 from 1000 m^2 area with cost: benefit ratio of 1:5.65.

Soil solarization with transparent LDPE (400 gauze) for 40-45 days was proved ideal pre-sowing treatment for the management of nematodes and other soil-borne diseases in cardamom nurseries. Mean soil temperature during the solarization period increased by 8.7°C in solarized beds and the germination was enhanced by 25.5%, while weed growth was suppressed by 82%. *Pythium vexans* was totally eliminated, while populations of *Rhizoctonia solani, Phyllosticta elettaria* and nematodes were suppressed to varying levels. Solarization was also found to stimulate the growth of cardamom seedlings (Eapen, 1995).

White polyethylene mulching for 15 days on beds prepared for planting betel vine revealed high reduction in nematode populations due to increased soil temperature to 44.1°C (Sivakumar and Marimuthu, 1987).

Soil solarization using thin transparent polyethylene mulches for six weeks caused as much as 72.6 and 88.5% decrease in the population densities of *Tylenchorhynchus vulgaris* and *Hoplolaimus indicus*, respectively in nursery beds (Table 25.2). Application of mahua cake before solarization usually caused slightly greater reduction in nematode population (Kamra and Gaur, 1998).

Table 25.2. Management of nematodes using soil solarization

Crop	Disease/Nematode	Duration	Results
Citrus	Root-knot nematodes, *Meloidogyne javanica, M. indica*	7-10 days	Suppress nematode population
Tomato nursery	Root-knot nematodes, *Meloidogyne* spp.	15 days	Decreased root-knot disease and weeds by 66% and 93%
French bean	Root-knot nematodes, *Meloidogyne* spp.	4-8 weeks	Nematodes effectively controlled
Cardamom nursery	Root-knot nematodes, *Meloidogyne* spp.	40-45 days	Nematodes suppressed
Betel vine	Root-knot nematode, *Meloidogyne incognita*	15 days	High reduction in nematode population
Turmeric	Root-knot nematodes, *Meloidogyne* spp.	40 days	Reduce nematode population

25.3.1. Soil solarization to manage root-knot nematodes

The effectiveness of solarization, a soil disinfection technique that uses passive solar heating, to control the incidence of root-knot nematode under greenhouse condition was accomplished by the application of 0.05-0.06 mm clear polyethylene sheets to moist soil for 4-week during the hot season.

In closed plastic greenhouses, the mulching of moistened soil with transparent polyethylene sheets for 4 weeks during the hot season, induce an increase of temperature. Maximum soil temperatures achieved under solarization were 51°C in the alluvial soil and 52.4°C in the heavy clay soil type at 20 cm. depth. In the non-solarization unheated greenhouse soil temperature achieved was 39°C (Fig. 25.5).

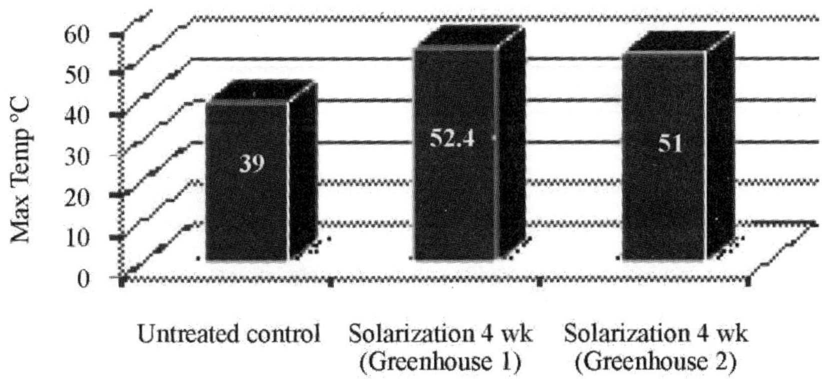

Fig. 25.5. Maximal temperature during soil solarization

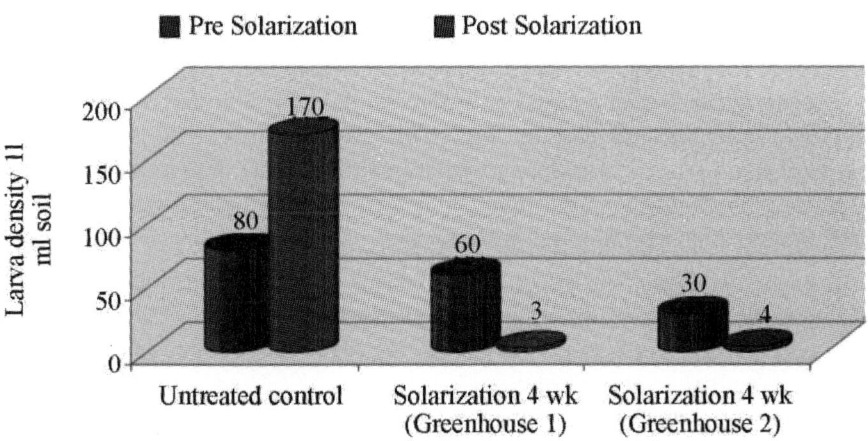

Fig. 25.6. Effect of soil solarization on the population densities of root-knot nematode in soil

The effect of soil solarization treatments on the population densities of the root-knot nematode, *M. incognita* was very clear. The larval density per 100 ml of soil was markedly reduced compared with the untreated control (Fig. 25.6). The soil solarization method had a significant influence to reduce the severity of root damage. Root galling index is the most important symptom for the root-knot nematode. This index was dramatically reduced in both solarized greenhouses compared with untreated control where the highest root galling index was found (Fig. 25.7).

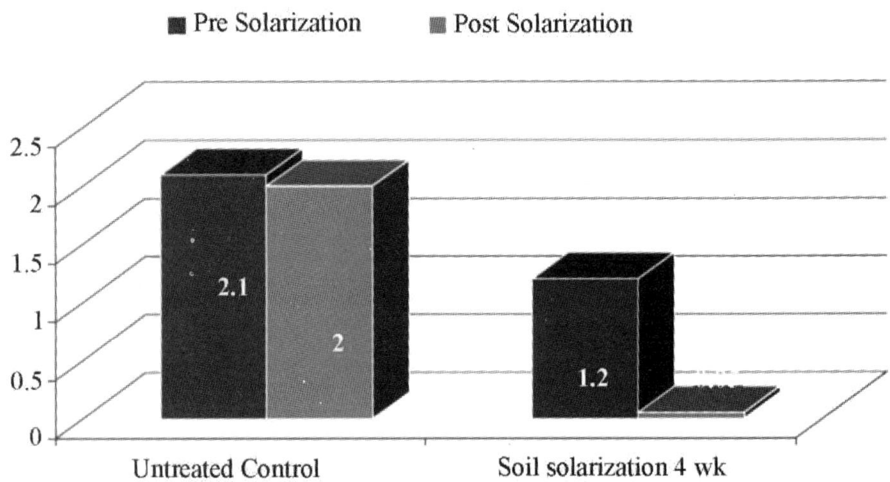

Fig. 25.7. Effect of soil solarization on root gall index (X - axis) by root-knot nematode

The growth of tomato plants, as indicated by visual assessment and plant height measurements was significantly improved by the soil solarization. Tomato fruit yields in solarized greenhouses were four-fold higher than those obtained in untreated control (Fig. 25.8). Such significant increases were consistent with effective control of nematode provided by solarization.

Solarization is also known to improve plant growth and yield through the release of nutrient induced by high soil temperatures. The soil solarization treatment also reduced nematode populations more than 90% in lettuce. The yield of lettuce was about two fold higher in solarized soil as compared with untreated control in several experimental plots.

Soil solarization should be done during the hottest period of the year. It offers a satisfactory and environmentally friendly solution for the control of root-knot nematode. This method is easily adopted in organic, conventional and integrated control growing system.

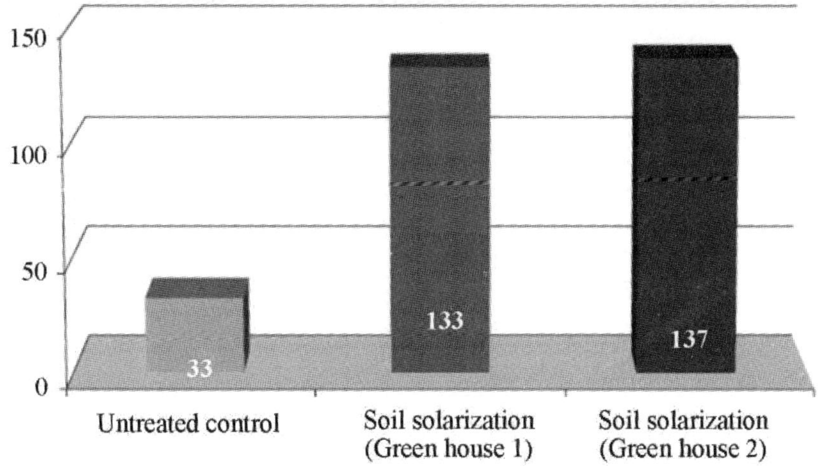

Fig. 25.8. Effect of soil solarization on tomato yield in tons/ha (X – axis)

25.4. IRRADIATION

There is good evidence that nematodes are killed or seriously affected by irradiation. Fassuliotis and Sparrow (1955) demonstrated that irradiation of potato tubers with X-rays inhibited sprouting and also development of *Globodera rostochiensis.* Cysts of this nematode exposed to 20,000 γ contained only brown and dead eggs and at 40,000 γ the eggs lost their contents completely. Wood and Goodey (1957) found that r-rays in between 48,000 to 96,000 γ inactivated *Ditylenchus myceliophagus* in mushroom compost. Irradiation of nematodes in the field is impracticable because of the length of time needed to irradiate even small areas and damage to plant roots. Ultraviolet light will also kill nematodes. These have not proved to be practical for nematode control in commercial agriculture.

25.5. OSMOTIC PRESSURE

Feder (1960) claimed that up to 100 per cent of the nematodes were killed when sucrose or dextrose was added to nematode infested soil at the rate of 1 to 5 per cent by weight. Feder and his colleagues (1962) showed that the osmotic effect was enhanced by the addition of 1 ton per acre of mono-sodium lauryl sulfate detergent to sugar soil mixtures. The application of 20 to 30 tons per acre of sugar to soil is impracticable and uneconomical.

25.6. WASHING PROCESS

Plant parasitic nematodes are often spread in soil adhering to potato tubers, bulbs and other planting materials and such spread is especially important with cyst forming species. Careful washing can considerably reduce the risk but, in practice, it is difficult to wash with sufficient thoroughness to remove all soil. Commercial vegetable, sugar beet and potato tuber washing apparatus have been developed.

25.7. SEED CLEANING

The seed gall nematode, *Anguina tritici* which causes 'ear cockles' of wheat has been recorded from most parts of the world where wheat is grown. Modern mechanical seed cleaning methods have been developed to remove the galls or cockles along with weed seeds and other foreign matter.

■■■■

Chapter 26

CULTURAL METHODS

26.1. INTRODUCTION

The direct control of nematodes by chemical or physical means is not often economical or practicable, especially on low acre value crops and indirect or cultural methods have to be employed. Cultural methods are attempts to adapt crop husbandry practices so as to minimize the effects of nematodes. Usually the operation will be a normal one carried out at little or no extra expense.

26.2. CROP ROTATION

Crop rotation attempts to keep nematode populations to a level at which crop damage is reduced to a minimum. By growing non-host crops, the numbers of plant nematodes are reduced, the number of years for this to occur depending on the initial population and the rate of population decrease. Rotations must also provide economically useful crops. In choosing rotations one must be careful not to promote a new set of pathogens in place of the ones to be controlled.

Continuous cropping with potatoes increases the cyst populations of *Globodera rostochiensis.* More than three year rotations with wheat, strawberry, cabbage, cauliflower, peas, maize and beans reduces the nematode population to a safe level. Likewise the cereal cyst nematode, *Heterodera avenae* can also be managed by following more than three year rotations with carrot, fenugreek and gram.

Some plants appear to be antagonistic to nematodes for they reduce populations to a greater extent than non-hosts. Osche and Brewton (1954) found that *Crotalaria spectabilis* and *C. striata* root systems have a toxic effect on root-knot nematodes and that, planted in advance of a crop, they may reduce subsequent damage. Meijneke and Oostenbrink (1958) showed that, *Tagetes* spp. reduce populations of *Pratylenchus, Tylenchorhynchus, Paratylenchus* and *Rotylenchus robustus* giving better growth of crops. Hence, these antagonistic plants can also be used in crop rotations to control certain plant parasitic nematodes.

Some of the cyst nematodes and certain species of root-knot nematodes attack only limited number crops. This permits a wider choice of unfavorable host crops for the rotation. For example, the principal host for the soybean cyst nematode (*Heterodera glycines*) is soybean and the planting of maize, cotton, groundnut, tobacco and many other crops in the rotation will result in decreasing the population each year. Soybeans following two years of maize in a field infested with *H. glycines* gave yield of 1615 kg per ha versus 538 kg per ha for the three successive years of soybean. In the fourth year, plots of soybeans following three years of maize yielded 2018 kg per ha, compared with 269 kg per ha for the continuous soybean plots.

Similarly, the root-knot nematode *Meloidogyne hapla,* a serious pest of groundnut, soybean and other crops, can be effectively controlled by rotations with maize or cotton, since these crops are non-hosts for this species. The planting of cotton, a non-host crop, for one year prior to planting groundnut, resulted in a yield difference of 1060 kg per ha and an increase in value per ha of $ 242 over that of groundnut following soybean, a favorable host for *M. hapla.*

Fig. 26.1. Effect of previous break-crop on root-knot nematode damage in the subsequent tomato crop. *Top* - Forage sorghum cv. Jumbo (background) and Lablab (foreground). *Bottom* - Roots of tomato crop grown after the nematode resistant sorghum break-crop (two plant roots on the left) and the nematode susceptible break crop Lablab (two plant roots on the right)

Table 26.1. Resistance ratings of potential break crop and standard control plant cultivars to species of root-knot nematode

Plant	Cultivar	Species of *Meloidogyne*		
		javanica	*hapla*	*incognita*
Lablab, *Lablab purpureus*	Highworth	SS	R	SS
	Rongi	SS	R	SS
Velvet bean, *Mucuna pruriens*	Florida	MS	HR	HR
Soybean, *Glycine max*	Frazer	SS	R	HR
Cowpea, *Vigna unguiculata*	Caloona	SS	HR	R
	Ebony	SS	HR	HR
Rhodes grass, *Chloris gayana*	Callide	SS	R/HR	R
	Katambora	R	HR	R
Pearl millet, *Pennisetum glaucum*		SS	R	SS
Forage sorghum, *Sorghum bicolor x sudanense*	Jumbo	R	HR	HR
	Octane	R	HR	HR
	Sprint	R	HR	HR
Maize, *Zea mays*	Hycorn	HR	HR	SS
Brassica	Bouncer	SS	SS	R
	Subzero	SS	SS	R
Brassica campestris	BFTap	SS	SS	R
Brassica carinata	BFBca	R	R	R
Brassica napus	BFStr	SS	SS	SS
	Goliath	SS	SS	R
Clover	Arrowleaf	MS	SS	SS
Clover - strawberry	Palestine	MS	SS	SS
Lucerne	Aurora	SS	R	HR

HR (highly resistant, multiplication rate = < 0.1), R (resistant, multiplication rate = 0.1 to < 1.0), SS (slightly susceptible, multiplication rate = 1.0 to < 10.0), MS (moderately susceptible, multiplication rate = 10.0 to < 100)

To control *M. incognita* and *M. javanica*, a tomato crop can be followed by groundnut without risk of damage to the groundnuts. While the groundnuts are growing, the nematodes cannot reproduce. Instead many of the larvae in the soil die or become non-infective because of starvation and the attacks of predators, fungi and diseases. If the population is reduced sufficiently, tomatoes can be grown again after groundnuts without serious injury (Taylor and Sasser, 1978).

In tropical and sub-tropical areas, forage sorghum has proved a most useful resistant break-crop because it can be planted at any time of the year and many cultivars are highly resistant to the warm-climate species of *Meloidogyne*. Also the crop is easy to grow, competes readily with weeds and produces large amounts of biomass in 2-3 months (Fig. 26.1). A well-managed break-crop usually reduces nematode populations to levels that will not damage the following vegetable crop.

Lists of potential break-crops resistant to particular root-knot nematode species have been presented in Table 26.1.

26.3. TRAP CROPPING

Trap crops (Fig. 26.2) are plants grown before or with the main crop into a smaller area. They are the more preferred host when grown with the main crop. Trap crops can increase the efficiency of control by concentrating the nematode pests in one location and by applying a chemical treatment without treating the main crop, or by destroying the trap crops and associated nematode pests through uprooting and burning.

Cowpea causes the root-knot nematode eggs to hatch; the larvae enter the roots and develop to immobile stage. Then the crop is destroyed before the nematodes mature. *Crotalaria* is highly susceptible to invasion by the root-knot nematode but is resistant to the development of larvae into adults. Specific resistance to nematodes which operates on a gene for gene basis, may involve a hypersensitive reaction to prevent invasion as it occurs in sun hemp (*Crotalaria spectabilis*), which survives invasion by *Meloidogyne* juveniles by failing to create the giant cells vital for the nematode survival (Cook and Baker, 1983).

Sree Bhadra, a high yielding variety of sweet potato released by Central Tuber Crops Research Institute, Trivandrum, is identified as a resistant trap crop for the root-knot nematode. The nematodes were able to penetrate the root but giant cell formation was not induced and as a result the nematodes die inside the roots. Planting this variety in root-knot nematode infested field helped in clearing the field free of nematodes and giving good tuber yield. Subsequently growing of susceptible crops like okra, coleus and yam escaped nematode damage (Mohandas, 2001).

Growing of trap crop like French marigold, *Tagetes patula* in alternate rows with potato reduced the root-knot nematode larval population in soil, root galling and tuber infestation while the yields increased up to 123% over control.

Solanum sisymbriifolium acts as a trap crop and can be used as sustainable and effective alternative to chemical control of potato cyst nematode (PCN). The crop acts as a completely risk-free trap crop for PCN by triggering hatch in the soil but preventing completion of the pest's life cycle, however long the crop is left in the ground.

Fig. 26.2. Trap cover crops - Sun hemp, *Crotalaria juncea* (monocrotarine); French marigold, *Tagetes patula* (α-terthinyl), Sorghum-Sudan grass (Dhurrin), Oil radish, *Raphanus sativus* (glucosinolate)

Tomato nursery beds previously planted with trap crop (marigold) gave maximum reduction in root-knot nematode population in soil which also increased the germination of tomato seeds and production of more healthy (nematode-free) seedlings (Rangaswamy *et al.*, 1999).

Growing of trap crop like knol-khol with brinjal but removed and destroyed before new generation of nematode could emerge helps in reducing the root-knot nematode population (Ayyar, 1926).

26.4. FALLOWING

Many nematode pests can be controlled by depriving them of plants on which to feed. From the practical stand point, however, keeping land fallow for the necessary length of time is not economically feasible or can even be dangerous because of soil erosion by wind or water. Nevertheless, crops that have been damaged by root infecting nematodes of any kind should not be allowed to stand in the field for a long time after harvest but should be destroyed immediately, preferably by plowing to expose the roots to the drying action of wind and sun. Weeds which are susceptible to nematodes should not be allowed to grow during fallow.

26.5. SELECTION OF HEALTHY PROPAGATING MATERIAL

With plants that are propagated vegetatively, it is often possible to eliminate nematodes by selecting for propagation uninfested plants or plant parts. Thus the citrus nematode (*Tylenchulus semipenetrans*), the golden nematode of potato (*Globodera rostochiensis*), the burrowing nematode of banana (*Radopholus similis*) and the coffee lesion nematode (*Pratylenchus coffeae*) can be eliminated by selecting nematode-free planting material. The wheat seed gall nematode (*Anguina tritici*) and the rice white tip nematode (*Aphelenchoides besseyi*) can also be controlled by planting nematode-free seeds.

26.6. REMOVAL OR DESTRUCTION OF INFECTED PLANTS

In tobacco, if the stalks are cut soon after harvest and the root system of the plant is then turned out and exposed, the population of root-knot nematodes is reduced through the drying action of the sun and wind. The same principle can also be utilized for the control of *Ditylenchus angustus* on rice.

26.7. INFLUENCE OF MANURING

Kirkpatrick *et al.* (1959a) showed that populations of *Xiphinema* and *Pratylenchus* were significantly lower under cherry trees receiving high rates of potash (2 and 6 kg K_2SO_4 per tree). They also showed (1959b) that the numbers of these nematodes were correlated negatively with leaf potash. Application of $CaCN_2$, NaCN and urea cyanamide is effective in controlling root-knot nematodes. This may be due to the release of HCN or NH_3 which are toxic to the nematodes.

26.8. FLOODING

Keeping the land submerged in water decreases oxygen content of the soil and kills the nematodes by asphyxiation. Chemicals lethal to nematodes, such as butyric and propionic acids, H_2S and NH_3, often develop in flooded soils. This method of control is effective against root-knot and stunt nematodes on vegetables.

26.9. COVER/GREEN MANURE CROPPING

Cover/green manure crops are planted between main crops which are valuable tools for managing crop nematode pests; which positively impact yields over time. The warm season legume cover crops are effective in reducing populations of certain plant parasitic nematodes by breaking their life cycles (Potter *et al.*, 1998; Vargas-Ayala *et al.*, 2000). The populations of sting (*Belonolaimus longicaudatus*) and root-knot (*Meloidogyne incognita*) nematodes were effectively reduced in cash crops by hairy indigo and joint vetch cover crops combined with mulching of cowpea clippings (Rhoades and Forbes, 1986). Likewise, the population densities of several root-knot nematode species were lowered (present simultaneously) by the cover crop velvet bean in greenhouse and field tests (Rodriguez-Kabana *et al.*, 1992).

26.10. ANTAGONISTIC PLANTS

26.10.1. Mustard

Morgan (1925) accidentally discovered that potatoes grown with white mustard in a pot of infested soil were less heavily attacked by nematodes than potatoes growing alone. Triffit (1930) showed that potato root diffusate was ineffective in the presence of leachings from the roots of mustard seedlings. Ellenby (1951) showed that, in field trials, mustard oils increased the yield of potatoes by reducing the severity of nematode attack. The active principle involved in mustard is allyl isothiocyanate which is toxic to the nematodes.

26.10.2. Pangola grass

Another reported naturally occurring nematicide is pyrocatechol. This compound accumulates in high concentration in roots of Pangola grass *(Eragrostis curvula*) and is a potent nematicide against *Meloidogyne javanica, M. hapla, M. arenaria thamsei* and *M. incognita acrita.*

26.10.3. Marigold

Meijneke and Oostenbrink (1958) showed that, by growing marigold, *Tagetes erecta* and *T. patula,* the populations of *Pratylenchus, Paratylenchus, Tylenchorhynchus* and *Rotylenchus* could be considerably reduced. *Tagetes* spp. also reduced *Pratylenchus* and *Meloidogyne* populations on tea. The active principle in *Tagetes* spp. is α-terthienyl which is toxic to these nematodes.

Some researchers believe that marigold root exudates prevent the nematodes from developing and their eggs from hatching. However, the nematode species and growth stages suppressed by marigold vary with the marigold species. For example, *Tagetes patula* 'Single Gold' (synonym: 'Groundcover') (Fig. 26.3) is an extremely poor host of the root-knot nematode, and there is limited penetration and development by this nematode on this plant. However, another marigold species, *T. erecta* 'Cracker Jack' (Fig. 26.4), behaved as a trap crop: root-knot nematodes were attracted to and entered its roots, but the development of their off-spring was impeded. In other cases, marigolds may behave as a trap crop by allowing penetration of nematodes but inhibiting their subsequent development and reproduction.

Fig. 26.3. A patch of marigold, *Tagetes patula* 'Single Gold' at the early flowering stage. Behind it is a border row of a sorghum Sudan grass hybrid (Photo: C.R.R. Hooks)

Fig. 26.4. Flowering marigold *Tagetes erecta* 'Cracker Jack' interplanted with cauliflower.

26.10.4. Asparagus

Rohde and Jenkins (1958) showed that *Asparagus officinalis* would not support populations of *Trichodorus christiei* for more than 40 to 50 days. Tomato, normally a good host of this nematode, supported only a low population when asparagus was growing in the same pots. A glycoside (asparagusic acid) is the active principle involved which is toxic to *T. christiei* and several other nematode species.

26.10.5. Crotalaria

Osche and Brewton (1954) have suggested that *Crotalaria* root system exerts a toxic effect on *Meloidogyne* populations. Barrons (1940) demonstrated that *Meloidogyne* larvae freely entered the roots of *Crotalaria* but failed to survive. Populations of *Xiphinema americanum* decreased under *Crotalaria spectabilis*.

26.10.6. Sesame

Atwal and Mangar (1969) showed that root exudates from sesame (*Sesamum orientale*) have nematicidal properties against *Meloidogyne incognita.* When Okra was grown in *M. incognita* infested soil it was only lightly attacked and there were fewer nematodes compared with when okra was grown in the absence of sesame.

26.10.7. Bitter cucumber

Cucurbitacins are triterpenoids that accumulate as aglycones or glycosides in cucumbers carrying a dominant gene for bitterness. Extracts from bitter cucumber attracted fewer *Meloidogyne incognita* larvae than did a comparable extract from non-bitter cucumbers which do not accumulate cucurbitacin. The cucurbitacins are repellent to *M. incognita.* Once larvae penetrate the roots it is possible that cucurbitacin (an antibiotic) acts as a toxin that either kills the invading larvae or repels them from the root.

26.10.8. *Carthamus tinctorius*

Two isomeric polyacetylene compounds extracted from flowers and roots of *Carthamus tinctorius* showed potent nematicidal activity against *Aphelenchoides besseyi*.

26.10.9. Neem (*Azadirachta indica*)

Neem is reported to contain at least two nematicidal compounds - nimbidin and thionemone. In a comparison test, neem root exudates were found to be more toxic to *Hoplolaimus indicus, Helicotylenchus indicus, Rotylenchulus reniformis, Tylenchorhynchus brassicae, Tylenchus filiformis* and *M. incognita* than marigold root exudates.

26.11. TIME OF PLANTING

In some cases, crops may be planted in winter when soil temperatures are still below necessary for activation of nematodes. Both early potatoes and sugar beets grow in soils too cold for cyst nematode activity. By the time the soil warms sufficiently for nematodes to become active, the plants are advanced enough to withstand a massive invasion of larvae. They continue to grow and eventually produce a crop.

26.12. BIOFUMIGATION

Stapleton *et al.* (1998) defined biofumigation as an agronomic practice that release volatile biotoxic compounds into the soil atmosphere during the decomposition of organic amendments. The isothiocyanates (ITCs) (related to the active ingredient in the commercial fumigants metham sodium and dazomet) are the most common volatiles produced during the breakdown of Brassicas that are highly toxic to nematodes (Fig. 26.5) (Delaquis and Mazza, 1995).

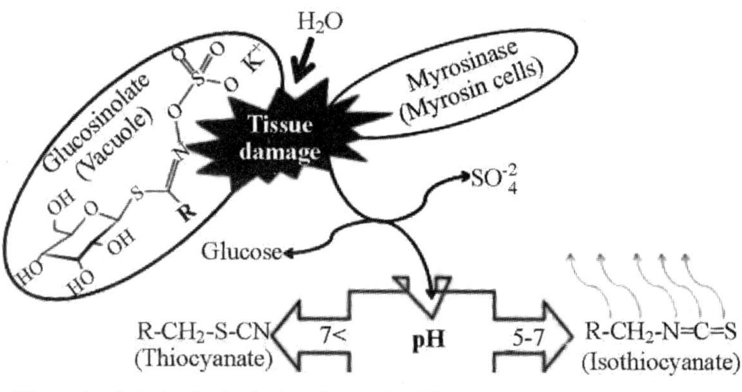

Fig. 26.5. Glucosinolate hydrolysis to release isothiocyanate.

The volatile chemicals (allelochemicals) released while decomposing *Brassica* tissues such as biologically-active green manures are utilized to manage soil-borne pathogens. The myrosinase enzymes at neutral pH hydrolyze glucosinolates (GSLs) to release ITCs following tissue damage. Most of the researchers believe that GSLs [sulfur containing chemicals (thioglucosides)] that are produced as secondary metabolites by Brassicas are responsible for providing the resistance against pathogens. The substantial quantities of ITC can be produced using black mustard (*B. nigra*) and Indian mustard (*B. juncea*) (Tollsten and Bergström, 1988) and could be utilized in a biofumigation cropping system.

The soil-borne plant parasitic nematodes are generally suppressed by cultivation and incorporation of cover/rotation crops, particularly Brassicaceae plants. Even though glucosinolates are thought to play an important role in nematode suppression, various studies have shown that nematode control efficacy and nematode reproduction in treated soils are not correlated with glucosinolate contents in plants (Potter *et al.,* 1998; McLeod and Steel, 1999). Other factors like non-glucosinolate compounds (other sulfur-containing compounds) (Bending and Lincoln, 1999), as well as biological and physiological factors, may also be involved in nematode suppression (Mazzola *et al.,* 2001). The biofumigation works better in sandier soils with low organic matter content.

Integration of soil incorporation of Brassicaceae plants (biofumigation) combined with soil tarping using plastic film in hot seasons enhances the nematode control efficacy by elevating soil temperatures due to soil solarization effect and preventing rapid emission of volatile nematicidal compounds from the soil to the atmosphere (Fig. 26.6) (Gamliel and Stapleton, 1993; Ploeg and Stapleton, 2001). The synergistic effect on nematode control efficacy has been noticed by combinations of sub-lethal soil temperatures (30–38.8°C) which render nematodes more sensitive to toxic compounds or to antagonistic microorganisms, and the biofumigation effects that release toxic volatile compounds. The nematode control efficacy can also be enhanced by the combination of biofumigation with non-Brassicaceae plants (such as pepper plant residues) and soil solarization.

Fig. 26.6. Nematode infested soil was amended with 1% (w/w) residues of oil radish (*Raphanus sativus*) and compared to unamended control

Cultivation of different mustard species (e.g. *Brassica juncea* var. *integrifolia* or *B. juncea* var. *juncea*) in nematode infested fields and their soil incorporation at flowering gives effective management of nematodes. The nematicidal compounds released during the decomposition of incorporated plant parts in a moist soil kill nematodes. The new crop can be planted or sown two weeks after incorporating plant material into the soil (it takes about two weeks for the plant material to decompose and stop releasing phytotoxic substances i.e., chemicals poisonous to plants).

■■■■

Chapter 27

CHEMICAL METHODS

27.1. INTRODUCTION

The primary advantage of chemical control over other methods is that the nematode population is reduced to a very low density within a matter of days after the chemical is applied, enabling the grower to plant a crop soon after treatment or in some cases at the time of treatment. Most crops are especially vulnerable to nematode attack during the seedling stage. Crops planted in treated soil develop good root systems, so that usually, in the case of annuals, the crop is mature before residual population of nematodes has increased to a level which would cause damage. Another important benefit resulting from chemical treatment is that in addition to nematodes; fungi, bacteria and weeds are also controlled. Weed control, alone, results in tremendous savings in labor costs in the production of vegetable transplants and ornamentals.

Damage caused by nematodes could not be fully appreciated until chemical nematicides were applied to soil in many parts of the world resulting in dramatic improvement of many different crops. Plant parasitic nematodes may be controlled by applying nematicides to the soil or to the potential host plant. Protecting the host plant is the more attractive method because it requires much less active ingredient per hectare and avoids the need to treat the great bulk of soil in which the crop will be grown, or is already growing. Control of phytonematodes by the use of nematicides is both difficult and expensive. Since nematicides are very costly, they should be used only when profitable i.e. when they increase the crop value many times over the cost of the chemical and of its application.

27.2. HISTORICAL

Kuhn (1881) first applied CS_2 to control sugar beet nematode in Germany, but results were not encouraging. In the United States, Bessey (1911) experimented with CS_2 for the control of root-knot nematodes in South Carolina, but the method proved impractical. Formaldehyde, cyanide, quick lime and numerous other chemicals were found to have nematicidal qualities, but all were too expensive to use even on high priced specialty crops.

The nematicidal qualities of Chloropicrin (tear gas) were discovered by Matthews (1919) in England. Godfrey, Oliveira and Hoshino (1934) established small field plots in Hawaii where they injected chloropicrin with a hand applicator which placed the chemical 6 to 8 inches deep at rates of 150 to 170 pounds per acre. To contain the gas, plots were covered with impermeable paper with the edges buried in soil. Results indicated a kill of 99 to 100 per cent of the nematodes.

Carter (1943) an entomologist of the Hawaiian Pineapple Research Institute, reported that a dichloropropene - dichloropropane (DD) mixture produced by the Shell Chemical Corporation was a promising new nematicide when applied at the rate of 250 pounds per acre.

Preliminary experiments with ethylene dibromide (EDB) were made in 1944 by Kagy and Youngson in California and by Christie in Florida. The same year the Dow Chemical Company introduced the chemical as a successful soil fumigant for the control of nematodes.

The discovery of these two practical nematicides (DD and EDB) marked the beginning of the new era in nematode control. These nematicides demonstrated the yield losses due to nematode infestations.

27.3. GROUPS OF NEMATICIDES

27.3.1. Halogenated hydrocarbons

Nematodes are sensitive to the halogens and most of our successful nematicides are halogenated hydrocarbons. Of these that have been tried, the five most effective compounds are – Chloropicrin, Methyl bromide, D-D, EDB and DBCP.

27.3.1.1. Chloropicrin (Tear gas)

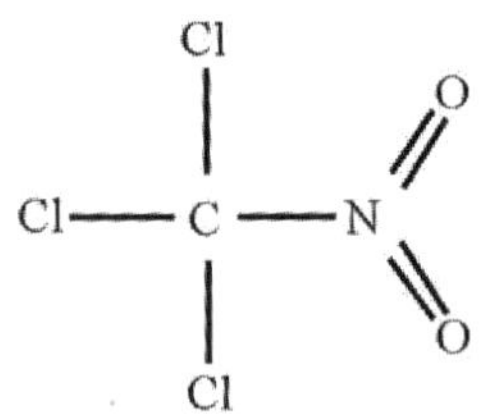

Chemical name	: Trichloronitromethane
Trade name	: Larvacide, Picfume
Dose	: 50 ml/m^2
Boiling point	: 112.4°C
Yellowish volatile liquid	

27.3.1.2. Methyl bromide

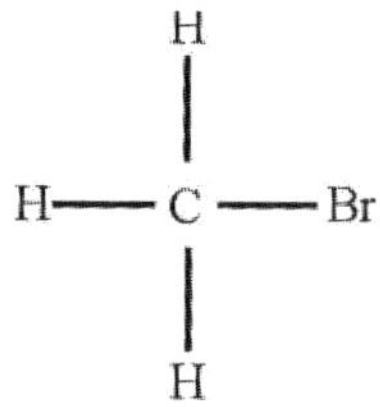

Chemical name : Bromomethane

Trade names : Embafume, Bed-Fume, Dowfume MC_2, Pest master

Dose : 30 ml/m^2

Boiling point : 4.4°C (Gas at ordinary temperature)

These two chemicals are biocides which kill nematodes, fungi, bacteria, weed seeds and insects that are present in the soil. These chemicals are used for fumigating potting soil or the soil in greenhouse benches, seed beds and plant beds. They are applied by injection under the plastic seal as they are highly volatile. After the treatment the seal should be allowed to remain for 2 days. After removing the seal, allow 2 days waiting period before planting is taken up.

Because methyl bromide is nearly odorless, chloropicrin may be added as a warning gas.

27.3.1.3. D-D

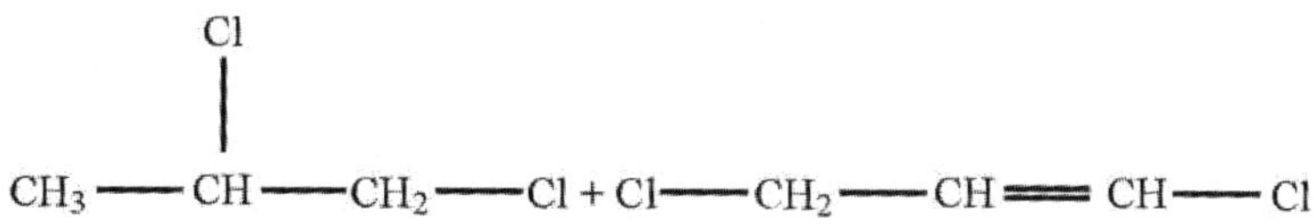

Chemical name : 1, 2- dichloropropane + 1, 3-dichloropropene

Trade name : D-D, Vidden D, Telone

Dose : 250 to 500 L/ha

Boiling point : 106-111°C

Dark brown volatile liquid.

27.3.1.4. EDB

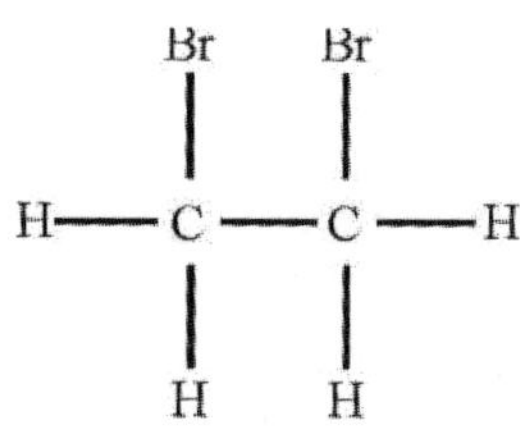

Chemical name : 1, 2-dibromoethane

Trade name : Dowfume W-40, Soilfume 40, Bromofume 40, Nemex 42 (containing 41% EDB), Dowfume W-85, Soilfume 85, Bromofume 85. Nemex 85 (containing 85% EDB)

Dose : 60 L/ha

Boiling point : 131ºC

Heavy, colorless, volatile liquid.

These 2 chemicals applied by injecting small quantities into the soil to a depth of 15 cm or deeper, with injections spaced usually at 30 cm intervals. They are highly effective nematicides suitable for field scale application and are used extensively and successfully as pre-planting treatments for annual crops. They may injure the roots of living plants and stunt their growth. To avoid injury to crops an adequate interval must elapse between application and planting, usually supplemented by tillage.

As compared with EDB, D-D is somewhat more effective for controlling cyst and lesion nematodes and about equally effective for controlling root-knot nematodes, when both are applied at comparable rates, EDB is reported to be more effective than D-D for controlling sting, stunt and potato rot nematodes. EDB cannot be used for certain bromine - sensitive plants like onions, garlic, lily, amaryllis, etc. D-D is more fungicidal than EDB and EDB is more insecticidal than D-D.

27.3.1.5. DBCP

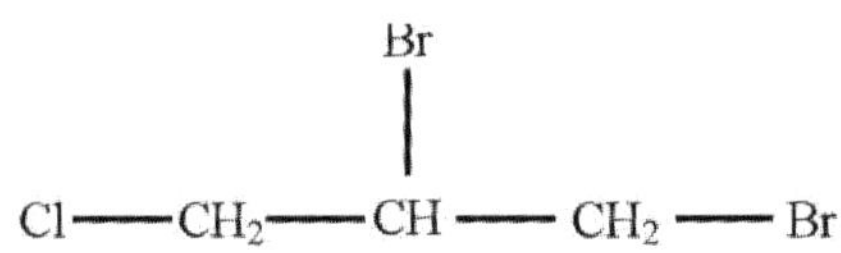

Chemical name : 1, 2-dibromo-3-chloropropane

Trade name : Nemagon, Fumazone

Dose : 25 to 50 L/ha

Boiling point : 195.6ºC

Heavy, straw-colored, volatile liquid.

It can be injected into the soil like other soil fumigants or an emulsifiable concentrate can be mixed with water and applied as a drench. Although slow acting, DBCP is a potent nematicide capable of giving a high degree of nematode control at comparatively low rates. It tends to linger in the soil and continue its nematicidal activity longer than other soil fumigants. DBCP appears to function more effectively than the other fumigants at high soil temperatures, probably because of its high boiling point. DBCP is not highly phytotoxic and can be applied around the roots of many kinds of living plants without appreciably affecting their growth. However, some kinds of plants are sensitive to it, e.g. onion, potato, tobacco, tomato, etc.

Recently, manufacturing restrictions have been imposed on fumigants containing 1, 2-dibromo-3-chloropropane because of their potential for causing cancer in some laboratory animals and sterility in man.

Nematicides having as an active ingredient one of these five halogenated hydrocarbons or a mixture of two of them are sold under many different trade names. A formulation containing both EDB and D-D is available under the trade name Dorlon.

Soil fumigation is an exacting procedure that must be done properly if satisfactory results are to be obtained. Following are a few of the more obvious considerations relating to application of soil fumigants:

- Before applying a fumigant, the land should be prepared to the consistency of a good seed bed. It should be smooth and free of lumps, clods and undecomposed crop residues. Heavy soils should be plowed, as disking alone will not loosen them to a sufficient depth.
- The soil temperature 15 cm below the surface should be between 10° and 30°C, preferably between 21° and 27°C.
- Fumigants are not fully effective in very dry or very wet soil. For best results in most soils the moisture content should approach but not exceed field capacity.
- After the fumigant is applied, holes, channels or furrows left by the applicator should be filled promptly and firmly and the land left smooth and firm at the surface. The land should remain undisturbed for one week, and then it should be deeply and thoroughly disked.
- When weather and soil conditions are ideal, 2 to 3 weeks' time should elapse between application and planting to avoid injury to the crop.

27.3.1.6. *Some effects of soil fumigation:* Soil fumigants may inhibit temporarily the action of nitrifying bacteria and retard the transformation of ammonia nitrogen to nitrate nitrogen. The resulting accumulation of NH_3 may be toxic to certain ammonia - sensitive plants like tomatoes, especially where such fertilizers as ammonium sulfate are applied before fumigation. It has been generally conceded that soil fumigation, when followed by a legume, is likely to reduce nodulation and may make inoculation of the seed necessary. However, Miller (1951) reports that peanuts grown on land following fumigation with EDB were more heavily nodulated than peanuts grown on comparable unfumigated land. It is logical to expect, therefore, that fumigation may, in some instances, increase nodulation.

27.3.2. Organophosphates

The need for a non-phytotoxic nematicide led to the investigation of this group of compounds in the hope of finding one that would control nematodes at rates not injurious to plants.

27.3.2.1. *Parathion*

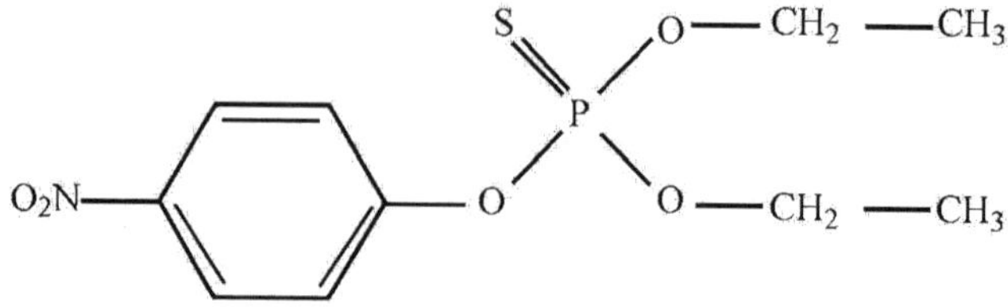

Chemical name : O, O-diethyl O-(p-nitrophenyl)-phosphorothioate

Trade name : Ekatex, Folidol, Thiophos

Formulation : 50% EC

Dose : 0.1% Spray

Dimock and Ford (1950) reported good control of the chrysanthemum foliar nematode with repeated applications of parathion spray. Raski and Allen (1948) greatly reduced the number of bud nematodes, *Aphelenchoides fragariae* per strawberry bud with two applications of parathion spray but failed to eradicate the infection.

Eggs laid by the nematode in the plant tissue are not destroyed by parathion. Therefore, the compound must be sprayed several times in order to kill the larvae as they hatch from the eggs.

27.3.2.2. *Dichlofenthion*

Chemical name : O-(2, 4-dichlorophyenl) O, O-diethyl phosphorothioate

Trade name : VC-13, Hexanema

Formulation : 75% EC or 5% granules

Dose : 8-12 kg a.i. /ha

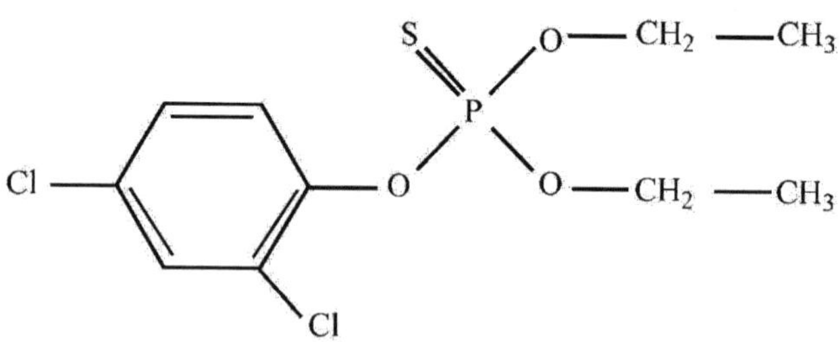

This compound is not volatile enough to function effectively as a soil fumigant; hence usually it is emulsified with water and applied as a drench. Dichlofenthion is less phytotoxic and will kill many kinds of nematodes at rates that are not injurious to living plants. It will control root-knot on living plants apparently by killing the larvae as they hatch and prevent reinfestation. It is being used with considerable success for arresting nematode damage to turf and many ornamentals.

27.3.2.3. *Thionazin*

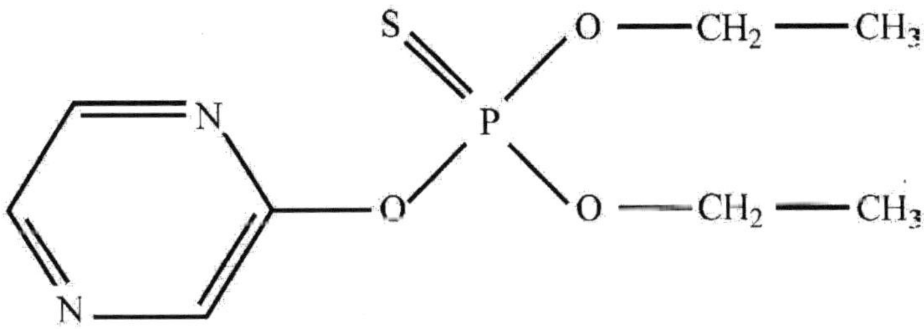

Chemical name : O, O-diethyl-O-2-pyrazinyl-phosphorothioate

Trade name : Zinophos, Nemaphos

Formulation : 10% granules

Dose : 4 to 8 kg a i/ha

This chemical is most effective against the sting nematode of maize at 1 kg a.i. per ha as row treatment. It is also effective against the root-knot nematode on vegetables. It can also be used at 600 to 800 ppm as seedling bare root dip treatment for plants infected with nematodes.

On carnation, excellent protection against *Meloidogyne incognita* was obtained by spraying the foliage at three weekly intervals.

Glasshouse experiments have shown that thionazin is systemic in action and is translocated both upwards and downwards in the plants and protection against nematode invasion persists for 18 to 20 days.

27.3.2.4. *Fensulfothion*

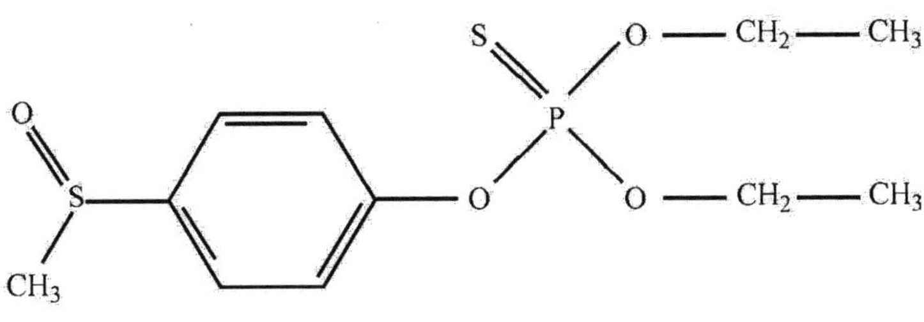

Chemical name : O, O-diethyl-O-p (methyl sulfinyl)-phosphorothioate

Trade name : Terracur - P, Dasanit

Formulation ; 5 to 10% granules

Dose : 4 to 8 kg a i/ha

This chemical is used to control root-knot nematodes both under nursery and field conditions. It is also effective against turf nematodes.

27.3.2.5. *Fenamiphos*

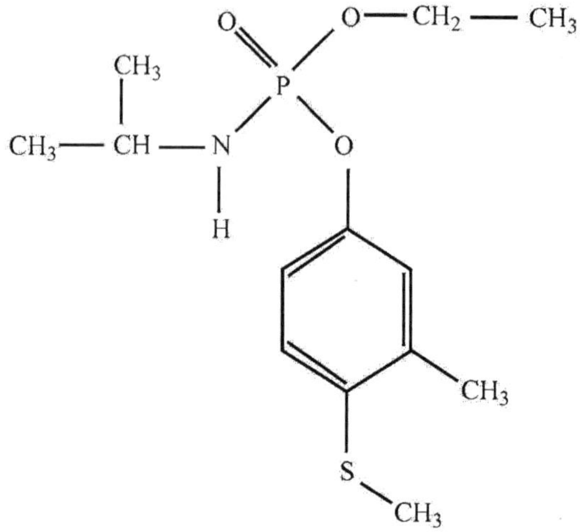

Chemical name : Ethyl-4-(methylthio) -m-tolyl isopropyl phosphoramidate

Trade name : Nemacur

Formulation : 10% granules

Dose : 4 to 8 kg a i/ha

It is a systemic nematicide/insecticide/miticide, with slight fungicidal action. The granular formulation is excellent against root-knot nematodes and because of its long persistence, even in warm soils. Pre-planting applications on annual crops have given good yield increases.

27.3.2.6. *Ethoprophos*

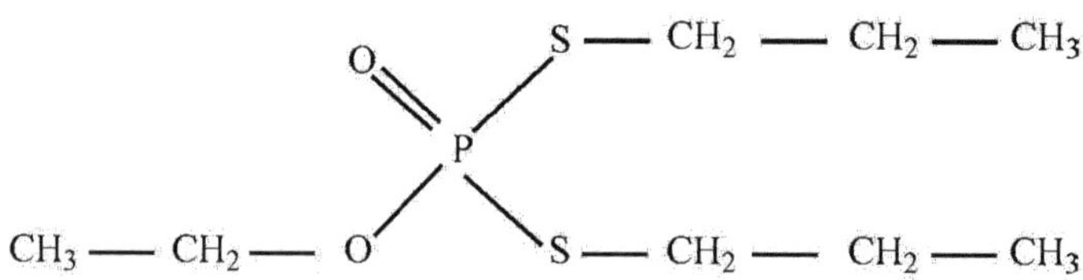

Chemical name : O-ethyl S, S-dipropyl-phosphorodithioate

Trade name : Mocap

Formulation : 10% granules

Dose : 4 to 8 kg a i/ha

It is a non-systemic nematicide with very little persistence. To obtain satisfactory results with root-knot nematodes, it is therefore necessary to apply the chemical when juveniles are at peak populations in the soil.

27.3.2.7. *Phorate*

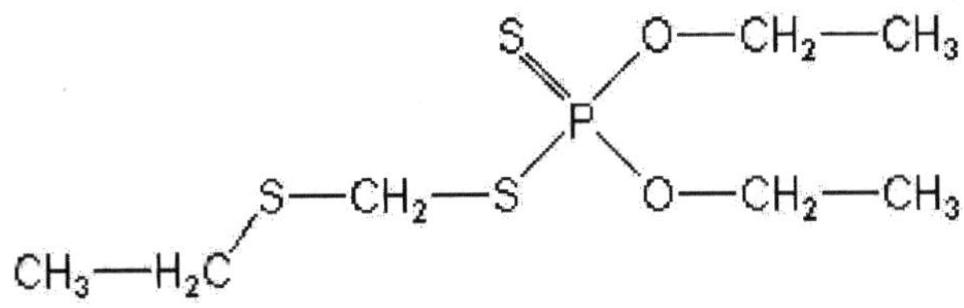

Chemical name	: O, O-diethyl S-(ethylthiomethyl)-phosphorodithioate
Trade name	: Phorate, Thimet
Formulation	: 10% granules
Dose	: 4 to 8 kg a i/ha

This is used to control stunt, root lesion and pin nematodes on ornamental plants.

Organophosphates can also be used as bare root dips for the control of plant nematodes (Table 27.1).

Table 27.1. Control of plant nematodes by dipping planting material in nematicide solutions

Plant	Nematode	Nematicide	Conc. ppm	Time min
Gladiolus corms	*Meloidogyne incognita*	Fensulfothion	600	15
		Ethoprop	900	15
Peach roots	*M. incognita*	Fensulfothion	1000	30
Rose roots	*M. hapla*	Ethoprop	1000	30
		Fenamiphos	1000	30

27.3.3. Dithiocarbomates

Certain chemicals of this group are effective in killing nematodes.

27.3.3.1. *Metham sodium*

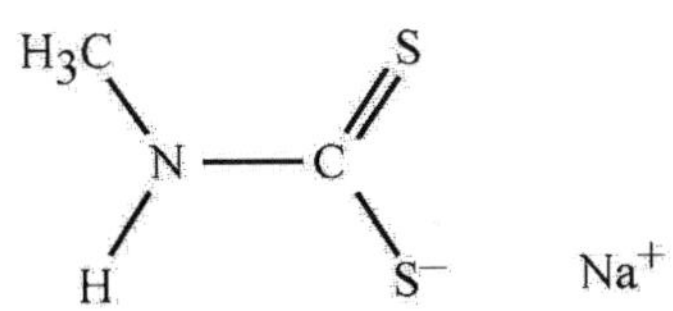

Chemical name	: Sodium N-methyldithiocarbamate
Trade name	: Vapam, Sistan, Vitafume, Unifume
Formulation	: 32-35% aqueous solution
Dose	: 100-200 ml/m^2

This solution is readily miscible with water and can be applied as a drench in irrigation water or by injection into the soil. After application, metham sodium produces a gas and distribution through the soil may result from dispersion in water or gaseous diffusion. It is used as a pre-plating treatment and is reported to control insects, fungi, nematodes, weeds and other pests. Because it is highly phytotoxic, an interval of two weeks or more must be allowed between application and planting.

27.3.3.2. *Aldicarb*

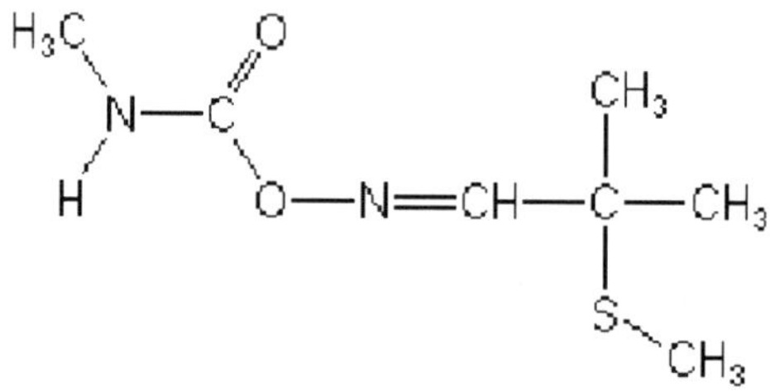

Chemical name : 2-methyl - 2- (methylthio) propionaldehyde-O-(methylcarbamoyl) oxime

Trade name : Temik

Formulation : 10% granules

Dose : 2 to 4 kg a i/ha

Aldicarb is effective in controlling root-knot nematodes on vegetables and tobacco both in nursery and field conditions. It can also be used as seedling bare root dip treatment for plants infected with nematodes. It is a systemic nematicide.

27.3.3.3. *Carbofuran*

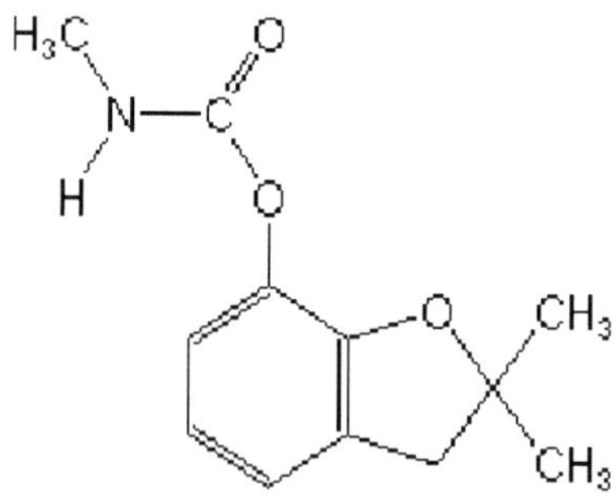

Chemical name : 2, 3-dihydro-2, 2-dimethyl 7-benzofuranyl methyl carbamate

Trade name : Furadan

Formulation : 3% granules and 50% WP

Dose : 4 to 8 kg a.i. /ha

It can be used to control root-knot nematodes on vegetables. It is also used for seed treatment to control the nematodes during the early stages of plant growth. It is a systemic nematicide.

There is good evidence for increased nematode tolerance to carbofuran in fields of corn that were treated with this compound for four to five years.

27.3.3.4. *Methomyl*

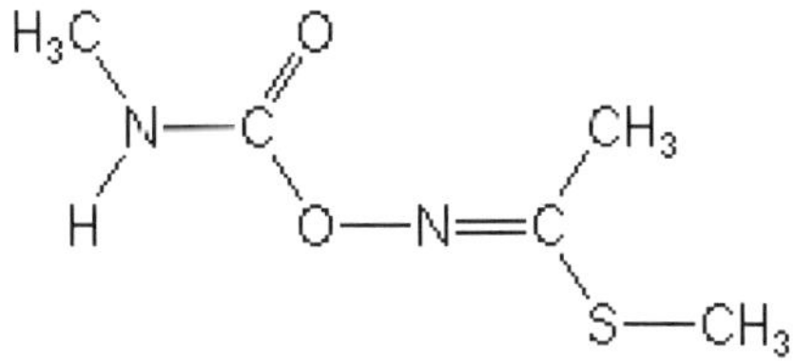

Chemical name : S-methyl N-[(methylcarbamoyl) oxy] thioacetamidate

Trade name : Lannate

Formulation : 90% WP

Dose : 4 to 8 kg a.i. /ha

It is a systemic nematicide

27.3.3.5. *Oxamyl*

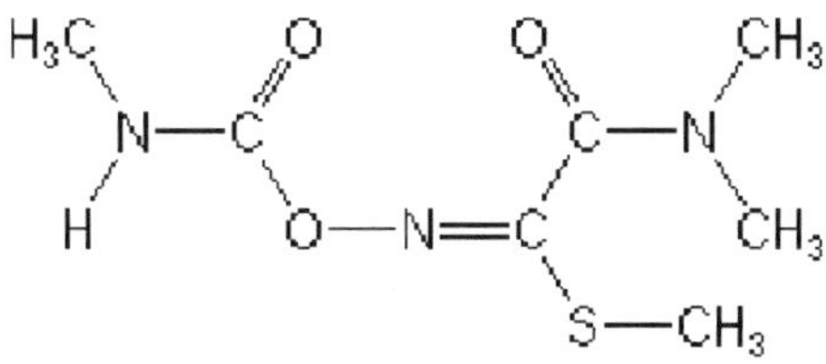

Chemical name : Methyl N, N'-dimethyl- N-[(methylcarbamoyl) oxy]-1-thioxamimidate

Trade name : Vydate, D-1410

Formulation : 40% EC

Dose : 4 to 8 L a.i. /ha (Fig. 27.1)

Fig. 27.1. Carrot field treated with oxamyl (left) for control of root-nematodes (right side untreated).

It is a systemic nematicide. It can be sprayed to the foliage to control the nematodes on roots. It also controls foliar nematodes.

27.3.4. Advantages of organophosphates and dithiocarbamates

- Effective against insects also (aphids, wire worms, white grubs and other soil insects).
- Much more effective at lower dosages.
- Easy to handle and apply.
- Much less phytotoxic and can be used in standing crop.
- Systemic and slow release in action.
- Can be used as seedling bare root dip treatment.
- Relatively non-volatile.
- Application often could be accomplished along with seeding making additional mechanical applications and the waiting period after treatment unnecessary.

27.3.5. Avermectins

Excitement is being generated by the results of a new class of insecticide - nematicide which have demonstrated a broad spectrum of anthelminthic activity. Promising results for the control of root-knot nematodes have been reported.

The aqueous solution of avermectins (250 ml of 0.001%/m^2 nursery bed) significantly reduced root galls of tomato seedlings raised in root-knot infested nursery beds as compared to carbofuran (10 g a.i. /m^2 nursery bed) or neem cake (1 kg/m^2 nursery bed). The treated tomato nursery beds yielded robust and healthy seedlings with no root-knot nematode infection (Fig 27.2) (Parvatha Reddy and Nagesh, 2002).

Fig 27.2. Tomato seedlings from avermectin treated and untreated nursery beds.

27.4. PLACEMENT OF NEMATICIDES

A few of the economically important plant nematodes infect stems, leaves or buds. In such cases, the parts of the plants on or in which the nematodes are located can be reached by sprays. A few compounds are systemic (e.g. oxamyl). They may be applied to leaves and become distributed throughout the plant to affect nematodes at the root surface as well as within the tissues.

Most of our important plant nematodes are inhabitants of the soil. There are three methods whereby a chemical can be brought into contact with soil-inhabiting nematodes, (a) by diffusion into the soil as a gas, (b) by dispersion through the soil in water, and (c) by mechanical mixing.

27.4.1. Gaseous diffusion

Most of the successful control of nematodes with chemicals has been achieved by soil fumigation, wherein the chemical is brought into contact with the nematodes by gaseous diffusion. For fumigating small areas, a hand applicator can be used. The area is divided lengthwise and crosswise with marks 30 cm apart and an injection is made where the marks intersect. Somewhat better coverage can be obtained by staggering the injection, i.e. making one row of injections where the marks intersect and the next row of injections midway between these points. For field scale applications of fumigants, power driven equipment is used that injects the fumigant in parallel streams 30 cm apart (Fig. 27.3).

Fig. 27.3. A tractor equipped for applying soil nematicide.

27.4.2. Dispersion in water

The practice of mixing a nematicide with water and applying the mixture as a drench has come into use as a means of controlling nematodes in turfs and ornamentals, such as gardenias and similar shrubs and perennials. Control procedures of this kind have been made possible by the introduction of such relatively non-phytotoxic nematicides as DBCP and metham sodium. Liquids are applied to furrows in the soil in carefully regulated amounts. The furrows are closed immediately and the soil is compacted. Nematicides are also added to irrigation water for treatment of standing citrus trees.

27.4.3. Mechanical mixing

Experiments have demonstrated that nematodes can be controlled by mixing granular chemicals with the soil mechanically. This method might enable the grower to prepare the land to a good tilth and incorporate both fertilizer and nematicide in one operation. Thorough mixing can be achieved in field scale either by disking or rotary tillage.

27.5. TYPES OF TREATMENT

27.5.1. Pre-plant treatment

As most of the halogenated hydrocarbons are phytotoxic, they must be applied to the soil before planting. About 3 to 4 weeks waiting period must elapse between application and planting.

27.5.2. Treatment at planting time

Most of the organophosphates and dithiocarbamates are much less phytotoxic and can be applied at the time of planting.

27.5.3. Post-plant treatment

DBCP, organophosphate and dithiocarbamates can be applied in the standing crop as they are less phytotoxic. This type of treatment is generally given in case of perennial crops.

27.5.4. Bare root dip treatment

Thionazin and aldicarb at 500 ppm can be used as seedling bare root dip treatment for 15 minutes to control nematodes on transplantable seedlings.

27.5.5. Seed treatment

Carbofuran can be used as seed treatment for the control of nematodes during the early period of plant growth.

27.5.6. Foliar treatment

Some systemic compounds (e.g. oxamyl) may be sprayed on leaves and translocated throughout the plant to affect nematodes at the root surface or within the tissues.

27.6. TYPES OF APPLICATION

27.6.1. Over all (Broadcast) application

Applying, the nematicide to the entire area is referred to as broadcast, solid, area or overall application. This method of application is used when crops are to be planted in rows spaced less than 60 cm apart.

27.6.2. Row application

An important way to reduce the cost on nematicides is by row rather than overall application. For crops grown in rows more than 60 cm apart, row application is widely practiced. The nematicide is applied only to the rows where the plants are to grow e.g. tobacco, cotton, vegetables. Depending on the spacing, the cost of the chemical is from one half to one tenth of that for broadcast application.

27.6.3. Spot application

For annual crops grown in widely spaced hills, spot application is practiced e.g. tobacco, cotton, vegetables. This type of application is more economical than row application.

27.6.4. Strip application

A strip pre-plant treatment can also be used for some perennial crops, such as grape and citrus, saving as much as 50 per cent of the chemical as compared with overall application. In this method only planting strip is treated with the nematicide. A disadvantage of this method is that recontamination from cultivation or irrigation is almost certain.

27.6.5. Site application

The site application is commonly used for replant problems in established orchards or vineyards. Experimentally, it has been demonstrated to be feasible with wide-spaced annual crops such as melons. For melons, a treating site of 60 sq. cm is sufficient, while tree crops require a minimum area of 2 to 2.5 m. The planting site method reduces the amount of chemical used considerably.

27.7. EVALUATION OF RESULTS

For easy interpretation of results, nematicidal trials are best done in soils where nematodes are the dominant pathogens and are uniformly distributed.

Direct soil sampling before and after treatment and during various stages of crop growth gives information on changes in nematode numbers. Control of root-knot nematodes can be assessed by counting the galls formed on the roots of indicator plants pot-grown in soil samples from treated plots. Control of cyst nematodes can be assessed by counting the number of larvae in the roots of potato indicator plants pot-grown for 6 weeks in soil samples taken from treated plots.

Organophosphates and dithiocarbamates may not reduce the nematode counts after treatment. They may affect the nematode sensitivity in finding the plant roots. Hence the evaluation should be based on host response.

The evaluation of results should also be based on symptoms and yield.

27.8. WAYS BY WHICH THE COST OF NEMATICIDES CAN BE REDUCED

- Seedling bare-root dip treatment in systemic nematicides like thionazin, carbofuran, oxamyl and aldicarb at 500 ppm for 15 minutes. This treatment completely eliminates root-knot infection and protects the seedlings for 6 to 8 weeks from nematode infestation after planting.
- Seed treatment with carbofuran prevents the nematode infection during the early period of seedling growth.
- Production of nematode-free seedlings by treatment of nursery beds.
- Row and spot application of nematicides in broad spaced crops.
- Indigenous manufacture of nematicides.
- Reduction of customs duty on nematicides which are imported.
- Sale of nematicides at subsidized price by Government agencies.

27.9. MODE OF ACTION

The mode of action of nematicides is not very well understood and the limited information available is supplemented by knowledge of insecticidal action that may apply to nematicides.

27.9.1. Organic halides

One theory that has been advanced to explain the toxicity of organic halides is believed to be caused primarily by the inhibition of some essential enzyme system in the nematode by chemical reaction of relative halide with some required nucleophilic (basic) center, such as sulfydryl, amino or dihydroxyl groups present in enzymes, and death ensues. The reactive halide may also react chemically with biologically important proteins or peptides or some regulator of cellular metabolism, inhibiting its function.

$$\text{Enzyme or protein} + \text{R: H} \quad \rightarrow \quad (\text{Enzyme or protein})\text{: R} + \text{H}$$

Another possible mode of intoxication of nematodes by alkyl halides may involve the interaction of halo-organic nematicide with an iron center in the respiratory chain of the nematode.

$$Fe^{++} \quad \rightarrow \quad Fe^{+++}$$

Compounds like metham sodium and dazomet which produce toxic volatile decomposition products such as methylisothiocyanate are highly reactive toward nucleophilic reagents and probably have the same mode of action as the reactive organic halides.

27.9.2. Organophosphates and carbamates

Even though the exact evidence for the biochemical mode of action of organophosphates and carbamates in nematodes is not clearly available, many similarities with their effects in insects and mammals, however, support the hypothesis that these nematicides kill animals by inhibiting cholinesterase with consequent disruption of nervous activity caused by accumulation of acetylcholine at the nerve endings.

The reactions involved in cholinesterase inhibition are: first a complex between acetyl-choline and the phosphate or carbamate is formed; secondly, this complex reacts to deliver the phosphorylated or carbamylated enzyme; and thirdly, the original enzyme could be formed by a hydrolysis reaction called dephosphorylation or decarbamylation. Organic phosphates in general cause irreversible enzyme inhibition and carbamates are reversible inhibitors.

Cholinesterases are found in many nematode species. Kamfe and Dietze (1972) were able to show the inhibition of cholinesterase located in the central nervous system.

Other toxic mechanisms cannot, however, be excluded. More exact information about the chemical and biochemical processes in poisoned and healthy nematodes are needed. Perhaps other enzymes than those studied in insects and mammals may also be involved, because organophosphate and carbamate compounds inhibit a wide range of enzymes.

27.9.3. Avermectins

Avermectins, produced by the soil actinomycete, *Streptomyces avermitilis* block the transmission of electrical potentials between interneurons and excitatory motoneurons in the ventral nerve cord. They inhibit transmission between inhibitory motoneurons and muscle by opening the chloride channels on the membranes but have little effect on excitatory neuromuscular junctions. The block of interneuron to excitatory motoneuron transmission can be reversed by picrotoxin, a known gammaminobutyric acid antagonist.

The different modes of action of nematicides have been summarized in Table 27.2.

Table 27.2. Common nematicides and probable modes of action

Nematicide	Mode of action
Fumigants	
Water: (required to activate toxicity in soil)	
MIT (methyl isothiocyanate) Metham (Vapam) Dazomet (Mylone)	MIT may react with amino acids, oxidases and nucleophilic sites on proteins; metham and dazomet form MIT.
Air: (required for movement in soil)	
1, 3-D (1, 3-dichloropropene) EDB (ethylene dibromide) DBCP (Nemagon) MB (methyl bromide) Chloropicrin	Alkyl halides may react with nucleophilic sites on proteins and oxidize iron porphyrins and hemeproteins.
Non Fumigants	
Organophosphates	
Ethoprop (Mocap) Fensulfothion (Dasanit) Fenamiphos (Nemacur)	Irreversible binding of acetyl cholinesterase, esterase inhibition and various pharmacologic actions.
Carbamates	
Carbofuran (Furadan) Aldicarb (Temik) Oxamyl (Vydate)	Reversible binding of acetyl cholinesterase, esterase inhibition and various pharmacologic actions.

Chapter 28

BIOLOGICAL METHODS

28.1. INTRODUCTION

Growing recognition that pesticides must be used cautiously to avoid damage to the environment has led to intensive search for other methods to control nematodes. Biological control aims at increasing the parasites, predators and pathogens of nematodes in the soil, in order to increase the mortality of plant nematodes. This can be done by changing the environment, by addition of organic amendments or by introducing other organisms directly. This should enhance the biotic component of environmental resistance. In practice this has not yet been possible on a large scale.

28.2. ORGANIC AMENDMENTS

Linford *et al.* (1938) observed that a significant reduction in root-knot nematode galls on cowpea occurred during the decomposition of organic matter. His work sparked an interest in various organic materials as possible soil amendments for the control of plant nematode diseases. Duddington *et al.* (1956) treated micro plots of beet heavily infested with *Heterodera schachtii* with organic matter (bran) which resulted in significant reduction in the final cyst population and increased the yield of beet. Duddington *et al.* (1961) decreased the number of *Heterodera avenae* larvae invading oat seedlings by green manure treatments of chopped cabbage leaves, and showed that this was due to the stimulation of the predacious fungi. Mankau and Minteer (1962) showed that castor pomace, a byproduct of castor oil extraction, eliminated all the larvae of *Tylenchulus semipenetrans* from soil. Johnson *et al.* (1967) found that amendments reduced incidence of root-knot on tomato. Singh and Sitaramaiah (1966) applied finely divided oil cakes to root-knot infested soil and noted a reduction in disease incidence in okra and tomato. Later, they observed that saw dust amendments in combination with urea were also effective.

Explanations for the possible mechanisms involved in the partial control of nematodes by the use of organic amendments include:

- The volatile fatty acids (formic, acetic, propionic and butyric acids), ammonia and hydrogen sulfide that are released during the microbial decomposition of organic amendments are directly toxic to the nematodes.

- Microbivorous nematodes rapidly reproduce on the abundant bacteria and decomposition products.
- Changes in the physical and chemical conditions of the soil may alter the host nematode relationship.
- Organic amendments improve soil conditions for more rapid root growth. This enhances the utilization of soil nutrients and masks the effects of nematode damage.
- Amino acids released during the decomposition of organic amendments may be responsible for reducing the nematode population in the soil (Parvatha Reddy *et al.,* 1975b). Yamashita and Akiya (1968) have shown that the amino acids are released from the organic amendments applied to the soil. Application of amino acids has been shown to control plant parasitic nematodes in the soil (Parvatha Reddy *et al.,* 1975a).

Unfortunately, the large quantities (50 to 150 tons per ha) of these amendments that are required often make their application commercially impractical.

28.3. PREDATORS

28.3.1. Predacious nematodes

These are characterized by either a large open stoma usually armed with teeth used in seizing and ripping open prey or swallowing them whole, or a pharynx equipped with an onchiostyle or stomatostyle for piercing the prey (Esser and Sobers, 1964) (Fig. 28.1 and 28.2).

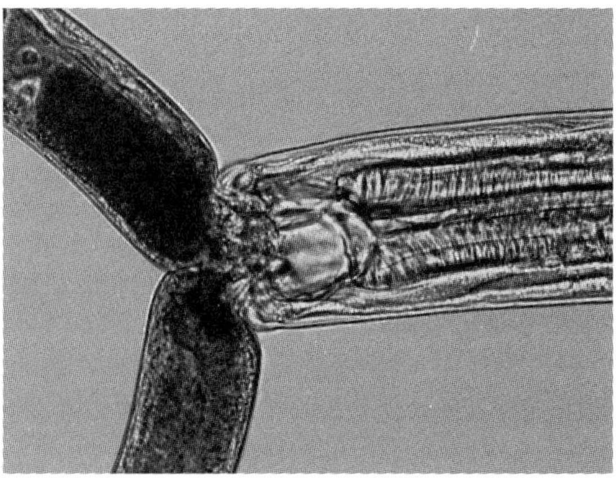

Fig. 28.1. Predatory nematode, Mononchus (Mononchida - Adenophorea)

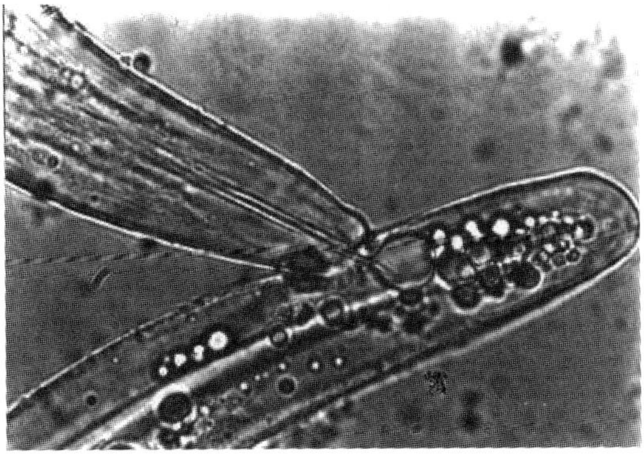

Fig. 28.2. *Aphelenchoides* (Aphelenchina-Tylenchida, Secernentea) feeding on nematode.

In the group of nematodes with open stoma, *Mononchus* spp. have been suggested for use in control of plant nematode diseases. Thorne (1927) observed that *Mononchus papillatus* feeding on *Heterodera schachtii* larvae and males. Yeates (1969) found that *Mononchoides potohikus* attacking a number of different nematodes. Other genera of predacious nematodes with open stoma are *Butlerius, Anatonchus, Diplogaster* and *Tripyla.*

The predacious, stylet-bearing nematode, *Seinura tenuicaudata,* is representative of those nematodes having stomatostyles. Linford and Oliveira (1937) and Hechler (1963) reported that this nematode attacked several different nematode species which appeared to become paralyzed soon after the stylet penetrated the cuticle. During feeding, the body contents of the prey were partially digested before being ingested. Two species of *Aphelenchoides,* attacking *Pratylenchus pratensis* and root-knot larvae, appeared to paralyse the prey, presumably by the injection of saliva.

Linford and Oliveira (1937) observed ten species of *Dorylaimus,* two of *Discolaimus* and one of *Actinolaimus* feeding on other nematodes and they considered that predation by these hollow stylet (Onchiostyles) nematodes was more important than that of *Mononchus.* Boosalis and Mankau (1965) found that under certain conditions the predation of *Thornia* and of other organisms on *Tylenchulus semipenetrans* may significantly decrease the population of this citrus pest.

28.3.2. Predacious fungi

While most predacious fungi are within the order Moniliales of the class Fungi Imperfecti, some belong to the Zoopagales, an order of Phycomycetes.

There are two types of predacious activities among these fungi: the nematode trappers and the endozooic parasites. The nematode trappers have adhesive networks, sticky knobs attached to the mycelium by short lateral branches and hyphal rings that in some species may also constrict to capture prey. The endozoic parasites attack their prey through the production of sticky spores that on contact adhere to the cuticle and germinate, forming tubes which penetrate the nematode. In other endozooic parasites, spores are ingested by the nematode, and then penetration usually occurs internally through the esophagus.

The trapping devices may be grouped as follows;

***28.3.2.1. Sticky branches*:** Short lateral branches often only a few cells long may anastomose to form loops, but the loops never form complex networks (Fig. 28.3).

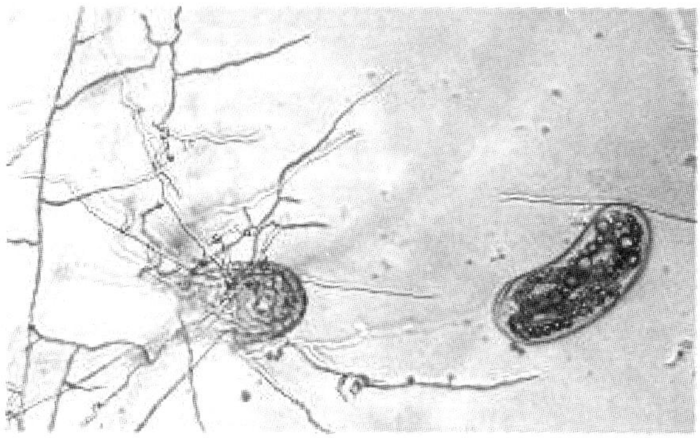

Fig. 28.3. *Dactylella oviparasitica* parasitizing the eggs of sugar beet cyst nematode.

28.3.2.2. Sticky networks: The mycelium curl around and anastamoses with similar branches or with parent hyphae. These loops produce complex three dimensional structures. The adhesive surface of the network helps to hold the prey e.g. *Arthrobotrys* sp. (Fig. 28.4).

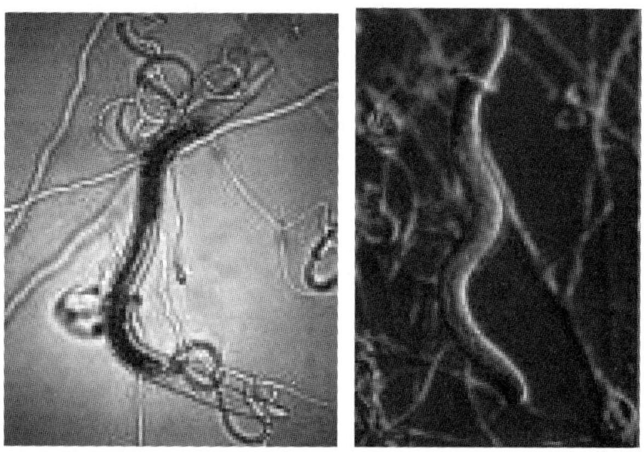

Fig. 28.4. Sticky nets Arthrobotrys oligospora.

28.3.2.3. Sticky knobs: Small spherical to sub-spherical lobes are present on one or two celled lateral hyphae. Only the terminal cell is sticky e.g. *Arthrobotrys candida* (Fig. 28.5), *Monacrosporium ellipsospora.*

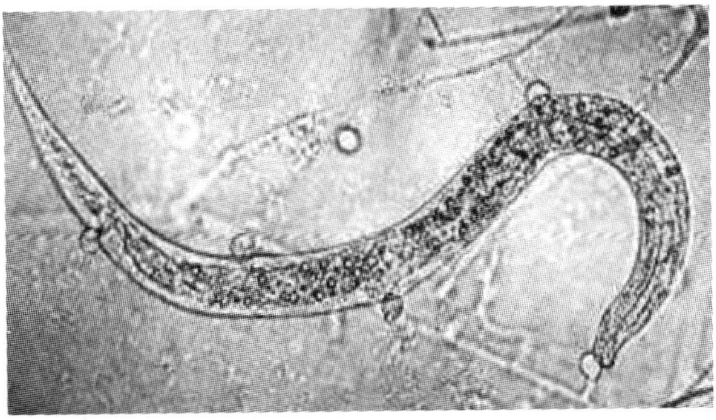

Fig. 28.5. Sticky knobs of *Arthrobotrys candida*

28.3.2.4. Constricting rings: The trap is formed when a short hyphal branch curls back on itself and anastamoses, forming a three or rarely, four celled ring. The ring is borne on one or two celled lateral hypha. When a nematode enters and contacts the inner walls of the ring cells, the cell bulge inward, filling the lumen of the ring and holding the nematode fast e.g. *Arthrobotrys dactyloides* (Fig. 28.6), *Monacrosporium bembicoides, Dactylaria brochopaga.*

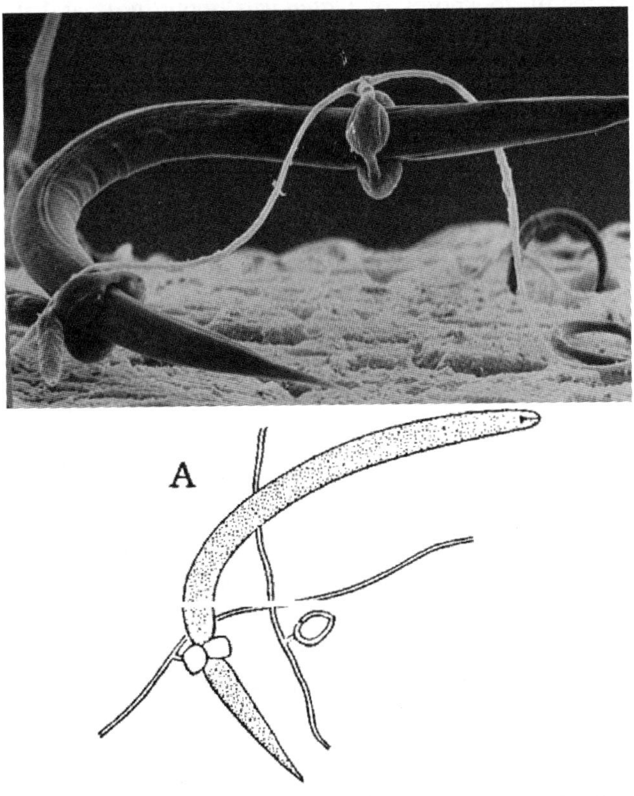

Fig. 28.6. A nematode is entrapped by two constricting rings of *Arthrobotrys dactyloides*

28.3.2.5. Non-constricting rings: The trap is formed similar to the constricting ring. Apparently, it is a non-adhesive trap and only after the nematode has struggled to escape and is firmly ensnared does the trap become an infective structure e.g. *Dactylaria candida.*

In addition to traps and adhesive products, predacious fungi may also produce a toxin, accounting for the death of some nematodes before the fungus actually penetrates the cuticle.

The endoparasitic fungi usually enter the nematode by a germ tube that penetrates the cuticle from a sticky spore. Once inside the nematode, the hyphae ramify throughout the body, absorb the contents and spore-bearing hyphae, and then emerge from the nematode cadaver. *Catenaria vermicola* frequently attacks sugarcane nematode pests. Mobile zoospores adhere to the nematode cuticle near the mouth, anus or vulva and, if not dislodged within a few hours, produce germ tubes that penetrate the nematode's body.

The fungus *Phialophora heteroderae* enters the cyst of *Globodera rostochiensis* via the oral or vulval openings and then penetrates the egg shell, so decreasing the number of viable larvae.

Sayre and Keeley (1969) showed in culture a linear relationship between increasing number of zoospores of *Catenaria anguillulae* and incidence of the fungus disease in the stem nematode, *Ditylenchus dipsaci.* However, massive doses of zoospores would be necessary in the field to obtain a mortality equivalent to that which occurred in the laboratory.

Numerous attempts have been made to use predacious fungi to control phytonematodes. Basically, the fungi have been used in three ways:

- Organic amendments have been added to soil to create conditions more favorable for the fungi already present in the soil.
- Specific nematode-trapping fungi were added to the soil.
- Both practices were done together.

These attempts have in a few instances given promise of success, but mostly the results have been inconclusive.

28.3.3. Tardigrades

They have been reported as possible biocontrol agents of phytonematodes on the basis of observed predation of plant parasitic nematodes (Linford and Oliveira, 1938). Sayre (1969) demonstrated that the tardigrade, *Hypsibius myrops,* fed on phytonematodes (*Meloidogyne incognita* and *Ditylenchus dipsaci)* and could reduce their populations on a substrate of moist sphagnum.

28.3.4. Turbellarians

A small rhabdocoel flatworm, *Adenoplea* sp., isolated from greenhouse soil, was found preying on larvae of *Meloidogyne incognita.* In greenhouse pot tests, when flatworms were added to soil, there was a reduction in the incidence of galling on tomato roots, but it was not sufficient to be of practical value.

28.3.5. Collembola

Murphy and Doncaster (1957) reported that *Onchyiurus armatus* was by far the most voracious of the predators on *Heterodera cruciferae*. Once the cyst wall was ruptured, as many as four or five insects joined in devouring the nematode.

28.3.6. Mites

Linford and Oliveira (1938) observed six kinds of mites feeding on root-knot nematodes. Murphy and Donscaster (1957) reported injury to *Heterodera* cysts by a mite. An oribatid, *Pergalumna* sp. is reported to eat *Tylenchorhynchus martini* in large numbers.

28.3.7. Enchytraeids

Schaerffenberg and Tendl (1951) reported control of the sugar beet nematode, *Heterodera schachtii*, when 150-200 enchytraeids were added initially to each pot containing a nematode infested sugar beet plant.

28.3.8. Predacious protozoa

A large amoeboid organism, *Theratromyxa weberi,* was observed preying on larvae of *Globodera rostochiensis* and on several other nematodes by Weber *et al.* (1951). A similar organism was found attacking *G. rostochiensis* in England and *H. schachtii* in Canada.

28.4. PARASITES

28.4.1. Parasitic fungi

Paecilomyces lilacinus is an effective parasite of *Meloidogyne* eggs. Egg parasites are more dramatic in reducing the nematode population. Nematode eggs of the group Heteroderidae and those deposited in a gelatinous matrix are more vulnerable to attack by these organisms than are those of migratory parasites. Once in contact with the cysts or egg masses, the fungus grows rapidly and eventually parasitizes all the eggs that are in the early embryonic developmental stages (Fig. 28.7).

Results on substrate study showed that the fungus grows abundantly in sterilized mashed potato, chopped water lily plants, ipil ipil (*Leucaena glauca*) leaves and corn grits.

This fungus is by far the most promising and practicable biological control agent for management of root-knot and cyst nematodes.

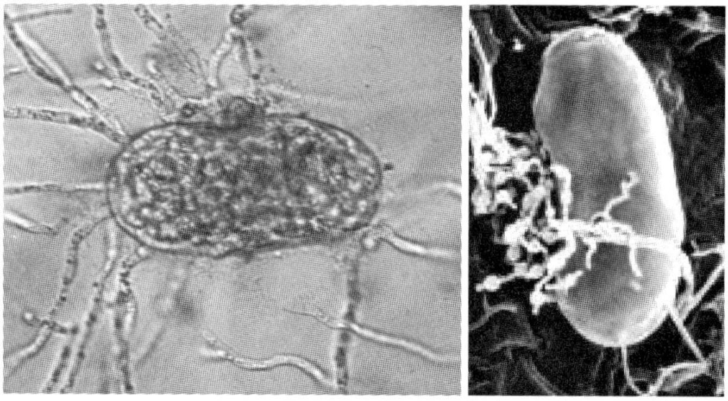

Fig. 28.7. *Paecilomyces lilacinus* (left) and *Pochonia chlamydosporia* (right) infecting nematode eggs.

28.4.1.1. Potato: *P. lilacinus* infected eggs of *Meloidogyne incognita* had destroyed the embryos within 5 days in laboratory tests. In greenhouse tests, a reduction of up to 70% root galls and egg masses was observed in the plants from fungus treated tubers. Since up to 60% of the eggs within egg masses were infected with the fungus, the size of subsequent nematode generations was reduced. In field tests, the fungus significantly reduced the tuber and root galling due to *Meloidogyne* infection. 86% of the egg masses collected from the roots of fungus treated plots were infected with *P. lilacinus* and over 54% of the eggs within these egg masses were destroyed. Yield of marketable potatoes (non-infected tubers) was significantly higher in the fungus treated plots. The fungus spread rapidly in the field which is a desirable characteristic of a successful and competitive biological control agent.

The fungus treated potato plots had higher yield and there were significantly lower numbers of *Globodera rostochiensis* cysts.

28.4.1.2. Tomato: Control of *M. incognita* was achieved by *P. lilacinus*. The fungus treated plots had higher yield than did control plots and those treated with carbofuran.

28.4.1.3. Citrus: The fruit diameter of sweet orange cv. `Valencia' from fungus treated trees was greater than that from trees treated with nematicides. The number of nematodes (*Tylenchulus semipenetrans*) in roots of and soil around citrus trees inoculated with *P. lilacinus* was significantly lower than that in the non-treated or the nematicide-treated trees.

28.4.2. Bacteria

A bacterial (*Pseudomonas denitrficans*) infection of *Xiphinema americanum* was reported by Adams and Eichenmuller (1963). Cell free extracts of a *Pseudomonas* sp. caused a high mortality of *Meloidogyne* sp. *Pasteuria penetrans* attacks species of *Meloidogyne* (Fig. 28.8) and *Pratylenchus*. Spores attach to the cuticle and multiply within. Infected larvae are less active than healthy ones; infected adult females of *Meloidogyne* are filled with spores and produce few or no eggs. The purified thermostable exotoxin of *Bacillus thuringiensis* was reported to be toxic to *Meloidogyne* spp. eggs and larvae *in vitro* at 1: 10 concentration.

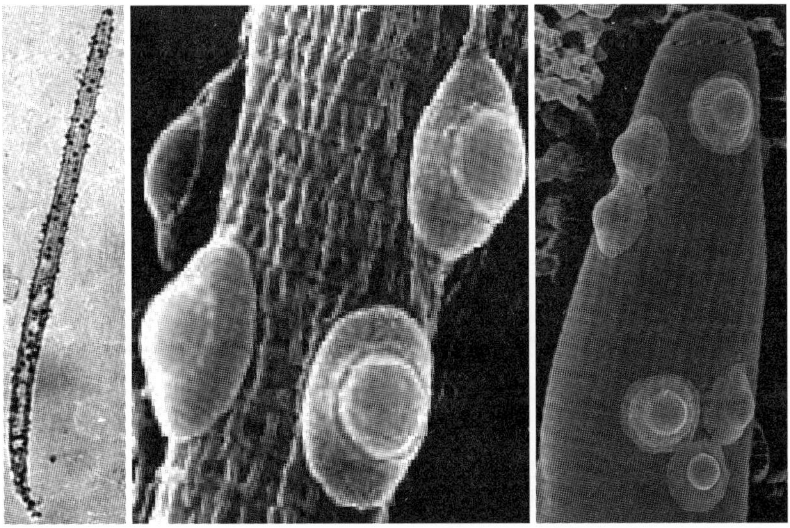

Fig. 28.8. Endospores of *Pasteuria penetrans* attached to the root-knot nematode larva.

A large number of plant growth promoting rhizobacteria (PGPR) are known to reduce nematode populations and important genera include *Agrobacterium*, *Alcaligenes*, *Bacillus*, *Clostridium*, *Desulfovibrio*, *Pseudomonas*, *Serratia* and *Streptomyces*. PGPR also induce systemic resistance against nematode pests. The use of PGPR as biological control agents of plant parasitic nematodes especially for potato cyst nematode has been reported as a successful strategy in the management of these nematodes (Sikora, 1992).

28.4.3. Viruses

There are only two reports of abnormal behavior of nematodes in which viruses were thought to be involved. Loewenberg *et al.* (1959) claimed that an agent which passed through a bacteriological filter caused sluggishness in *Meloidogyne incognita* larvae and an inability to form galls. The condition was believed to be caused by a virus. However, neither the virus particles were not observed in diseased nematodes, nor the possible involvement of a mycoplasma eliminated.

The second report concerns the phenomenon of swarming (aggregation of nematodes in masses). Electron microscope studies of the cuticle of *Tylenchorhynchus martini* swarmers revealed abnormal and disruptive biological changes in the cuticle layers that suggested a diseased condition. Virus-like particles were observed in the hypodermis and muscle layers, digestive and reproductive systems and on the surface of the cuticle and were isolated from homogenate of swarming nematodes.

It is difficult to recognize any real biological control implications from the facts presented to date regarding either of the above viral diseases of nematodes.

28.4.4. Parasitic protozoa

Williams (1960) reported that *Caliculosporidum penetrans* caused a disease incidence of 34 per cent in populations of root-knot nematodes. Later he found the disease occurring in *Xiphinema* sp. A parasite infecting *Meloidogyne javanica* and resembling *C. penetrans* was discovered by Prasad and Mankau (1969). *C. penetrans* has also been reported to parasitize *Pratylenchus brachyurus* in the United States, *Meloidogyne javanica* in Japan and several species of plant parasitic nematodes in the Netherlands. Esser and Sobers (1964) reported a population of *Helicotylenchus microlobus* to be heavily infected with *Caliculosporidum* sp. About 25 per cent of the infected nematodes were dead and filled with spores or had them adhering to the cuticle.

28.4.5. Rickettsias

Intracellular bacterium like microorganisms tentatively identified as rickettsias were found in cyst nematode species *Heterodera glycines, H. goettingiana* and *Globodera rostochiensis*. The microorganisms appear to be passed to off-spring transovarially from generation to generation. They occurred both in reproductive and non-reproductive tissues. Only minor pathological effects on host cells were observed.

The observation of dividing stages of rickettsia like organisms in *Xiphinema index* associated with the roots of diseased grapevines containing the same organisms may indicate a possible vector role for the nematode.

28.5. ARBASCULAR MYCORRHIZAL FUNGI (AMF)

The mycorrhizae are symbiotic associations between plant roots and certain species of fungi. In practical agricultural terms, mycorrhizas are traditionally ectomycorrhizas and arbuscular mycorrhizal fungi as reported by Ikram (1990).

28.5.1. Ectomycorrhizae

The ectomycorrhizas grows around the root surface and are easily seen with the naked eye while endomycorrhizas grow inside roots. The ectomycorrhizas are common to many temperate forest tree species and this group is characterized by a sheath of visibly-swollen fungal tissue (mantle) which encloses the ultimate rootlets of the root system, together with intercellular infection of the epidermis and cortex (Fig. 28.9). Some species of ectomycorrhizas have been shown to reduce ectoparasitic nematode damage of forest tree roots, but are not effective against root-knot nematodes.

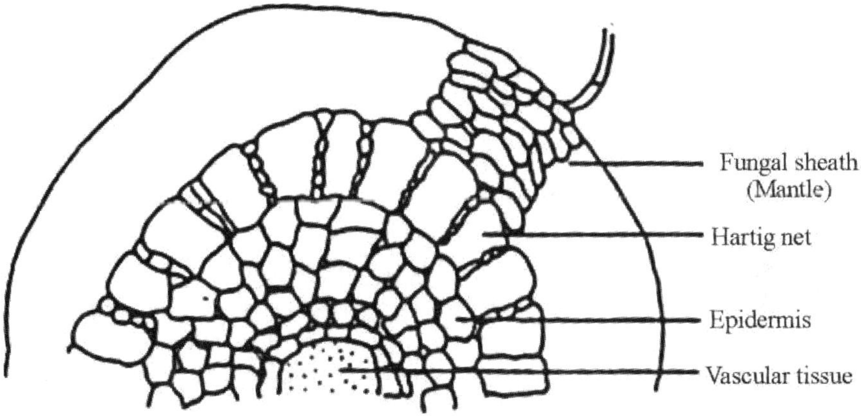

Fig. 28.9. Diagrammatic representation of ectomycorrhizae (Ikram, 1990)

28.5.2. Endomycorrhizae

Endomycorrhizae are the most spectacular and widespread type of mycorrhizas present in nearly all major agronomic crops (vegetables, fruit crops, flowers, forest trees and plantation crops). The invading endomycorrhizal hyphae grow on the root surface forms an appressorium and penetrates into the cortex. Hyphal growth in root cortical cells are both inter-and intracellular where they form the two diagnostic structures, vesicles and arbuscules (Fig. 28.10). Many species do not form vesicles, therefore, the new name arbuscular mycorrhizal fungi.

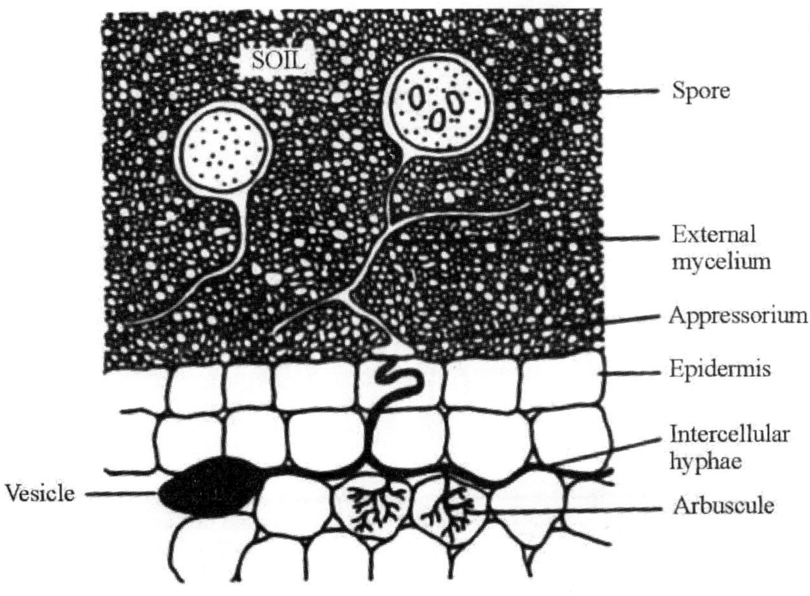

Fig. 28.10. Diagrammatic representation of endomycorrhizae (Ikram, 1990)

28.5.3. Mode of action

A plant with a well-established symbiont density is stronger because it has:

- Increased resistance to nematode parasites and root pathogens
- Increased tolerance to drought, salt and other abiotic stress factors
- Improved phosphate uptake
- Less root injury following transplanting

The colonization of plants with endomycorrhizal fungi apart from providing plants with nutrients also has a depressive effect on root-knot nematodes. The obligate symbiotic endomycorrhizal fungi protect their host against root-knot nematode attack by competition as reported by Sikora (1978) and Hussey and Roncadori (1982). Sikora (1978) found that penetration and development of *M. incognita* in tomato was significantly reduced by the endomycorrhizal fungus, *Glomus mosseae* under glasshouse conditions. Because these fungi occur commonly together with plant parasitic nematodes in the roots or rhizosphere of the same plants, they interact with both host plant and nematodes (Stirling, 1991).

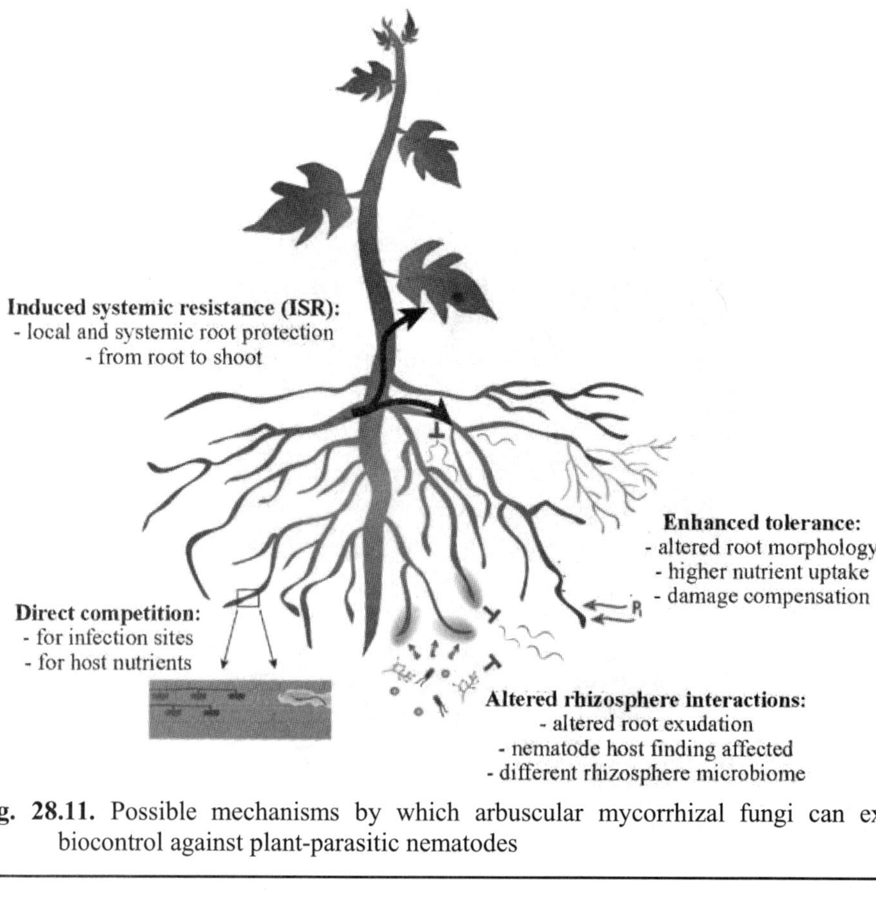

Fig. 28.11. Possible mechanisms by which arbuscular mycorrhizal fungi can exert biocontrol against plant-parasitic nematodes

Smith (1987) explained possible hypotheses for the beneficial effects of endomycorrhizal fungi on plants parasitized by nematodes. He postulated that the symbiotic mycorrhizae: (i) reduce or alter root exudates consequently affecting egg hatch or nematode attraction, (ii) retard nematode development or reproduction within root tissue, and (iii) parasitize female nematodes and their eggs.

The possible mechanisms by which arbuscular mycorrhizal fungi can exert biocontrol against plant-parasitic nematodes are summarized in Fig. 28.11:

- Induced systemic resistance
    - Local and systemic protection from root to shoot
- Enhanced tolerance
    - Altered root morphology
    - Higher nutrient uptake
    - Damage compensation
- Direct competition
    - For infection sites
    - For host nutrients
- Altered rhizosphere interactions
    - Altered root exudation
    - Nematode host finding affected
    - Different rhizosphere microbiome

Chapter 29

HOST RESISTANCE

29.1. INTRODUCTION

Plant resistance is regarded as an extremely feasible method for controlling nematodes. It is an effective, economical and environmentally safe means of reducing losses from nematode pests. Use of resistant plants enables the grower to control the parasitic nematodes without increasing production costs associated with the purchase of expensive chemicals, applicators and in the numerous mechanical operations that go into the production of a crop. Resistance is especially valuable in controlling nematodes in low value crops and minor crops which can increase crop yields equal to that obtained by soil fumigation.

The use of plant varieties resistant to nematodes may be the only method of control available, which is economic on a field scale, for certain nematodes. At present, largely because of limited research in this area, a few nematode - resistant varieties of crop plants are available to the commercial grower. To produce crop plants resistant to parasitic nematodes involves several stages in research:

- The nematode must be defined as clearly as possible because different races of species may produce different symptoms in the same species of plant. It is always advisable to assess resistance with several nematode populations from different localities to see if racial differences occur.
- It is necessary to ensure that laboratory or greenhouse techniques for the assessment of resistance provide maximum opportunity for the plants to react to the nematodes. Where tests have to be done initially on field scale, well randomized and statistically designed plot experiments should be conducted.
- Valid criteria for the assessment of resistance must be devised to indicate whether the plant is immune, tolerant or resistant.
- Different varieties or closely related species of the plant being studied are then tested for resistance and some idea is obtained for the genetics of heritable resistance, so that a plant breeding program can be started.
- Crosses and back crosses are made with potentially useful plant material until a variety is obtained which appears to fulfill the requirements of a

nematode resistant plant and which still has commercially desirable qualities.

- The resistant plant is tested under field conditions in different localities and soil types to assess the influence of environment on resistance.

29.2. DEVELOPMENT OF NEMATODE RESISTANT VARIETIES

Although a few nematode resistant plant varieties have been developed by selection from commercial varieties, the most common method of obtaining resistant varieties is to cross plants having desirable commercial characters with those possessing nematode resistance. An original cross between a commercial variety and a nematode resistant source, followed by repeated back crosses to commercial varieties, is generally used in breeding programs to incorporate resistance into a commercially acceptable plant variety.

Roots infected with root-knot nematodes are commonly used as inoculum in breeding experiments. Infected roots containing many egg masses are chopped and equal quantities are added to the soil in which the test plants are to be grown. Inoculum of the cyst nematodes is obtained by washing the cyst from soil by flotation and sieving. These cysts are freed of accompanying debris, proportioned by counting, and added to the soil. Large quantities of stem nematodes for use in breeding programs are grown routinely on host tissues. As these nematodes are generally found inside stem or bulbs, their recovery is relatively easy. Infected roots are often used as sources of inoculum of migratory endoparasitic nematodes. Large quantities of some species, primarily lesion nematodes, are grown on callus tissue or excised root cultures. Inoculum of ectoparasitic root feeders is particularly difficult to obtain because of the time required to extract these nematodes from soil and to separate the test species from other species. For most ectoparasitic root feeders the only practical method is to test resistance in the field or greenhouse in soil infested by more than one nematode species.

Plants are categorized as being immune, resistant, susceptible or tolerant to nematode diseases. An immune plant is one that is able to prevent infection with no disease expression, i.e. absolute resistance. A susceptible plant is one that cannot overcome or withstand the injurious effects of the nematode and permits it to develop to its fullest capacity. A resistant plant is one that cannot prevent entry of the parasite but is able to prevent, restrict or retard its development. Tolerance refers to the ability of a plant of endure invasion by the nematode without appreciable symptom expression or damage.

Generally, nematode resistance or tolerance is tested by measuring plant performance and rating disease symptoms, such as root galls. Because growers are primarily interested in yield and quality of product, these are very important criteria. However, the rate of nematode reproduction also should be determined.

Soil temperature is probably the most important environmental factor in the expression of resistance. Holtzmann (1965) found that tomatoes were less resistant at 30 and 34.5°C than at 20 and 25°C. The detailed study by Dropkin

(1969) showed that the resistant tomato cv. Nematex grown at 28ºC and below was highly resistant to *M. incognita acrita*. Similarly, as temperature increases, nematode galling on root-knot resistant tomato cv. Floradel becomes more severe, causing maximum damage at approximately 35°C (Fig. 29.1). However, resistance diminished with each degree above 29ºC and at 33ºC the seedlings were fully susceptible. Similar temperature effects were observed on beans and sweet potato. However, not all resistant plants are affected similarly by increased temperatures. Dropkin (1969) also observed that the resistance of the African horned cucumber, *Cucumis metuliferus* increased as the temperature rose from 28 to 32ºC. He also found that resistance reactions differed between cultivars within the same species.

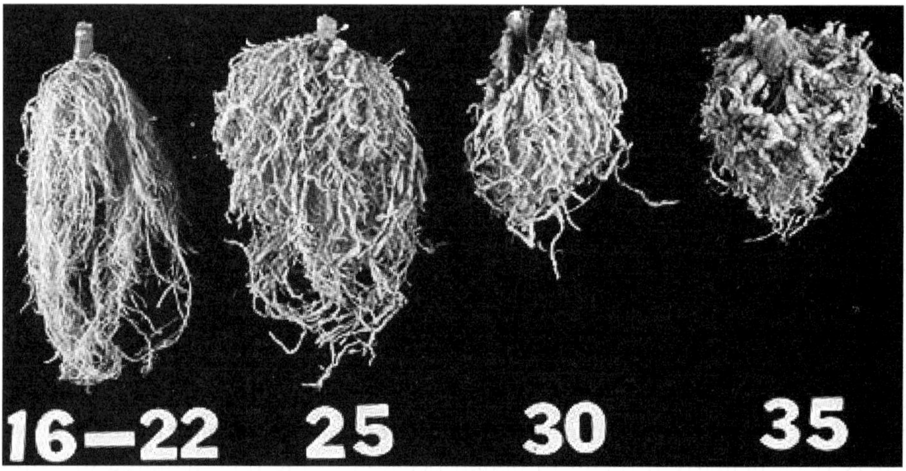

Fig. 29.1. The influence of temperature on *Meloidogyne incognita* galling on tomato cv. Floradel (Courtesy G.S. Abawi)

29.3. NATURE OF RESISTANCE

Nematode resistance could obviously be of more than one type. First there might be resistance to invasion. Rohde (1990) suggests that failure to produce a hatching factor by plants may confer resistance in the presence of cyst nematodes. He also considers that repulsion or lack of attraction by plants would increase resistance. This type of resistance may be associated with root exudates that either repels the infective stage larvae from the roots or one toxic to them.

Secondly, there might be resistance after invasion resulting in the larvae not being able to complete their development and reproduce. Most nematode resistance appears to be of this type. Nematodes enter roots of most resistant plants, but often in smaller numbers than they enter roots of susceptible plants. Williams (1958) found that fewer *Globodera rostochiensis* larvae invaded the more resistant species of *Solanum*. The root systems of resistant tomatoes were invaded by fewer larvae of *Meloidogyne incognita*, usually one half or less, than those of susceptible varieties. However, they found no difference between the numbers of larvae entering resistant and susceptible sweet potato varieties. Nematodes in resistant plant roots may die and disappear quickly, especially where there is a hypersensitive (necrotic) reaction to invasion. Furthermore, it is possible that nematodes may invade and then leave a resistant root whereas in a susceptible root they may tend to stay. Hence, it is necessary to make nematode counts within a day or two of inoculation otherwise death and decay of larvae in resistant plants may be interpreted as reduced invasion. Plant resistance may cause not only slower nematode development but also production of fewer eggs by females. In most cases, development of females does not proceed further than the third stage, while males may mature normally.

Thirdly, there could be tolerance of nematode attack, i.e., although invaded by the larvae which develop normally, the plants show relatively little loss of yield. Mountain and Patrick (1959) found that peach seedlings of the variety "Shalil" supported high root populations of *Pratylenchus penetrans* without discernible stunting.

Plant resistance may be related to the failure of the host to supply some nutrient necessary for rapid reproduction or even for survival of the nematode. The formation of few or no giant cells at feeding sites of maturing female nematodes is associated with plant's resistance to cyst and root-knot nematodes. Infection by stem nematodes causes conspicuous enlargement and separation of parenchymatous cells in susceptible plants but only slight cell enlargement and separation in resistant plants. In two wild species of *Cucumis,* Fassuliotis (1970) found that resistance was expressed by the development of small giant cells which impeded the development of females and shifted the sex ratio towards maleness.

Nematode populations around the roots of resistant plants sometimes decline at a more rapid rate than can be explained by starvation, and it is presumed that toxins of plant origin are responsible. Oostenbrink *et al.* (1957) showed that the cultivation of *Tagetes patula* between the rows or around the foot of other plants reduced the *Pratylenchus* populations in their roots as well as in the soil, giving better growth of perennials in the second year or in the succeeding crop. *Tylenchorhynchus dubius* was also suppressed but *Rotylenchus robustus* and a number of Tylenchida were unaffected. The active nematicidal principle from *Tagetes* roots was found to be α-terthienyl.

Rohde and Jenkins (1958) found that the asparagus variety `Mary Washington' was resistant to *Trichodorus christiei.* Populations of this nematode declined more rapidly in soil containing asparagus roots than in soil without roots and nematode numbers did not increase on a suitable host if its roots were intermingled with those of asparagus. A glycoside with a low molecular weight, aglycone, was present in the rhizosphere of asparagus which either killed the nematodes or inhibited their reproduction. This toxic compound secreted from asparagus roots was also toxic to *Paratylenchus projectus, Tylenchorhynchus claytoni, Helicotylenchus nannus, Belonolaimus gracilis*, *Tylencholaimus stiratus* and *Diphtherophora* species. Rohde (1960) showed that the toxicity of this compound was related to interference with acetylcholinesterase in the nervous system of the nematode.

Phytoalexins formed only after infection of plant tissue by the nematode are also involved in plant resistance to nematodes. The resistance factors are localized within the individual plant cells (of a chemical nature) attacked by the nematode and are not generally translocated in the plant.

Death of cells hypersensitive to nematode feeding and browning of tissues around infecting nematodes occurs in many resistant plants in response to infection by several different nematode species. Hypersensitivity may determine resistance in chrysanthemums infested with *Aphelenchoides ritzemabosi.* Leaves of resistant varieties brown quickly and there is no nematode multiplication, so that initial infestation is isolated and does not spread to the other leaves of the plant. This was attributed to the absence of a nutritional factor in leaves of resistant varieties. Hypersensitive reactions have been postulated for resistance of red clover, alfalfa and pea to the stem nematode; citrus to the citrus nematode; lima bean to the lesion nematode and soybean to root-knot nematodes. Browning of plant tissues as a result of feeding of a lesion nematode (*Pratylenchus penetrans)* appears to be caused by accumulation and oxidation of phenolic compounds. In apple and peach rootstocks, lesion formation as a result of feeding of lesion nematodes is roughly correlated with the amygdalin content of rootstocks. Sometimes death of hypersensitive cells may cause extensive damage to a particular plant species even though the nematode species causing the damage cannot reproduce on the plant.

Formation of wound periderm or corky layers, which wall off and retard the development of citrus nematode larvae in roots of resistant citrus plants and root-knot nematode larvae in old yam tubers, has been reported.

Webster (1969) favors the hypothesis that a gene in the plant cell controls either the type of protein bound to a plant growth regulator or the occurrence of a particular enzyme in the plant. The response in susceptible plants depends on the appropriate proteolytic enzymes breaking the specific peptide bonds and so releasing tryptophan and gibberellic acid which produce an actively growing or modified tissue that provides the nematode with adequate nutrition. Presumably, the plant is resistant when any of the requirements in this casual chain is deficient.

Wilski and Giebel (1966) propose that resistance of potatoes to *Globodera rostochiensis* depend on the presence of phenolic glucosides that are hydrolyzed in resistant plants by β -glucosidases secreted by the nematode. Consequently, free polyphenols are released which cause necrosis. Later experiments indicated that in both susceptible and resistant potatoes, β - glucosidase secreted by the nematode released IAA, kinetin and phenolic compounds from complexes. By the interaction of IAA oxidases and the phenolics, the IAA was destroyed. Consequently there is an increase of IAA in susceptible plants and a decrease in resistant plants. Whether syncytia or necrotic cells are formed is then determined by the balance between IAA and kinetin; a preponderance of kinetin leads to a necrotic reaction. In resistant potatoes there is a higher ratio of mono-to polyphenols that inhibits IAA destruction only slightly, i.e. IAA decreases.

29.4. PROBLEMS ASSOCIATED WITH RESISTANCE

- Unrelated plant species, especially those with different chromosome numbers, are difficult to hybridize.
- Undesirable agronomic characters are often linked to nematode resistance, making transfer of resistance without these characters difficult.
- The isolation and identification of enzymes secreted by nematodes during feeding and their influence on plant tissues.
- Elucidation of the biochemical differences between resistant and susceptible varieties, associated with differences in cellular response to nematode feeding.
- Identification of the chemical in plant cells which promote and inhibit nematode development and reproduction.

Table 29.1 lists nematode-resistant cultivars of crop plants.

Table 29.1. Nematodes resistant crop cultivars

Crop	Nematode	Resistant cultivars	Reference(s)
Coffee (*Coffee* spp.)	*M. exigua*	Amfillo 1141-2, Bangelan, Barbuk Sudan 1171-26, Bukobensis, Dalle Mixed 1150-2, Kawisari Cols 6 and 8, Laurenttii Col 10, N 39, Robusta Cols 2 and 10, Tafara Kela 1161-9	
Cotton (*Gossypium* spp.)	*Meloidogyne* spp.	Auburn 56, Bayou, Clevewitt (Fig. 29.2)	
Banana	*Radopholus similis*	Kadali, Pedalimoongil, Ayiramkapoovan, Peykunnan, Kunnan, Pisang Seribu, Tongat, Vennettu Kunnan, Anaikomban	Charles *et al.*, 1983 Rajendran *et al.*, 1979
Citrus	*Tylenchulus semipenetrans*	Trifoliate orange, Swingle citrumelo	Parvatha Reddy, 1983
Grapevine	*Meloidogyne incognita*	Black Champa, Dogridge, 1613, Salt Creek, Cardinal, Banquabad	Parvatha Reddy and Singh, 1984 Darekar & Patil, 1982

Crop	Nematode	Resistant cultivars	Reference(s)
	Rotylenchulus reniformis	Dakshi, Joazbeli, Mukchilani	
Peach (*Prunus persica*)	*M. incognita*	Nemaguard, Okinawa, Rancho Resistant. S - 37, Shalil, Yunnan	
	M. javanica	For Valley 234-1, Nemaguard, Okinawa	
Plum (*Prunus* spp.)	*M. incognita* and *M. javanica*	Marianna 2524 and 2623, Myrolaban 29, 29C, 29D and 29G	
Rice (*Oryza sativa*)	*M. incognita* and *M. javanica*	International	
Soybean (*Glycine max*)	*M. arenaria*	Bragg, Dyer	
	M. hapla	Bragg (moderately resistant), Ilbroy	
	M. incognita	Bragg, Delmar, Dyer, Hampton, Hardec, Hill, Hutton, Jackson, Kahala, Kaikoo, Kailua, Laredo, Makapa Summer, Peking (resistant); Amredo, Bethel, Blackhawk, Dare, habard, Hood, Illsoy, Laredo, Monroe, Palmerto, Roanoke; York (moderately resistant)	
	M. javanica	Hampton (moderately resistant)	
Tobacco (*Nicotiana tabacum*)	*M. incognita*	Coker 86, 254, 258, and 347, Florida 22, Ga 1469, NC 79, 88, 95, 98 and 2512, Speight G-23, G-28 G-33, G-41, Virginia 080 and 770	Parvatha Reddy, 1981
Tomato	*Meloidogyne* spp.	Nematox, SL-120, NTR-1, SL-12, Patriot, VFN-8, VFN Bush, Piersol, Radiant, Nemared, Ronita, Anahu, Bresch, Helani, Campbell-25, Punuui, Arka Vardan, Pelican, Hawaii-7746, Hawaii-7747, Hisar Lalit	Rao & Tikoo, 1975 Patel *et al.*, 1979 Jain *et al.*, 1983 Kanwar & Bhatti, 1990
Brinjal	*M. incognita*	Giant of Banaras, Black Beauty, Gola	Alam *et al.,* 1974
Chilli	*M. javanica*	579, CAP-63, Pusa Jwala	Jain *et al.*, 1983
Bell Pepper (*Capsicum*	*Meloidogyne arenaria*	Burlington, California Wonder Special, Nemaheart, Oakview,	

Crop	Nematode	Resistant cultivars	Reference(s)
annum)		Wonder, Red Chile, Ruby King, Santanka x S	
	M. incognita	Nemaheart, Red Chile, Santanka x S	
	M. javanica	California Wonder Special, Early California Wonder, Nemaheart, Oakview Wonder, Red Chile, Santanka x S	
Potato	*M. incognita*	Kufri Dewa	Raj & Gill, 1983
	Globodera rostochiensis	Kufri Swarna	
Cowpea	*M. incognita*	Barsati Mutant, Iron, New Selection, G-152, 92-1-B, IC 9642-B, TVU 2430-P	Sharma & Sethi, 1976 Darekar & Patil, 1981 Singh & Parvatha Reddy, 1982a
French bean	*M. incognita*	Banat, Blue Lake Stringless, Bountiful Flat, Brown Beauty, Cambridge Countess, Gallaroy, Kenya-3, Pinto W5-114, Seafarer, Suttan's Masterpiece	Singh *et al*., 1981
Muskmelon	*incognita*	Scarsol	Khan *et al*., 1971
	M. javanica	S-445	Jain *et al*., 1983
Watermelon	*M. incognita*	Shehjanpuri	Khan *et al*., 1971
Ridgegourd	*M. incognita*	Panipati, Meerut Special	Khan *et al*., 1971
Ashgourd	*M. incognita*	Jaipuri, Agra	Khan *et al*., 1971
Pumpkin	*M. incognita*	Jaipuri, Dasna	Khan *et al*., 1971
Lettuce (*Lactuca sativa*)	*M. hapla*	Salinas (Fig. 29.3)	
China aster	*M. incognita*	Shashank-Resistant Poornima-Mod. Resistant	Nagesh *et al*., 1995b
Tuberose	*M. incognita*	Sringar-Resistant Suvasini-Tolerant	Nagesh *et al*., 1995a

Fig. 29.2. S-1046 test fields showed improved growth of cotton lines with resistance to reniform nematodes (*right*) compared to susceptible lines (*left*) at mid-season and harvest.

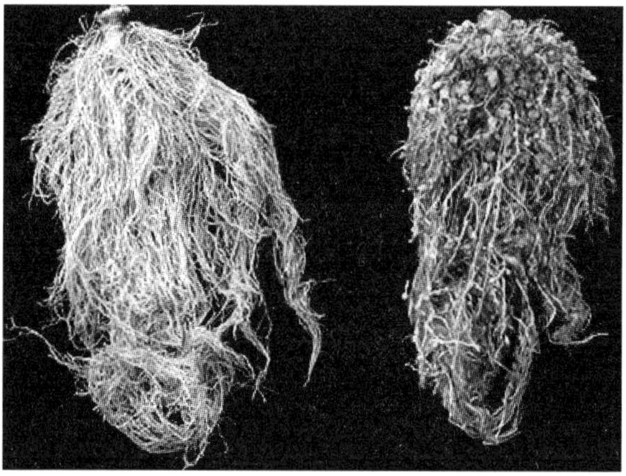

Fig. 29.3. Meloidogyne hapla resistant lettuce cv. Salinas (*left*) and susceptible lettuce cv. Ithaca (*right*). (Courtesy N.A. Mitkowski)

29.5. BIOTECHNOLOGICAL APPROACHES FOR NEMATODE RESISTANCE

Biotechnology has a role to play in incorporation of resistance against nematodes and biological control of plant nematodes. A number of genes that mediate nematode resistance have now been or soon will be cloned from a variety of plant species. Nematode resistance genes are present in several crop species and are important components of many breeding programs including those for tomato, potato, soybeans and cereals (Trudgill, 1991). Several resistance genes have been mapped for chromosomal locations or linkage groups and some of them have been cloned. The first nematode resistance gene to be cloned was *Hs1pro-1*, a gene from a wild relative of sugar beet conferring resistance to *Heterodera schachtii* (Cai *et al.,* 1997). The cDNA, under the control of the CaMV 35S promoter, was able to confer nematode resistance to sugar beets transformed with *Agrobacterium rhizogenes* in an *in vitro* assay (Cai *et al.,* 1997). The *Mi* gene from tomato conferred resistance against a root-knot nematode and an aphid in transgenic potato (Rossi *et al.,* 1998). The gene *Mi* is a true *R* gene, characterized by the presence of NBS and LRR domains. *Gpa2* gene that confers resistance against some isolates of the potato cyst nematode *Globodera pallida* was identified. This gene shares extensive homology with the Rx1 gene that confers resistance to potato virus X suggesting a similarity in function (Van Der Vossen *et al.,* 2000).

Biotechnology offers sustainable solutions to the problem of plant parasitic nematode control. There are several possible approaches for developing transgenic plants with improved nematode resistance; these include anti-invasion and migration strategies, feeding-cell attenuation, and antinematode feeding and development strategies. The essential elements of an effective control strategy are (*a*) genes that encode an antinematode effector protein, peptide or interfering RNA and (*b*) promoters that direct a specific pattern of expression for that effector. There are essentially three approaches for engineering resistance: transgenic expression of natural resistance genes in heterologous species; targeting and disruption of the nematode; and feeding site attenuation (Thomas and Cottage, 2006).

27.6. CONCLUSION

Resistance in plants is dependent upon physiological differences which render plants unable to meet the metabolic requirements of the parasite or they may inactivate the stimulatory substances released by the parasite. Resistant plants secrete toxic substances which slow down all host parasite interactions. Resistance often depends upon one or few enzymes and the metabolic compounds they control.

The biochemical basis of resistance in plants to nematodes has been little studied. Research in this area should aid in developing resistant varieties of plants. The study of the nature or mechanisms of resistance may be helpful in two ways:

- Breeding of resistant varieties might be speeded up by laboratory analysis of plants during intervals between tests on growing plants. Cytological studies could be made to locate the characteristics which determine resistance and help to incorporate these in desirable agricultural varieties.
- New chemicals may be found in resistant plants which would be useful in the discovery of new chemical means of nematode control.

Chapter 30

INTEGRATED MANAGEMENT

30.1. INTRODUCTION

Plant parasitic nematodes cause significant economic losses in crop plants. Root-knot nematodes (*Meloidogyne* spp.) are among the most significant of all plant parasitic nematodes. To alleviate nematode problems, integrated nematode management (INM) procedures are implemented based on the principles of prevention, population reduction and tolerance. INM seeks to stabilize populations of target nematodes at acceptable levels resulting in favorable long term socio-economic and environmental consequences. In practice, integration of different methods of nematode control is the rule rather than the exception. In most cases, practical control of a nematode disease involves integration of several diverse control measures.

30.2. INTEGRATED NEMATODE MANAGEMENT

The grower combines several measures to prevent the growth of damaging nematode populations. Where there is an advisory service equipped to monitor field populations, these control measures can be utilized with minimum amounts of nematicides. Without advisory services, integrated control of nematodes is hardly possible and relatively heavy dosages of chemical nematicides are employed.

INM can be defined as a systems approach to reduce nematode damage to tolerable levels through a variety of techniques, including predators and parasites, genetically resistant hosts, natural environmental modifications and when necessary and appropriate, chemical nematicides (Fig. 30.1).

Decisions to implement INM programs are based on the perceived value of the potential crop loss and the cost of the management procedure. Theoretically this involves determining an economic threshold. Management procedures should usually be implemented when the marginal revenue derived from the management input is equal to or exceeds the marginal cost.

The design, development and implementation of INM systems for plant protection require extensive interaction of nematologists with scientists in complementary disciplines.

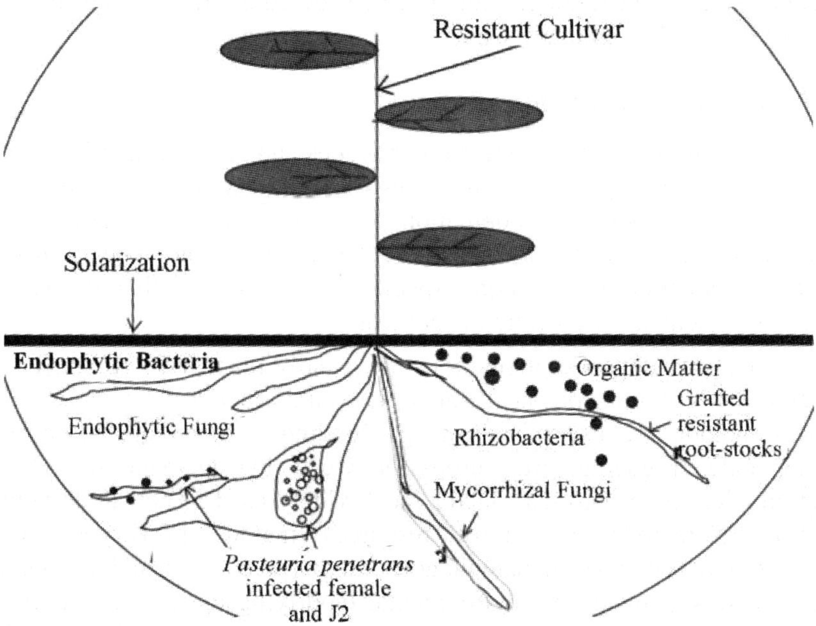

Fig. 30.1. INM methods that can be used to manage plant nematodes

INM includes the research, development, technology transfer and implementation needed to integrate two or more control procedures to manage one or more nematode species. The components of INM systems include biological and environmental monitoring, agricultural production systems (APS) and pest-crop ecosystem models and system design and implementation. The temporal elements of the process include APS problem analysis (detection and diagnosis), system evaluation (APS and INM option), INM system design, recommendation, implementation and feedback. The procedures available for INM system design and implementation are based on the principles of exclusion, population reduction and tolerance. The spatial aspects of INM include production system, institutional and geographical components.

30.2.1. Exclusion

Crop losses caused by plant parasitic nematodes can be avoided through preventing the introduction of specific nematodes or nematode problems in areas where the species do not exist. The focal point of exclusion is the target nematode species. Exclusion procedures should be used as first order defenses to prevent dissemination and establishment.

Exclusion procedures include sanitation, certified plant material, nematode-free soil or planting media, population reduction or eradication procedures, APS design manipulations and regulatory activities. Quarantines are used to prevent or slow the spread of plant parasitic nematodes. Certified plant material and nematode-free planting media or equipment are used for nematode exclusion. All available exclusion procedures should be considered in the system evaluation and design of the INM program.

30.2.2. Population reduction

Crop losses caused by plant parasitic nematodes can be alleviated by procedures designed to decrease population densities of the target species to an acceptable level. Whenever possible, population reduction strategies should stabilize nematode population densities below the damage threshold. The focal point of population reduction is the target species. It is assumed that the species of concern is present and eradication is not feasible.

Established nematode populations are usually managed through population reduction by cultural, physical, biological and chemical methods. Cultural methods include the use of fallowing, flooding, cover crops, crop rotation, planting date, rogueing, trap crops or weed management for population reduction. Heat is the physical factor most widely used in nematode population reduction. Nematode bio-control resources include prokaryotes (viruses, rickettsias and bacteria), nematophagous fungi (predacious and endoparasitic), and predacious, parasitic or environmental modifying invertebrates (nematodes, protozoa, mites, etc.) and antagonistic plants (*Tagetes* spp.). Nematicides can directly or indirectly reduce populations associated with nematode infested materials and can be used for soil application, root dips, foliar applicants or seed treatments.

30.2.3. Tolerance

Crop losses caused by plant parasitic nematodes can be reduced by manipulating the host in relation to its environment. The objective of tolerance is to elevate the damage threshold. The focal point is the host crop.

Procedures that protect or increase plant tolerance have excellent potential for INM. These include cultural manipulation, chemical application, mycorrhizal colonization and resistant cultivars.

30.3. FUTURE STRATEGIES

- Soil treatment before planting or between crops
    - Soil drying
    - Soil solarization in hot season
    - Rotation with non-host crops
- Biological enhancement during seedling production
    - Arbuscular mycorrhizal fungi
    - Mutualistic fungal or bacterial endophytes
    - *Pasteuria penetrans*

    - Resistant cultivars
- Biological enhancement of seedlings prior to transplanting
    - Plant health-promoting rhizobacteria e.g. *Bacillus cereus* S18 and *Rhizobium etli* G12
    - Grafting resistant/tolerant rootstocks
- Stimulation of natural antagonistic potential
    - Incorporation of organic composts e.g. chicken manure
    - Planting and incorporation of green manure e.g. caster, sesame
- Management methods during crop growth
    - Systemic nematicides
    - Optimum plant fertilization and irrigation
- Control methods at harvest
    - Trap cropping using hosts and degree day monitoring
    - Antagonistic cropping with e.g. Tagetes or Crotalaria
    - Removal or composting roots to kill nematodes
- Biofumigation with plant residues or green manure crops

In conclusion, the logical use of the methods listed above can lead to effective INM of root-knot over time in the field. However, research both on basic and applied aspects is still needed to make such approaches available to the farmer under practical conditions. Government support for small scale industry may be needed in order to produce biocontrol agents on a large scale for market use by resource poor farmers. In addition, training of extension experts is required to teach them how to optimize INM combined with biological control for root-knot nematodes management.

30.4. CASE DESCRIPTIONS

Potato tomato, brinjal, okra, sweet potato, carrot, tobacco and strawberry APS's are presented as case descriptions of INM systems.

30.4.1. Potato

The integrated control of the potato cyst nematode, *Globodera rostochiensis* can be achieved by a rational combination of crop rotation, resistant cultivars and nematicides (Jones, 1969) (Table 30.1).

The control of potato rot nematode, *Ditylenchus destructor* was achieved by the combination of disease escape; hygiene, in the form of seed certification and crop rotation (Winslow and Willis, 1972).

- Healthy seed, planted late, harvested early and stored as cool and dry as possible.

- Proper rotation-potatoes not more frequently than once in 3 or 4 years and avoid other susceptible crops.
- Field hygiene (removal of old infested tubers and weed control).

Table 30.1. Integration of control methods for *Globodera rostochiensis*

Sl. No.	Control method (s)	Resulting population (% initial population)	Kill (%)	Population after growing and harvesting a susceptible cultivar, calculated at two assumed nematode multiplication rates* (% initial population)	
				30 x	70x
1	4 years without potatoes	3	97	90	<100
2	1 year with resistant potatoes	20	80	>100	>100
3	Nematicide treatment (fumigant)	25	75	>100	>100
4	1 and 2	0.6	99.4	18	42
5	1 and 3	0.75	99.25	22.5	52.5
6	2 and 3	5	95	>100	>100
7	All three methods	0.15	99.85	4.5	10.5

*The observed maximum reproductive rate is thought to lie between 25 and 75 times but could be greater.

30.4.2. Tomato

Meloidogyne species have the potential for causing serious problems in tomato production. Proper site selection and preparation for transplant productions are essential components of INM. Fallowing, crop rotation, cover crops and nematicides can be used. The transplants are inspected at the site of origin for root-knot nematode galls. Certification and inspection should be followed. *Meloidogyne* spp. infections avoiding detection are usually below the damage thresholds (Bird, 1981).

New variations on these involve combinations of cultural (deep plowing to a depth of 20 cm) and chemical methods (nursery bed treatment with aldicarb at 0.4 g a.i. per m^2 and main field treatment with aldicarb at 1 kg a.i. per ha) for the control of root-knot nematodes in tomato which also registered maximum yield.

Red gram (*Cajanus cajan*) and *Crotalaria spectabilis* were used in conjunction with clean fallow and a nematicide (fensulfothion) for managing root-knot nematode populations in tomato transplants.

30.4.3. Brinjal

Deep plowing (20 cm) together with spot application of aldicarb at 1.0 kg a. i. per ha increased growth of brinjal plants and significantly reduced the root-knot nematodes.

Nursery treatment with metham sodium at 25 ml per sq. m. along with deep plowing (20 cm) in the main field gave significant reduction in root-knot nematode population and better brinjal yields.

30.4.4. Okra

Deep plowing (20 cm) together with seed treatment (carbofuran at 3% w/w) was found to be effective for the control of root-knot nematodes in okra.

Deep plowing (20 cm) followed by fallowing for one month, fallowing for one month after weeding, integration of aldicarb application at 0.8 kg a i per ha at sowing after either of the cultural practices or deep plowing (20 cm) together with carbofuran or aldicarb seed treatment resulted in the control of the reniform nematode (*Rotylenchulus reniformis*) and better yield of okra (Lakshmanan and Sivakumar, 1981).

30.4.5. Sweet potato

Control of root-knot damage to sweet potatoes involves integration of at least three methods:

- The selection of seed roots which are either free of nematodes or freed of nematodes by hot water treatment.
- Planting of the seed roots into beds of sand or coarse textured soil that is either nematode - free or, if infested, is pre-plant fumigated.
- Transplanting of the clean `slips' into nematode-free soil or pre-plant fumigated soil.

30.4.6. Carrot

Meloidogyne hapla is a severe pathogen of carrots. The damage threshold is very low and infection by one *M. hapla* can make a carrot unsuitable for the fresh market. A highly commercial INM system has been evolved using a 5 year rotation in conjunction with a cover crop, soil fumigation and tolerant cultivars. A soil fumigant is applied the fall before the first of two consecutive carrot crops, followed by 2 years of onion production and 1 year of the cover crop sudex. Onions are relatively poor hosts for *M. hapla* and have a high economic threshold. In fields where nematode problems are detected after the first growing season, a low rate of a soil fumigant may be applied before the second carrot crop, or the rotation design may be altered to delete the second carrot crop. Tolerance to *M. hapla* exists in commercial cultivars of carrots (Bird, 1981).

30.4.7. Tobacco

The development of an integrated program for the control of *Meloidogyne incognita* in tobacco has been of immense value to the state of North Carolina (Taylor and Sasser, 1978). On problem fields where the average yearly income without control is $ 1,047 per hectare:

- Sanitary procedures, such as plowing up tobacco roots at the end of the season, add $ 711.
- Rotation with resistant crops, such as groundnut or cotton, adds $ 1,104.
- Use of resistant tobacco varieties, such as NC 95, adds $ 939.
- Use of nematicides adds another $ 1,378.

The total increase is $5,179 or almost five times as much as without nematode and disease control.

30.4.8. Strawberry

When a soil is heavily infested with *Pratylenchus penetrans,* intelligent combination of nematicides and crop rotation is desirable. Soil application of nematicides may reduce the nematode population to a very low level. An advisable cropping sequence could then be:

(1) Strawberry (which is susceptible but does not cause a large increase of the residual nematode population)

(2) Potato (moderately susceptible and causing moderate nematode reproduction)

(3) Oats or barley (not susceptible but causing nematode reproduction)

(4) Beet (not susceptible and suppressing the population)

After the above cropping sequence, moderately susceptible crops can be grown again successfully.

It is seen from the above discussion that the integration of the different methods offers effective control possibilities which are of great practical value in the suppression of harmful nematode population densities as well in preventing the buildup of measurable populations of nematodes.

SECTION VIII
NEMATODE PESTS OF CROP PLANTS

Chapter 31

HORTICULTURAL CROPS

31.1. CITRUS, *Citrus* spp.

Nematodes are one of the limiting factors in citrus production throughout the world. The first record of an association between a nematode and citrus appears to be that of Neal (1889), who found *Meloidogyne* sp. parasitizing citrus roots in Florida. But it was not until the discovery in 1912 of *Tylenchulus semipenetrans* on the roots of citrus trees in California that a nematode was found to cause a disease condition of citrus, later called 'slow decline'.

The major nematode problems on citrus include *Tylenchulus semipenetrans* (Cobb, 1914b), *Radopholus similis* (Suit and Du Charme, 1953), *Belonolaimus longicaudatus, Trichodorus christiei, Paratrichodorus porosus, Hemicycliophora arenaria, Meloidogyne* spp., *Pratylenchus brachyurus, P. coffeae* (Siddiqi, 1964), *Xiphinema brevicolle, X. index* and *Hoplolaimus indicus* (Gupta and Atwal, 1972).

31.1.1. Citrus nematode, *Tylenchulus semipenetrans*

The citrus nematode was discovered in 1912 in California on the roots of citrus trees that exhibited a `mottled' appearance. Cobb (1914b) published a more detailed account of its distribution, morphology and life history. Siddiqi (1961) reported *T. semipenetrans* for the first time from India in a paper presented before the Indian Science Congress and observed that about 80 per cent of the citrus trees at Aligarh, Uttar Pradesh, were infested. This nematode causes 'slow decline' and is considered to be one of the factors responsible for die-back disease of citrus trees in India. The annual reduction in the world citrus crop due to `slow decline' is estimated at 8.7 to 12.2 per cent.

31.1.1.1. Symptoms: *T semipenetrans* causes `slow decline' disease of citrus. The symptoms include a general reduction in tree growth and vigor, yellowing and shedding of leaves and production of under sized fruits. The decline symptoms are often more pronounced in the uppermost portion of the trees (Fig. 31.1). In Israel, the decline symptoms are associated with a critical nematode population of approximately 40,000 larvae and males per 10 grams of roots, above which the tree performance is markedly lowered.

Fig. 31.1. Citrus trees exhibiting symptoms of citrus decline caused by the citrus nematode, *Tylenchulus semipenetrans*. Note die-back symptoms (thinning canopy and bare branches) in the upper canopy.

Soil particles usually cling tightly, even after washing to the gelatinous egg masses which cover the protruding part of the nematode body (Plates 31.2 and 31.3). In heavily infested roots, the cortex separates readily from the vascular stele.

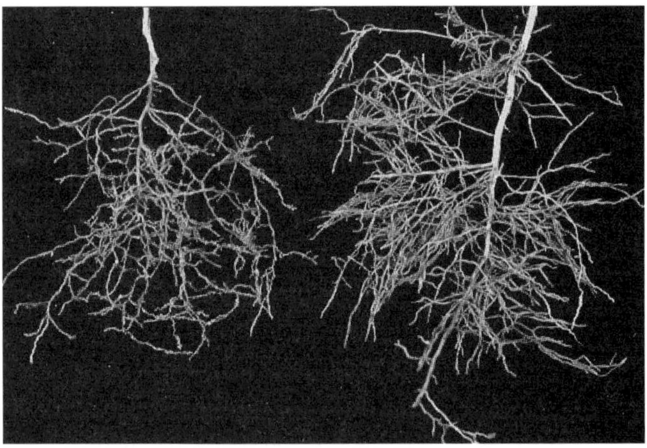

Plate 31.2. Left - citrus roots infected with *Tylenchulus semipenetrans,* Right - healthy roots.

31.1.1.2. Life cycle: Van Gundy (1958) studied the life cycle in detail. Eggs hatched in 12 to 14 days at about 24°C and sex differentiation was possible at the second stage. The second stage male larvae developed to maturity within 7 days without feeding. The second stage female larvae required 14 days to locate the root, feed on the epidermal cells and molt. The life cycle from egg to egg required 6 to 8 weeks.

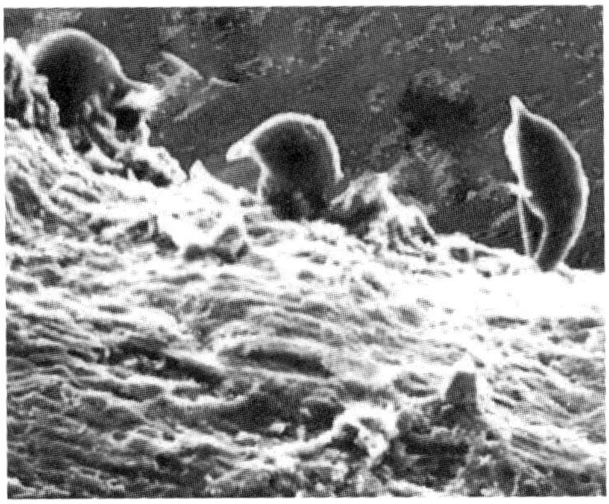

Plate 31.3. Citrus nematode females attached to the citrus root.

In experiments conducted in Israel, the minimum time required at 24°C for completion of the life cycle was 14 weeks on *Poncirus trifoliata*, 10 weeks on *Ruta bracteosa* and 7 weeks on *Citrus aurantium* and *C. limettoides* (Cohn, 1965).

31.1.1.3. Host-parasite relationship: Maximum numbers of larvae enter surface layers of roots which are 4 to 5 weeks old and feed on the more mature parts of the roots. The young females penetrate deeper and become established at a depth of several cortical cells. A feeding site around the head of the female is soon established. This consists of 6 to 10 cortical cells, the so called 'nurse cells' around a cavity that house the nematode head (Fig. 31.4). The nurse cells have thickened walls, enlarged nucleus and nucleolus and no vacuole. As the nematodes feeds and reproduces, a large proportion of the feeder roots in upper soil layers are destroyed, the uptake of water and minerals from soil is reduced and symptoms appear in the above ground tree parts.

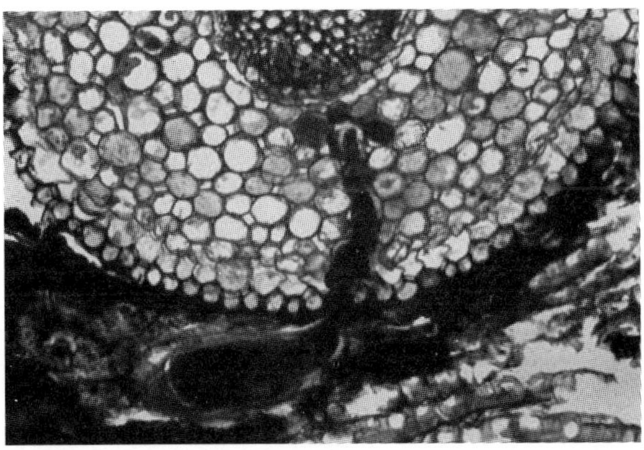

Fig. 31.4. Cross section of citrus root parasitized by *Tylenchulus semipenetrans*. Note the nematode head surrounded by the 'nurse cells'.

31.1.1.4. Host range: Vilardebo and Luc (1961) list 29 species of *Citrus*, 21 citrus hybrids and 11 other species of Rutaceae as hosts from 23 countries. The citrus nematode has developed strains or biotypes to attack grasses (*Andropogon rhizomatus, Panicum* spp.), Olive (*Olea* spp.), grapevines (*Vitis* spp.), Persimmon (*Diospyrous lotus*), Pear (*Pyrus communis*), *Calodendrum capense,* climbing hemp weed (*Mikania batatifolia)* and lilac (*Syringa vulgaris)* (Stokes, 1969). In India, *T. semipenetrans* is known to attack grapes and loquats (Parvatha Reddy and Singh, 1978a).

31.1.1.5. Interaction with other pathogens: In the presence of *T. semipenetrans*, significantly more citrus plants become infected by *Fusarium oxysporum* and *F. solani* and the combined activity of the nematode and *F. solani* is thought to be a factor in the debilitation of citrus.

31.1.1.6. Ecology: Optimal nematode reproduction occurs at soil temperatures of 28-31ºC. In California, optimum nematode reproduction occurred in soils with a clay content of 10-15%. A pH range of 5.6-7.6 has been found favorable for nematode reproduction.

Tree performance deteriorates when the nematode infestation attains a critical level of 40,000 larvae per 10 g of feeder roots which may be attained in orchards in 12-17 years after infested seedlings are planted in virgin soils.

31.1.1.7. Spread and survival: The widespread distribution of *T. semipenetrans* throughout the citrus growing regions of the world is mainly due to the transfer of infested citrus planting material. Movement of soil accounts for more local and short distance spread of the free living stages of the nematode. Agricultural implements, animals and man, wind and water are common agents in spreading the nematode in soil.

In soil, the nematode may remain viable in the absence of a host for as long as 10 years.

31.1.1.8. Physiologic races: Four races of *T. semipenetrans* that differ in pathogenicity to sweet orange, citrange, trifoliate orange, grapevine and lilac have been selected from different locations in California (Baines *et al.,* 1974). A 'grass strain', which parasites *Andropogon rhizomatus* but failed to infect citrus was reported by Stokes (1969).

31.1.1.9. ***Management***

(a) Preventive measures: The citrus nematode is easily transported to new orchards through infested seedlings and budded plants which are usually carried "balled" with infested soil. Therefore, nurseries should never be established on or near old citrus orchards and nursery soil should be sterilized before planting.

Care should also be taken not to spread infection through tools, machinery and irrigation water used in infested groves.

(b) Chemical methods

(i) Bare root dip treatment: The citrus nematode can be controlled by bare root dips in water at 45°C for 25 min or in suitable concentrations of fensulfothion, ethoprop or thionazin.

(ii) Soil treatment: The nematicides that have been successful in nematode control are ethoprop, carbofuran, aldicarb, phorate, dimethoate and fensulfothion. Metham sodium controlled *Phytophthora* at the same time as *T. semipenetrans.*

Chhabra *et al.* (1977) found that fensulfothion was very effective at 3 kg a.i. /ha in controlling the nematodes on pumello and increased the yield by 68-76%. Significant reduction in the citrus nematode population (76 to 84 per cent) in sweet orange and increase in yield (61 to 244 per cent) was observed with soil application of aldicarb and carbofuran both at 6 kg a.i. per ha (Baghel and Bhatti, 1983a).

(c) Biological control: Incorporation of organic soil amendments into the soil has reduced citrus nematode population. A *Mononchus* sp. was observed feeding on *T. semipenetrans* by Cobb (1914b). Several species of nematode trapping fungi of the genera *Arthrobotrys, Dactylella* and *Dactylaria* have been found in association with the citrus nematode. But biological control is not yet successfully put to use.

The number of nematodes in roots and soil around citrus trees inoculated with *Paecilomyces lilacinus* was significantly lower than that in the non-treated or the nematicide treated trees (Jatala, 1986). Application of *P. lilacinus* in the form of spore suspension checked the multiplication of *T. semipenetrans* by 71.7 to 97.8% (Mani and Murthy, 1989).

Arbuscular mycorrhizal fungus, *Glomus mosseae* when added to the soil with *T. semipenetrans* in *Citrus jambhiri,* limited the nematode development and partially neutralized the adverse effect of the nematode (Baghel *et al.,* 1990).

(d) Resistant rootstocks: Citrus rootstocks which are resistant or highly tolerant to the citrus nematode would offer practical and effective means of control. Trifoliate orange (*Poncirus trifoliata*), Troyer Citrange and Swingle Citrumello and its hybrids (Citrumello) were highly resistant to the citrus nematode (Parvatha Reddy and Singh, 1978b; Parvatha Reddy and Agarwal, 1987; Mani and Reddy, 1986).

Van Gundy and Kirkpatrick (1965) discovered that in citrus both a morphological reaction (a hypersensitive cells reaction to feeding and the formation of a wound periderm in the root tissue) and a possible toxic effect of plant secretion on the nematode, retard the development of the early stages. Cohn (1965) found a similar morphological retarding effect in *Severinia buxifolia* grown in soil.

(e) Integrated management: Integration of organic amendments (neem cake and castor cake) with biological control agents (*Paecilomyces lilacinus, Trichoderma harzianum, Verticillium lecanii, Pasteuria penetrans, Pseudomonas* *fluorescens*) and arbuscular mycorrhizal fungi (*Glomus fasciculatum*) was found to be effective for the management of the citrus nematode.

31.1.2. Lesion nematode, *Pratylenchus coffeae*

The nematode root rot of citrus in Uttar Pradesh is principally caused by *P. coffeae* (Siddiqi 1964). This nematode constitutes the dominant factor in the nematode complex of citrus in Aligarh.

31.1.2.1. Symptoms: *P. coffeae* infected citrus plants exhibit poor growth, general dieback and produce under-sized fruits.

31.1.2.2. Host-parasite relationship: Adults and larvae of *P. coffeae* are found in the cortical tissues of the feeder roots where a tiny, brownish black lesion is formed which gradually expand to girdle the rootlets (Fig. 31.5). Radewald *et al.* (1971) found that when the nematode invaded a root-tip, the meristem often was destroyed and lateral root initiation usually occurred near the destroyed root tip. The young plants of *Citrus limon* and *C. sinensis* show poor growth when infected with *P. coffeae* (Siddiqi, 1964). A marked reduction in the top as well as root weight of the infected seedlings was noticed. O'Bannon and Tomerlin (1969) reported that *C. limon* seedlings inoculated with 20 *P. coffeae* had peak populations at 36 weeks with 10,000 nematodes per g root extracted; shoot growth was reduced by 22 per cent.

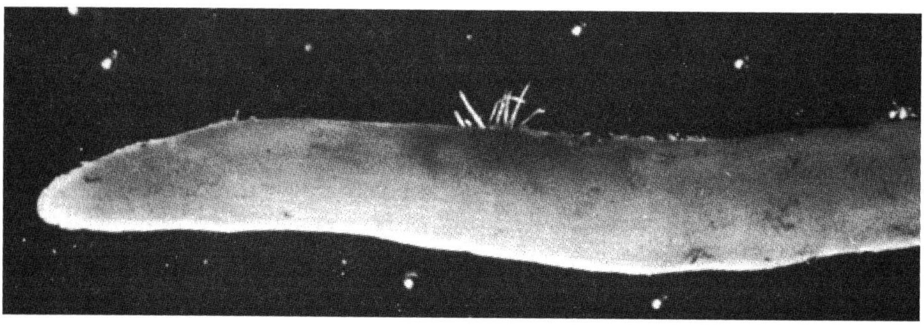

Fig. 31.5. Rough lemon roots infected with *Pratylenchus coffeae.*

31.1.2.3. Host range: *Citrus limon, C. sinensis, C. reticulata* and banana have been found to be hosts of *P. coffeae* in Uttar Pradesh (Siddiqi, 1964). The nematode is transferable from banana to citrus plants and from lemon to orange plants and *vice versa*. Radewald *et al.* (1971) reported this nematode on *C. jambhiri*.

31.1.2.4. Survival: *P. coffeae* survived best in excised infected roots of *C. jambhiri* stored in soils with initial moisture content near field capacity and at temperatures ranging from 10-32°C, and were infective after 4 months. The nematodes did not survive extended storage at temperatures over 38°C. The optimum temperature for reproduction was 29.5°C.

31.1.2.5. Management: Infestation in citrus seedlings of *P. coffeae* and *P. brachyurus* was eliminated by treating with fensulfothion and fenamifos at 4.4 kg a.i. per ha and the stem diameter of the treated seedlings increased by 44 and 46 per cent, respectively after 9 months. The population of *P. coffeae* can be effectively reduced by applying aldicarb or carbofuran at 4 kg a.i. per ha (Baghel and Bhatti, 1983a).

Four rootstock selections of *Microcitrus australis* x *M. australasica* hybrid and Rubiboux 70-A5, Trifoliate orange (*Poncirus trifoliata*) had nematode populations of less than 100/g root. They were considered to show some resistance to *P. coffeae* because of low numbers.

31.1.3. Root-knot nematodes, *Meloidogyne* spp.

Thirumula Rao (1956) reported that *Meloidogyne* sp. caused considerable damage to citrus in Andhra Pradesh when a susceptible crop like tobacco or okra was grown as an intercrop. This is the first report of nematode damage to citrus in India.

Whitehead (1968) reported *M. indica* on citrus from India. The root-knot nematode, *M. indica* on citrus has also been reported from Andhra Pradesh and Gujarat. *M. indica* infects *C. aurantium, C. sinensis* (Whitehead, 1968) and *C. aurantifolia* (Parvatha Reddy *et al.*, 1981). Acid lime orchards in North Gujarat, particularly in Banaskantha district were observed to be heavily infected by the root-knot nematode *M. indica*. *M. indica* infection was observed in Banaskantha, Mahesana and Anand districts of Gujarat.

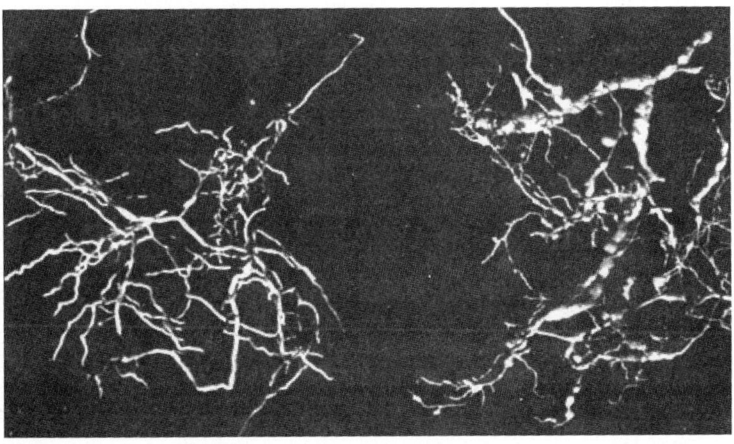

Plate 31.6. *Right*-citrus roots infected with *Meloidogyne indica, Left*- healthy roots

31.1.3.1. Symptoms: *M. indica* infected trees exhibited severe stunting, yellowing, poor in vigor, unthrifty in appearance, twig blight leading to citrus decline symptoms. The plants were showing yellowing of foliage in initial stage and were completely drying at later stage. On uprooting such infected plants, numerous small to very big sized galls/knots were noticed on the roots, resulting in very poorly developed root system (Patel *et al.*, 1999). They fail to flower and produce fruits even after several years of planting. The roots have conspicuous galls on pioneer and fibrous roots (Plate 31.6). In advanced stage, large cavities can be observed in place of galls. Egg masses can be seen as thin films spread over the root surface.

31.1.3.2. Management

(a) Chemical methods: The initial seedling dip treatment with carbosulfan at 500 ppm for 6 hr recorded more plant height and effectively controlled root-knot nematodes.

(b) Organic amendments: Among the soil applications, mustard cake, farm yard manure and poultry manure at 2.5 kg/plant were found more effective in increasing the plant growth and reducing root-knot index.

(c) Host resistance: Citrus rootstocks such as Rangpur lime 8784, Sour Orange Tirupati, Citrumello 4479, Rangpur 8748, Acid Lime PDKV, Sai Sarbati and Galgal were found resistant against *M. indica*. Rangpur lime Chethalli, Trifoliate Orange Chethalli, Nasnaran, Hazara Australia, Rangpur lime Kirumakki, Pramalini and Anand Selection were moderately resistant.

31.2. BANANA, *Musa* spp.

The burrowing nematode, *Radopholus similis,* is the most important banana pest. *Pratylenchus coffeae, Meloidogyne incognita* and *Helicotylenchus multicinctus* are the other nematode parasites of banana that are of some economic importance.

31.2.1. Burrowing nematode, *Radopholus similis*

Cobb (1893) first described the burrowing nematode from necrotic root lesions of banana growing in Fiji. The disease of bananas caused by *R. similis* is known throughout the world by different names, the most common of which are *Radopholus* root rot, blackhead, blackhead toppling disease and decline. Nair *et al.* (1966) first reported this nematode on banana from Kerala state, India. Yield increases of 30-60% have been obtained following the control of the nematode over a range of growing conditions.

31.2.1.1. Symptoms: Wounding of banana roots by the burrowing nematode usually induces reddish-brown cortical lesions which are diagnostic of the disease (Plate 31.7). These lesions are clearly seen when an affected root is split longitudinally and examined immediately. Root and rhizome necrosis is manifested by varying degrees of retarded growth, leaf yellowing and falling of mature plants.

With the increase in nematode population, feeding roots are invaded and destroyed as fast as they are formed. The resulting setback in the uptake of plant nutrients leads to debility and smaller fruit. The lesioning of the primary roots together with the girdling and death of those roots which anchor the plant to the ground makes the plant prone to tip over under wind pressure (Plate 31.7).

Plate 31.7. *Left* - Toppled banana plant due to poor root system caused by *Radopholus similis*. *Right* - Banana roots infected with *Radopholus similis* showing reddish-brown lesions

31.2.1.2. Life cycle: The burrowing nematode has a migratory endoparasitic habit and although the stages remain vermiform throughout, sexual dimorphism is apparent with adult males being somewhat degenerate and probably non-parasitic. Eggs are normally laid in infested tissue over 7-8 days at the rate of about 4 eggs per day. The life cycle from egg to egg extends over 20 to 25 days with eggs taking 8-10 days to hatch and the larvae 10-13 days to mature.

31.2.1.3. Host range: Poucher *et al.* (1967) list 244 hosts for the burrowing nematode. It is known to infect *Musa* spp., *Ipomoea batatas, Pueraria phaseoloides javanica,* pepper, areca nut, coconut, citrus, sugarcane, coffee, maize, avocado, tea, vegetables, ornamentals, trees grasses and weeds.

31.2.1.4. Host-parasitic relationship: The histopathology of banana roots infected by *R. similis* was studied by Blake (1966). On entering the root, the nematodes occupy an intercellular position in the cortical parenchyma where they feed on the cytoplasm of nearby cells, so destroying them and causing cavities to develop. These cavities coalesce and are continually enlarged by the nematodes feeding and tunneling laterally and towards the endodermis, producing the characteristic reddish brown lesions throughout the cortex. Three to four weeks after infection, when extensive cavities have formed, one or more deep cracks with raised margins appear on the root surface. The stele is not invaded. Eventually the root system may be reduced to a few short stubs.

31.2.1.5. Interaction with other pathogens: By wounding the host roots, nematodes may provide infection sites for other micro-organisms. Newhall (1958) showed that the incidence of Panama wilt caused by *Fusarium oxysporum* f. sp. *cubense* was doubled in the presence of *R. similis* during an experimental period of 3 months. When Gros Michel bananas infected with *R. similis* were inoculated with *F. oxysporum* f. sp. *cubense*, the onset of wilt was considerably shortened.

Risbeth (1960) suggested that nematodes caused the breakdown of resistance to Panama wilt in Lacatan bananas.

31.2.1.6. Spread and survival: The most important means whereby the burrowing nematode is spread from one locality to another is through planting material (corms). Trimming sets (paring), which is often done before they are planted, is usually not sufficiently rigorous to remove infection that extend deeply into a set.

Nematodes may be carried by water draining down a slope from an infested plantation above. *R. similis* may be disseminated by water that drains from infested areas and reenters either irrigation channels or streams from which water is being taken to irrigate bananas. Soil that adheres to implements, tyres of motor vehicles and shoes of plantation workers may also spread *R. similis.*

The nematode does not seem to have any special adaptations for survival in the absence of a susceptible host. Free living stages and eggs do not survive fallow or storage in plastic bags for more than 12 weeks. After a plantation is destroyed, the corms may remain succulent for several months and nematodes in these corms probably survive for as long as the corms remain succulent.

31.2.1.7. Physiological races: Du Charme and Birchfield (1956) distinguished 2 biotypes of *R. similis,* one that attacks citrus and the other that attacks bananas. The citrus biotype attacks many plant species belonging to diverse families including banana. The banana biotype seems to be restricted to *Musa* spp., sweet potato, *Pueraria phaseoloides javanica* and possibly to a few other plants but not citrus. Now the citrus biotype is recognized as a separate species *R. citrophilus*.

Venkitesan (1976) reported that population of *R similis* isolated from pepper roots did not infect *Citrus sinensis, C. reticulata, C aurantifolia, Coffea arabica,* coconut and areca nut; but infected banana only. Banana population infected pepper, coconut and areca nut; coconut population infected pepper and banana but not areca nut; while areca nut population infected pepper and banana but not coconut. He concluded that these four isolates are different `host types'.

31.2.1.8. Management

(a) Disinfecting banana sets

To avoid introducing inoculum into a new plantation, sets may be disinfected either by paring, heat therapy, and chemical treatment or by a combination of these methods.

(i) Paring and pralinage: The removal of infected tissue along with some of the surrounding healthy tissue will normally disinfect sets (Loos and Loos, 1960) (Fig. 31.8). Because pared sets are prone to fungal attack and to avoid recontamination by *R. similis*, Loos and Loos (1960) suggested dipping the pared sets in a Bordeaux mixture - DBCP paste (made by mixing 20 kg hydrated lime, 20 kg copper sulfate, 1288 ml 70% DBCP and 455 liters of water). A similar technique using carbofuran has been developed in India.

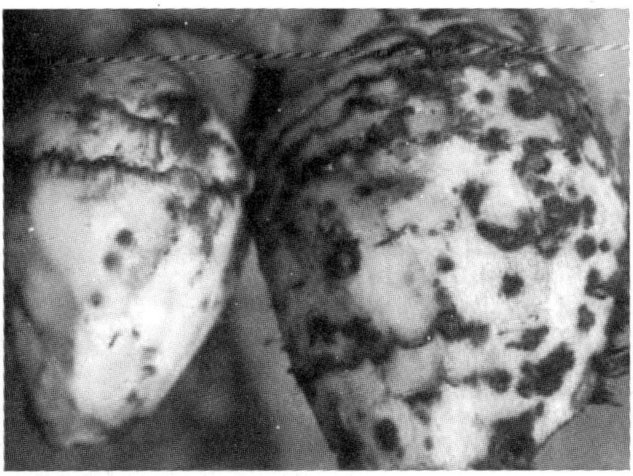

Fig. 31.8. Pared banana suckers

(ii) Heat therapy: The pared sets can be disinfected by dipping them in a hot water bath at 55°C for 25 minutes without causing significant set losses (Fig. 31.9). Fungicides applied after treatment would probably decrease any post-treatment losses caused by the fungi.

Fig. 31.9. Hot water treatment of planting material

(iii) Certified seeds: Certified nematode-free sets can be obtained by either paring/heat treatment or selection.

(b) Crop rotation and cultivation

The number of nematodes in soil may be decreased either by fallowing or crop rotation. The biotype of *R. similis* that attacks banana apparently has a narrow host range and limited survival in soil in the absence of a host. The destruction of an infested area either mechanically with a bulldozer or by injection of an herbicide into the pseudo stem and replanting the area with nematode free sets probably within 12 months would be a practical approach to the rehabilitation of infested plantations. After destroying an infested plantation, planting of cover crops like *Panicum maximum* var. *trichoglume,* marigold, and perennial legume (*Phaseolus atropurpureus*) did not support *R. similis* population (Fig. 31.10). Loos (1961) planted sugarcane in soil following the destruction of bananas infected with *R. similis* and found that *R. similis* was eliminated after 10 weeks.

Fig. 31.10. Intercropping with marigold and perennial legume

Flood fallowing for about 5 months destroys not only the *Fusarium* but also the burrowing nematodes (Loos, 1961).

(c) Chemical soil treatment: Soil application of fenamiphos at 3 kg a.i. per ha was effective in increasing plant height, plant girth, number of leaves and yield and in reducing the nematode population (Parvatha Reddy et al., 1991). Various chemicals (carbofuran, ethoprofos, oxamyl, and phenamifos) have also been found to be effective for the management of *R. similis* on banana under field conditions.

(d) Resistant varieties: Selections (tetraploids) from crosses between wild bananas (diploid male parent) and 'Gros Michel' (triploid female parent) show considerable resistance to *R. similis* as well as to Panama wilt and Sigatoka (*Mycosphaerella musicola*) diseases. A new variety Bodles Altaport, a cross between Gros Michel and Pisang Lilin, has been released for growing on the hill sides of Jamaica. This variety has an extensive root system, withstands wind and is more resistant to *R. similis* than Lacatan.

(e) Integrated management: Integration of organic amendments (FYM/ neem cake/karanj cake) with bioagents (*Trichoderma harzianum / T. viride / Pochonia chlamydosporia / Pastueria penetrans/Pseudomonas fluorescens*) or endomycorrhizae (*Glomus mosseae*) was found effective against the burrowing nematode.

The following INM measures are suggested for the effective management of the burrowing nematode:

- Mulching with black polythene at 20 per cent moisture can reduce the populations of nematodes.
- Intercropping with sun hemp, coriander, marigold or lucerne can reduce nematode population in the soil.
- Fallowing for a period of 3 months after banana can suppress the burrowing nematode population.
- Flood fallowing for about 5 months destroy not only the Panama wilt fungus - *Fusarium* but also the nematodes.
- Use of resistant/tolerant cultivars.
- Use of nematode-free certified planting material.
- Increased use of green manure and other organic amendments.
- Crop rotation with paddy, sugarcane and coriander.
- Application of neem cake at 800 grams per plant (two split applications - while planting and 1 month after planting) can reduce the population of the burrowing nematode and also increase the yield of banana.
- Paring and pralinage with carbofuran at 5 grams per sucker reduce the incidence of nematode.
- Dipping of banana suckers in 0.1 per cent aldicarb solution increase plant height, bunch weight, fingers per bunch, induce early flowering and effectively reduce the burrowing nematode population.
- Application of bio-control agents like *Paecilomyces lilacinus* and *Trichoderma harzianum* to the plants along with neem cake and farm yard manure (FYM) is beneficial. One kg of each of these bio-agents can be used to enrich 1 ton of farm yard manure. Farm yard manure added with these bio-agents should be left under shade for a period of 15 days and in between it is advisable to mix FYM thoroughly at an interval of 3 days to enrich FYM with the bio-agents. This enriched FYM can be added at 1 kg per pit at the time of planting and thrice after planting with an interval of 2 months at the same dosage.

31.2.2. Lesion nematode, *Pratylenchus coffeae*

P. coffeae has been reported to cause decline of banana plantations. The lesion nematode causes most damage to banana in association with *Fusarium oxysporum.* f. sp. *cubense* (Vadivelu *et al.*, 1987). In India, crop loss caused by the root-lesion nematode in banana cv. Nendran was reported to be 45%.

31.2.2.1. Symptoms: The symptoms caused by *P. coffeae* are very similar to those caused by *R. similis*. Both *R. similis* and *P. coffeae* are associated with the blackhead toppling diseases of bananas. *P. coffeae* enter the roots and feeds on cortical cells resulting in the production of dark lesions in the cortex (Fig. 31.11). The corm is similarly infected resulting in necrosis of the outer tissues.

Fig. 31.11. *Left* - Blackened and necrotic roots of banana caused by *Pratylenchus coffeae*. *Right* – Lesion nematodes in the cortical cells of banana root

31.2.2.2. Life cycle: The life cycle of *P. coffeae* is similar to that of *R. similis* except that *P. coffeae* males penetrate and feed on the underground plant tissues and cause decay. The egg to egg cycle is completed in about 27 days at 26-32°C. The optimum temperature for reproduction is 29.5°C.

31.2.2.3. Survival and spread: The lesion nematode can survive for 6 months in the soil, in the absence of its host. It has been observed infesting the corm, and hence can spread occur through planting material.

31.2.2.4. Management methods

(a) Cultural methods

(i) Fallowing: A fallow period of 6 months after destruction of all banana plants eliminated *P. coffeae*.

(ii) Intercropping: A significant reduction in *Pratylenchus* sp. population (85%) was observed in the banana field where *Tagetes erecta* was grown as intercrop and significantly increased the fruit yield (12 kg/plant) when intercropped compared with the untreated control (7 kg/plant) (Sundararaju *et al.*, 2002). Intercropping with sun hemp in banana field was found effective in reducing *P. coffeae* population by 38.4%, followed by marigold and cowpea which recorded a 29.0 and 22.3% reduction, respectively. Rice or green gram after banana suppressed the population of *P. coffeae*.

(b) Biological methods

(i) Antagonistic fungi: Application of *Trichoderma viride* at 20 g/plant at the time of planting and repeated 3 months after planting is effective in controlling *P. coffeae* as well as reducing the incidence of Panama wilt in banana.

Application of *Paecilomyces lilacinus* at 10 g along with FYM at 500 g or neem cake at 250 g/plant effectively suppressed the lesion nematode population by 60 and 63%, respectively (Poornima *et al.*, 2004).

(ii) Antagonistic bacteria: Soil application of *Pseudomonas fluorescens* at 20 g/plant (2.5 x 10^8 cfu/g) was effective in increasing the plant growth parameters (pseudo stem height and girth, number of leaves, root length and weight) and in reducing the lesion nematode population in soil (67.12 %) and roots (62.00%).

(c) Host resistance

Banana cvs./hybrids viz., Kunnan, Pey Kunnan, Then Kunnan, Nattu Poovan, Pidi Monthan, Chirapunji, Singhlal, Sakkarachayam, Malai Kala, Manik Champa, Kartobium Tham, Vennettukunnan and Hybrid-74 (Matti x Pisang Lilin), H-21, H-55, H-59, H-65, H-84, H-109, H-110 were found tolerant/ resistant to *P. coffeae* (Vadivelu *et al.*, 1987; Sundararaju *et al.*, 2004).

31.2.3. Spiral nematode, *Helicotylenchus multicinctus*

H. multicinctus is an important parasite of banana throughout the banana growing areas of the world. This nematode causes serious decline of bananas in the Jordan valley, Israel and was responsible for about 20% loss in yield.

31.2.3.1. Symptoms: The spiral nematode incites discrete relatively shallow, necrotic lesions on banana roots (Fig. 31.12). In Israel, *H. multicinctus* causes extensive root necrosis, dieback and dysfunction leading eventually to debility of the entire plant.

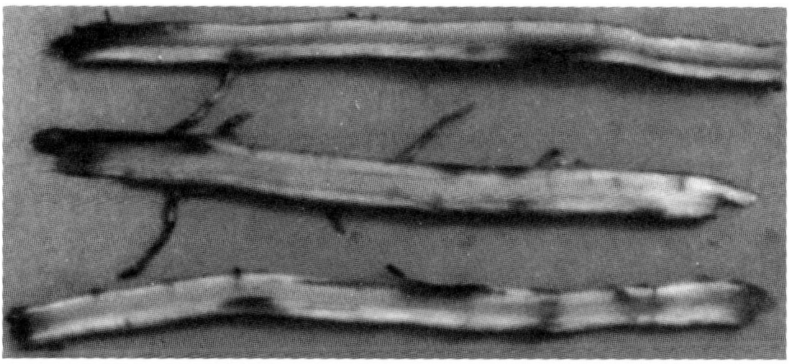

Plate 31.12. Banana roots infected with *Helicotylenchus multicinctus*

31.2.3.2. Host range: This nematode is known to infect banana, cocoa, sweet potato, citrus, sugarcane, cassava, coffee, maize, mango, oil palm, rice, rubber, tea, yam, avocado, Bermuda grass and grapevine.

31.2.3.3. Spread and survival: The nematodes easily get introduced into virgin land with banana soil and suckers (rhizomes), usually brought from old infested plantations. The nematodes survived 4 months without host plants.

31.2.3.4. Management

(a) Physical methods

The practice of peeling or paring of banana sets to remove necrotic tissue and dipping in hot water (53-55°C) for 20 min. is advocated.

(b) Cultural methods

(i) Crop rotation: Rice or green gram after banana suppressed the population of *H. multicinctus*.

(ii) Intercropping: Intercropping of banana with sun hemp was found effective and reduced *H. multicinctus* population by 38.4%, followed by marigold and cowpea which recorded a 29.0 and 22.3% reduction, respectively.

(c) Chemical methods

In recent years, emphasis has been shifted to use of granular nematicides (oxamyl/aldicarb/carbofuran/fensulfothion/ethoprophos/phenamiphos) for the management of the spiral nematode.

31.2.4. Root-knot nematodes, *Meloidogyne* spp.

Root-knot nematodes are present in all banana-growing regions and can be especially damaging to the cultivars of the Cavendish subgroup. The most frequent species associated with banana are *M. incognita, M. javanica, M. arenaria* and *M. hapla*.

31.2.4.1. Symptoms: In aerial plant parts, *Meloidogyne* can cause reduction in size and chlorosis of leaves, compromising plant development. Reduced yield and delay and unevenness in fruit ripening are also observed. Leaf symptoms include a progressive dieback of older leaves from the leaf tip. Heavily-infested plants are stunted with thin pseudo stems and discolored greenish-yellow bands apparent along the leaf blades.

The most distinctive symptom of *Meloidogyne* infestation is the appearance of galls on primary and secondary roots, which become swollen and distorted with heavy infestations. Galls vary in size, reaching up to 15 mm in diameter (Fig. 31.13). In bananas, *M. incognita* infection typically results in the formation of small galls on the roots or fine dilations in thicker fleshy roots. Different *Meloidogyne* species can occupy the same gall, as well as colonize the outer layers of the rhizome, where reddish-brown patches appear in the conducting vessels. Observation of longitudinal sections of infected root material with the naked eye can reveal females at various stages of maturation, together with associated blackened haloes known as "female's chamber".

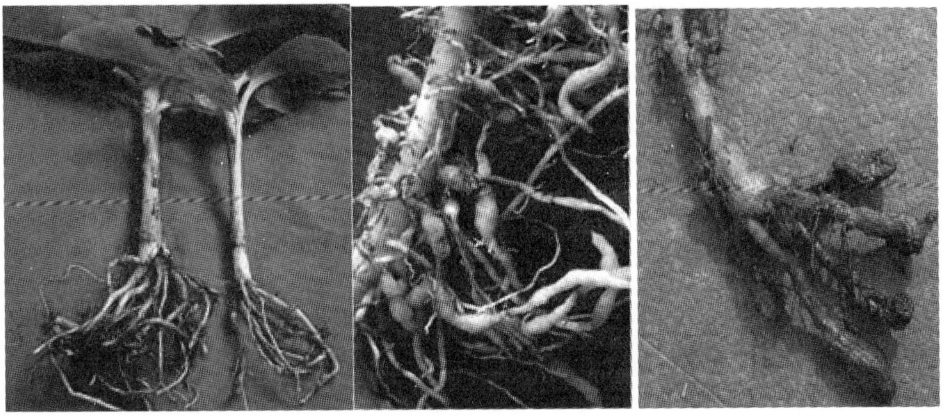

Fig. 31.13. Banana plants and roots infected with root-knot nematodes.

31.2.4.2. Management: Measures to prevent introduction and spread - such as use of certified planting material and clean farm machinery - are important management strategies.

Crop rotation is a possible management strategy. In the case of root-knot nematode species with broader host ranges - such as *M. incognita, M. javanica* and *M. arenaria* - effective crop rotation remains a challenge.

(a) Integrated management: Growing antagonistic plants namely *Tagetes erecta, T. patula* and *Crotalaria juncea* as intercrops in banana field drastically reduces the root-knot nematode population.

Infected corms can be disinfected of root-knot nematodes by paring the planting material followed by immersing it in hot water at 50-55°C for 30 minutes.

The biocontrol agent, *Pochonia chlamydospora* parasitized the eggs of root-knot nematode and prevent the hatching significantly.

The promising biological agents such as *Paecilomyces lilacinus, Trichoderma viride, Pochonia chlamydospora*, VA mycorrhiza, *Glomus fasciculatum* and bacterium, *Pasteuria penetrans* are recommended to control nematodes on banana.

Application of carbofuran at 20 g/plant or Rugby at 5 g/plant or Caldan at 5 g/plant applied twice once at planting and second application after three months of planting is known to reduce the nematode population and increase the yield significantly.

Application of neem cake at 250 g/plant twice (once at planting and second after four months) helps in suppressing the nematode population.

31.3. PAPAYA, *Carica papaya*

31.3.1. Root-knot nematodes, *Meloidogyne* spp.

31.3.1.1. Symptoms: Orchard show patches of poor growth with many plants missing in rows. Symptoms include poor growth, yellowing of foliage, dropping of leaves, reduction in leaf production, weak vigor and premature dropping of fruits. Roots exhibit galls of varying sizes (Fig. 31.14). In heavily infected old roots, adjacent galls join together and form large galls.

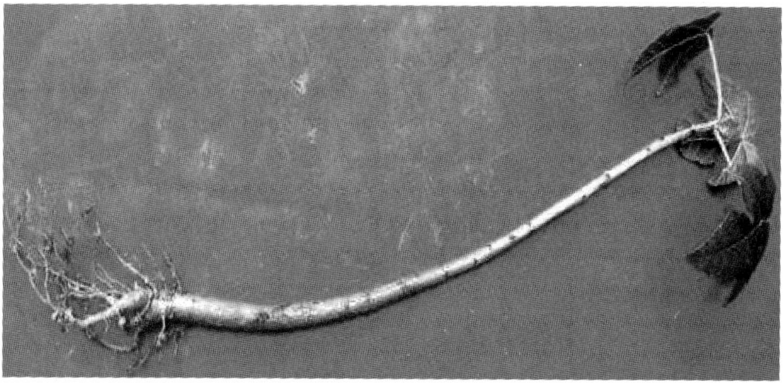

Fig. 31.14. Papaya infected with root-knot nematodes

31.3.1.2. Management

- Hot water treatment of papaya seedling roots at 50°C for 10 min.
- Neem cake at 2.5 t/ha gave reduction in nematode population in both soil and roots and increased yield by 260%.
- Application of carbofuran and fenamiphos each at 2 kg a.i. /ha.
- Nursery bed treatment with *P. lilacinus* and *T. harzianum* each at 1 g per plant.
- Use resistant papaya cvs. CO 2, Pusa 22-3, Pusa Delicious, Pusa Dwarf and Pusa Giant.

31.4. GRAPEVINE, *Vitis vinifera*

The root-knot (*Meloidogyne incognita*) and the reniform (*Rotylenchulus reniformis*) nematodes are the most important grapevine pests.

31.4.1. Root-knot nematode, *Meloidogyne incognita*

The root-knot nematode is responsible for 25 to 50 per cent yield loss in established vineyards of `Muscat Hamburg' in Tamil Nadu. Young, shallow rooted vines are often hit harder than are old deep rooted vines. Nematode damage is most severe when young vines are replanted in old vineyards.

Grapevine cultivars Bangalore Blue, Gulabi and Muscat are highly susceptible to the root-knot nematode.

31.4.1.1. Symptoms: The affected vines show decline in vigor with inadequate sprouting after pruning, reduction in leaf size, marginal drying of leaves and perceptible reduction in yield. The nematode produces small galls on the roots (Fig. 31.15). Heavy infections may completely destroy the root systems of young plants. Galling of roots blocks the flow of water and food, thus starving the plant. Those parts of the roots beyond the point of attack often die.

Fig. 31.15. *Left* – Grapevine plants stunted by root knot nematodes. *Right* - Grapevine roots infected with root-knot nematodes

31.4.1.2. Ecology: Grapevines are attacked and seriously injured by root-knot nematodes only when the vines are grown in light, sandy or sandy loam soils. Vines in fertile loams and heavier soils are seldom seriously damaged.

31.4.1.3. Management: A well planned control program should include the use of nematode-free planting material, nematode resistant rootstocks and the selection of the most effective soil treatment for the root-knot nematode infestation.

(a) Hot water treatment: Hot water treatment (51.7°C for 5 minutes) for root-knot nematode control is necessary if roots are lightly infested with nematodes but otherwise have made good growth. The roots are submerged in hot water maintained at 51.7°C for 5 minutes. The vines are then cooled quickly and planted immediately.

(b) Chemical control: Soil application of aldicarb or carbofuran at 6 kg a.i. per ha is effective and increased the fruit yield.

(c) Resistant rootstocks: In soils where the root-knot nematode infestations are heavy, it may be impossible to grow grapevines without the use of nematode resistant rootstocks.

The grapevine rootstocks 1613, Dogridge and Saltcreek are resistant, while 1616 and Talaki 5-A are moderately resistant to the root knot nematode (Fig. 31.16).

Fig. 31.16. Healthy and vigorous vines on either side are grafted on root-knot nematode resistant rootstocks, while the weak and missing positions are grafted on susceptible rootstocks.

31.4.2. Reniform nematode, *Rotylenchulus reniformis*

The reniform nematode is a serious problem on Anab-e-Shahi variety of grapevine.

31.4.2.1. Symptoms: Feed on and damage secondary and feeder roots (Fig. 31.17). Affected roots show brownish discoloration. As a result, nutrient uptake is affected and vine appears sick.

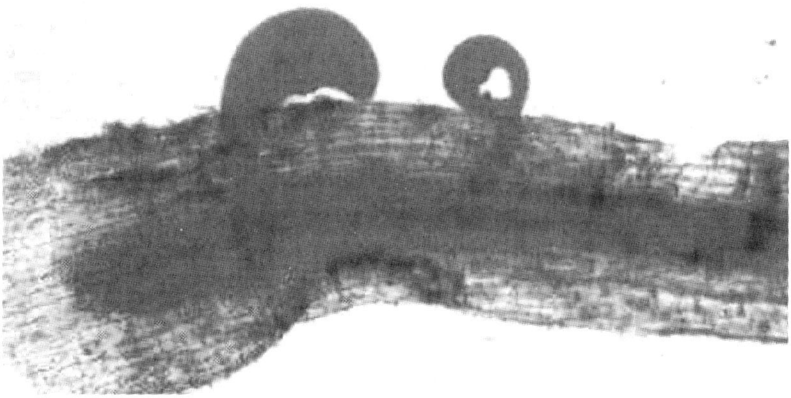

Fig. 31.17. Grapevine root infected with reniform nematode

31.4.2.2. Management

(a) Chemical methods: Soil application of Metham Sodium at 40 liters per ha is effective and increased the bunch weight by 63 per cent.

(b) Resistant cultivars: Grapevine cultivars such as Dakshi, Joazbeli and Mukchilani are resistant to the nematode.

31.4.3. Dagger nematode, *Xiphinema* spp.

31.4.3.1. Symptoms: The dagger nematode, *X. index,* feeds on root tips causing swelling and bending as well as many dead feeder roots. If population levels are high; multiple prolonged attacks can result in darkened, necrotic spots that spread through the root tip (Fig. 31.18). It is common to see poor growth and or stunting of the plant, yellowing or wilting of the foliage, and reduced root systems which can include root necrosis, lack of feeder or secondary roots, and occasional tufts of stubby rootlets.

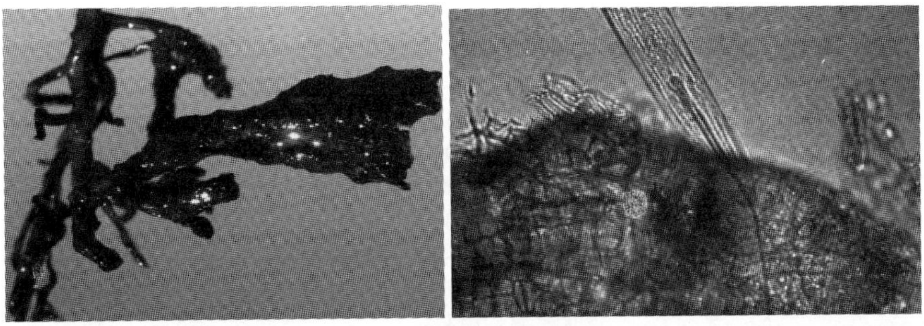

Fig. 31.18. *Left* - Terminal galling of grapevine root caused by a dagger nematode, *Xiphinema* sp. *Right* - A dagger nematode, *Xiphinema* sp. feeding on root tip.

Virus transmission by dagger nematode produces symptoms on leaves such as yellowing of veins, mosaic, and malformation with symptom expression less apparent among white varieties and in warmer regions. *Xiphinema index* is especially important because of its ability to transmit Grapevine fan leaf virus when feeding on grapevine roots.

31.4.3.2. Management: The most effective control of both *Xiphinema index* and *Meloidogyne javanica* was obtained with carbofuran at 0.2 g/liter. Populations of *Xiphinema index* and *Meloidogyne javanica* juveniles were highly susceptible to amendments composed of neem powder (at 4.08 g/pot) + Fertinemakil fertilizer (at 2.5g/pot) as well as marigold + Fertinemakil.

31.5. GUAVA, *PSIDIUM GUAJAVA*

31.5.1. Root-knot nematode, *Meloidogyne enterolobii* (= *M. mayaguensis*)

The worst nematode problem affecting guava is the root-knot nematode, *Meloidogyne enterolobii* which is recognized as a limiting factor in commercial guava production. In and around Ayakudi village, Palani taluk, Dindigul district, and Sirumugai village, near Mettupalyam, Coimbatore district, Tamil Nadu, young guava trees were exhibiting symptoms of yellowing, shedding of leaves, drying of branches, reduced vigor, less productivity etc. On examination of affected trees, root-knot nematode was found to be the causal agent for this menace. Further, the nematode was identified as *M. enterolobii,* a species of root-knot nematode so far not encountered in India. The unseen hidden enemies of root-knot nematode associated with guava are capable of killing the tress outright within three months after the onset of symptoms. It is considered that the nematode might have spread from other countries through saplings.

31.5.1.1. Symptoms: *Meloidogyne enterolobii* affects growth, yield, life span and tolerance to environmental stresses of infected plants. Young infected guava trees exhibit symptoms of yellowing, shedding of leaves, drying of branches, reduced vigor and less productivity. Severely infected trees decline rapidly culminating in the death of the plants. Moderate infestations are associated with general chlorosis, nutrient deficiency symptoms and reduced flowering and fruiting (Plate 31.19). Roots of infected trees show multiple galls and root necrosis, causing a drastic decrease in feeder roots (Cetintas *et al.*, 2007). The root-knot nematode infects the whole root system, from small rhizoids to more lignified roots (Plate 31.19). Overall, damage due to *M. enterolobii* may consist of reduced quantity and quality of yield. This syndrome is making cultivation of guava in the infested areas unfeasible, causing serious economic problems to growers and the economy of the region (Carneiro *et al.*, 2007).

Fig. 31.19. Symptoms of root-knot nematode on guava

The decline and death of guava trees parasitized by *M. enterolobii* is a complex disease involving other micro-organisms like *Fusarium solani* (Gomes *et al.*, 2010), *Verticillium dahliae, Pythium aphanidermatum,* and *Trichoderma* sp. (Avelar-Mejía *et al.*, 2001).

31.5.1.2. Management

(a) Prevention

- Use nematode-free planting material.
- Use sterilized potting media (e.g. sterilized by steam or roasting).
- Use clean implements and tools.

- Do not use previously infested fields for planting. Do proper land preparation. Deep plowing exposes nematodes to sunlight.
- Use recommended fertilizers to improve crop vigor.
- Use trap crops like marigold.
- Green manuring with repellent crop before planting sun hemp and mustard.
- Add burnt paddy husk to planting hole.
- Use resistant varieties and rootstocks such as Horana White, Horana Giant, Pubudu and Kanthi (all are moderately resistant).

(b) Monitoring

- Regularly observe nursery plant roots for the characteristic symptom of root galls.
- Sample the side roots of mature plants to observe for root galls.
- Observe above-ground symptoms like leaf yellowing, withering and scorching of leaf edges with purple patches followed by examination of root systems for galls.

(c) Direct control

- Avoid surface irrigation from infested fields to healthy fields.
- Add compost to the soil to reduce the effect of nematodes (after infestation).
- Remove weeds and infested crop residues to protect nearby healthy tree.
- Intercrop with Crucifers like mustard, cabbage and incorporate into soil (before flowering) to suppress nematodes through biofumigation.
- Add poultry manure, cow dung, goat dung, coconut and rubber pomace, neem cake and green manure (Adathoda, Tithonia, Gliricidia, neem and Kalawel) in planting pit as a soil amendment.
- Remove infested plants from the field.

(d) Intercropping with non-host cover crops: Use of non-host cover crops like peanut (cvs. IAC Oirã, IAC Poitã, IAC Tatuí and Cavalo vermelho), rye (cv. IPR 89), green gram, oat (cv. IAPAR 126), rape seed (cvs. CAN 420, CAN 401), foxtail millet, finger millet, cowpea, field bean, castor (cv. IAC 80), and velvet bean was found effective for the management of this nematode.

Fourteen fruit trees are non-hosts to *M. enterolobii* (atemoya, avocado, cashew, citrus rootstocks, coconut and mango) or poor hosts (assai palm, common mulberry, jaboticaba, grape rootstocks, passion fruit, sapodilla and soursop).

Growing non-host crops or black fallow are the most efficient methods for reducing *M. enterolobii* populations. Unfortunately, the list of non-host plants is very small, including cabbage, garlic, grapefruit, maize, peanut, sour orange and welsh onion.

(e) Chemical methods: Application of carbofuran at 15 g a.i./ m^2 in the rhizosphere region before flowering, pulverizing the soil in the basin area to mix the chemical in soil followed by irrigation was found effective for the management of guava root-knot nematode.

(f) Biological methods: Application of *Paecilomyces lilacinus/Psudomonas fluorescens*) at 20 g/m^2 (about 9 m^2 around the tree trunk) in the rhizosphere region at the time of flowering gave effective control of *M. enterolobii.*

(g) Host resistance: Only Costa Rican wild guava, *P. friedrichsthalianium* was considered resistant to *M enterolobii*. No nematodes were detected in the roots of the 'Criollo' and 'Arrayán' rootstocks, confirming the resistance of *P. friedrichsthalianium* (Bogantes-Arias and Newcomer, 2010).

The resistance was observed only in *P. guajava* accessions S3 K-10, A-06, J-16 and B-12 (Cassasa *et al.*, 1997; Milan, 2007, 2010).

Carneiro *et al.* (2012) reported that only one accession of *P. friedrichstalianium* (Costa Rican wild guava) was resistant and compatible as rootstock with *P. guajava* 'Paluma', in field conditions.

Milan (2007) found that only *P. guajava* Acc. B-12 has the potential to control the root- knot nematode and graft-compatible with commercial clone of guava because they are from the same species. *P. guajava* Acc. B-12 had the potential to control the nematode and could be used as a rootstock for guava commercial clones.

(h) Integrated management: The guava growers are advised to avoid nematode infected saplings, removal and destruction of nematode affected trees, application of farm yard manure at 100 kg, neem cake at 250g and nematode egg parasitic fungus *Paecilomyces lilacinus* at 25g and carbofuran at 60g/tree at early stage of nematode infestation to overcome the nematode problem and to enhance fruit yield.

Application of 1 kg neem cake enriched with bioagents + carbofuran at 7g a.i./ m^2 just before flowering was found effective against *M. enterolobii.*

31.6. POMOGRENATE, *PUNICA GRANATUM*

31.6.1. Root-knot nematode, *Meloidogyne incognita*

The root-knot nematode, *Meloidogyne incognita* Race 2 is a serious pest of pomegranate (in Maharashtra, Karnataka, Gujarat and Tamil Nadu) which primarily cause galls on root system that eventually are seriously hampered in their main functions of uptake and transport of water and nutrients, reduced plant growth and productivity of the pomegranate (Khan *et al.*, 2005; Nasira *et al.*, 2011).

31.6.1.1. Symptoms: Severely infested plants by root-knot nematode exhibited yellowing of foliage resulting in stunted plant growth. These plants produced less number of fruits or no fruits which might be due to the nematode induced nutritional deficiency. In severe cases the galls were predominantly found on the entire root system (Plate 31.20). The galls were white in color turned to light brown and hardy when they became old. The intensity of root-knot nematode damage increased with increase in age of the plant. In general, more than five year old plants were severely affected by root-knot nematode. Egg masses were observed inside as well as outside of the galls. More numbers of females were found in a single compound gall.

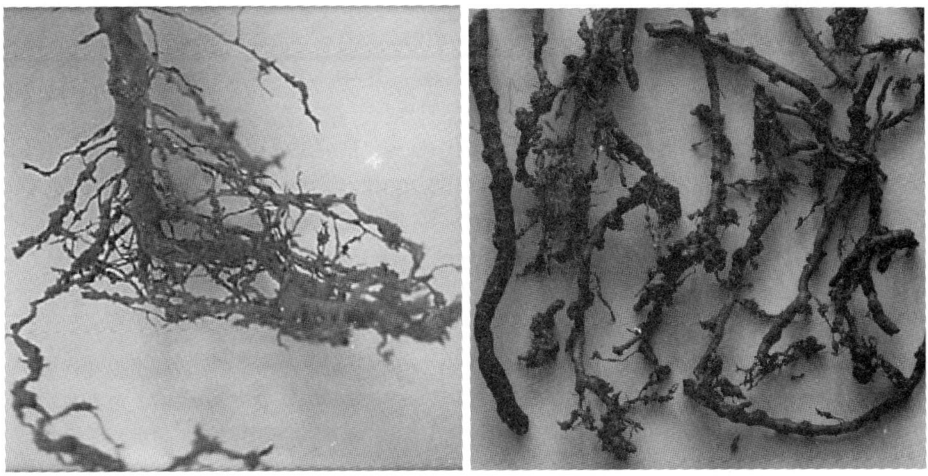

Fig. 31.20. Pomegranate roots infected with *Meloidogyne incognita.*

31.6.1.2. Management

(a) Intercropping: Intercropping with sun hemp, marigold, mustard, fenugreek, onion, and garlic during crop rest period was found effective against root-knot nematode on pomegranate.

(b) Botanicals: Application of marigold flowers gave the highest reduction percentage in root galling with value of 98.8%, followed by colocynth fruits extract which averaged 96.4% without significant differences between them. The lowest egg-mass index (EI) was obtained from plants receiving marigold extract (0.0) with reduction percentage value of 100 % (Nour El-Deen *et al.*, 2016).

(c) Biological methods: *Pseudomonas fluorescens* at 20 kg/ha was found to be most effective in reducing *M. incognita* population (31.28%) and number of root galls/5 g roots (29.28%) and increasing the fruit yield of pomegranate (18.99%) with 2.37 B:C ratio; followed by *Trichoderma viride* at 20 kg/ha (reduction in *M.*

incognita population - 28.3%, number of root galls/5 g roots -23.9%, increasing the fruit yield -16.78%, and B:C ratio - 2.33) (Pawar *et al.*, 2013).

The symbiotic bacterium of entomopathogenic nematode *Xenorhabdus* treatment significantly inhibited formation of galls on roots with reduction percentage of 97.9%, followed by those treated with *Photorhabdus* with value of 96%. Regarding egg-masses numbers, *M. incognita* was barely reproduced on plants treated with *Xenorhabdus* filtrate with reduction percentage value of 95.8%, followed by *Photorhabdus* with value of 91.7 % (Nour El-Deen *et al.*, 2016).

(d) Chemical methods: Application of carbofuran at 2 kg a.i./ha in nursery, and 4 kg a.i./ha in the main field at transplanting and at flowering gave effective management of *M. incognita* on pomegranate.

(e) Integrated management: Application of farmyard manure at 20 tons/ha, neem cake at 250 g/tree followed by bioagent, *Paecilomyces lilacinus* at 25 g/plant was found effective to manage root-knot nematode problem in pomegranate and to increase crop productivity.

Two applications of FYM enriched with *P. lilacinus* at 15-25 kg around each plant at 3 months interval along with application of liquid formulation of *P. lilacinus* through drip; showed recovery from symptoms, and putting forth new flushes. Saprophytic microbial population in soil has multiplied many fold, reduced egg masses and females in roots and juveniles of root-knot nematodes were not seen in soil.

The farmers of North Gujarat region cultivating pomegranate are advised to apply talc based formulation of *Pochonia chlamydosporia* (10^8 spores/g) or *Paecilomyces lilacinus* (10^8 spores/ g) at 100g/ plant along with mustard cake at 1.5, 2.0 and 2.25 kg/ plant in first, second and third year, respectively; two times i.e., first during November and second during April for the effective management of root- knot nematodes.

Application of bioagents (*T. viride, P. lilacinus, Ps. fluorescens*) at 10-20 g/m^2 in nursery, and 5 kg/plant of FYM/compst enriched with bioagents in the main field at transplanting and at flowering was found effective against *M. incognita*.

The following INM measures have been recommended for effective management of pomogrenate root-knot nematodes:

- Intercropping with sun hemp, marigold, mustard, fenugreek, onion, and garlic during crop rest period.
- Use of biocontrol agents such as *Pseudomonas fluorescens, Paecilomyces lilacinus, Pochonia chlamydosporia, Trichoderma harzianum, T. viride* at 20 kg/ha.
- Use of entomopathogenic nematodes (*Steinernema* spp., *Heterorhabditis* spp.) and their symbiotic bacteria (*Xenorhabdus, Photorhabdus*).

- INM using enriched neem cake/ mustard cake/FYM/compost/ vermi-compost with biocontrol agents (*Pseudomonas fluorescens, Paecilomyces lilacinus, Pochonia chlamydosporia, Trichoderma harzianum, T. viride*).
- Enforcement of domestic quarantine.

31.7. POTATO, *Solanum tuberosum*

The cyst nematodes (*Globodera rostochiensis* and *G. pallida*) and the root-knot nematodes *(Meloidogyne incognita, M. javanica* and *M. hapla*) are the major problems in India.

31.7.1. Cyst nematodes, *Globodera rostochiensis, G. pallida*

The potato cyst nematode (PCN) is the most important quarantine pest that is causing concern to potato production and trade throughout the world. In India, the potato cyst nematode, *Globodera rostochiensis* was first reported by Jones (1961) who detected the cysts from the roots of potato plants which exhibited the symptoms of yellowing of leaves from Nilgiris Hills of Tamil Nadu state. This report initiated a great awareness amongst administrators in the country resulting in establishment of Nematology Unit in Tamil Nadu as well as initiating an Indo-German Collaborative Program of intensive chemical application (Dasanit) in the infested area to avoid further nematode spread. The potential damage of this nematode to potato cultivation was such that the Destructive Insect Pest Act 1919 was amended by the Tamil Nadu Government in 1971 to ensure strict prohibition of movement of seed potatoes from infested area to any other parts of the state or the country.

The later surveys have indicated the presence of both species of potato cyst nematode (*Globodera rostochiensis* and *G. pallida*) restricted to potato growing regions of Nilgiris and Kodaikanal Hills of Tamil Nadu (Krishna Prasad, 1986) and Kerala (Ramana and Mohandas, 1988).

Ganguly *et al*. (2010) reported the occurrence of *Globodera rostochiensis* and *G. pallida* from the Potato Breeding Farms of Central Potato Research Institute located in Kufri, Shimla District of Himachal Pradesh. The potato fields at Lister House Farm and Main Farms were found to be heavily infested, supporting 58–548 cysts per 100 g soil. The reported occurrence of potato cyst nematodes in Himachal Pradesh, confirming the earlier report of *G. pallida* in this state by EPPO (European and Mediterranean Plant Protection Organization), is a cause of serious concern with ramifications on export and quarantine issues.

31.7.1.1. Symptoms: Under conditions of severe infestations, plant growth is stunted and wilting occurs during hot parts of the day (Fig. 31.21). The disease usually occurs in patches. Plants show tufting of leaves on the top as the outer leaves turn yellow and die. Root system is smothered, secondary roots are induced at the collar region and the plants can be easily pulled out. Tubers formed are less in number and reduced in size. In extreme cases tuber formation is arrested.

Fig. 31.21. Potato plants and roots infected with cyst nematodes

31.7.1.2. Management

(a) Quarantine: The infested area should be quarantined and cultivation of host plants to be discontinued indefinitely. Volunteer seedlings in the infested area should be destroyed. The seed material should not be allowed to move out of the infested area. The infested area can be put to non-agricultural use like construction of buildings, etc.

(b) Crop rotation: It is the most practical method of control. It is recommended to grow the host crop once in every 3 years if the infestation is confined to a few patches, but only once in 4 or 5 years if the infestation is extensive. In Nilgiris Hills, potato is rotated with wheat and strawberry to control the cyst nematode.

(c) Disease escape: Early varieties of potatoes like Irish Cobbler, usually suffer less than do late varieties. These start growth in spring at soil temperature too low for much nematode activity and may be harvested before many of the nematodes can reproduce.

(d) Resistant varieties: *Solanum vernei* (multiple gene resistance) and *S. tuberosum* s. sp. *andigena* (single gene resistance) were found to have marked resistance to the cyst nematode of potato. *S. andigena* roots produce a hatching factor and are readily invaded by the larvae, which, however, fail to mature and reproduce. The failure is because the root tissues are not induced to produce giant cells, but instead the cells near the nematode head become necrotic and there is inadequate nutrition for the female to develop.

In the United States, both the USDA and Cornell University scientists have evolved two commercial potato varieties *viz.*, `Peconic and `Wauseon' which were resistant to the cyst nematode. `Peconic' derives its resistance from the subspecies andigena. Wauseon is a selection from a cross between B 4159-8 and Katahdin.

(e) Chemical methods: The infested area can be treated with an eradication dosage of fesnulfothion or aldicarb both at 30 kg a.i. per ha in two doses at an interval of 15 to 20 days. Dazomet, which liberates methyl isothiocyanate, is more nematicidal in pearls and clay soils and often gives large yield increases.

(f) Integrated management: The combination of disease escape (planting early varieties) and the use of nematicides gave good control of the potato cyst nematode. Hygiene, in the form of seed certification, combined with crop rotation in seed growing areas is effective in controlling the cyst nematode. Use of a resistant variety followed by a nematicide would kill 99 per cent of the nematode population. The most effective measure combines crop rotation, use of nematicides and resistant varieties to keep the nematode at an economically acceptable level.

Fensulfothion applied at planting time at 30 kg a.i. per ha almost completely prevented nematode infestation in the next two seasons. Even a lower dose of 15 kg a.i. /ha for three seasons and 7.5 kg per ha in subsequent seasons is effective in keeping the nematode under check so as to obtain reasonable good yields. The object of integrated control is to bring down the infestation level to the minimum by the use of nematicides and manage the populations further by suitable crop rotations (potato x wheat x strawberry x maize x bean) or resistant varieties.

An integrated system utilizing resistant cultivars, non-host crops and susceptible cultivars with chemical treatment (Aldicarb) will successfully manage densities of *Globodera rostochiensis* at desired levels.

31.7.2. Root-knot nematodes, *Meloidogyne* spp.

Root-knot nematodes, *Meloidogyne* spp., are by far the most important nematode parasites of potatoes in India. Severe infestation of root-knot nematodes leading to crop failure have been noticed in Mahasu district of Himachal Pradesh and Hassan in Karnataka where farmers did not get even as much as planted by them.

31.7.2.1. Symptoms: Plants affected by root knot nematodes are generally stunted, slightly yellow and may wilt during hot weather. The diagnostic symptoms of a *Meloidogyne* attack on potatoes are the galls found on roots and tubers. Galling on tubers may render them unsalable (Plate 31.22).

31.7.2.2. Life cycle: Species of *Meloidogyne* are sedentary root and tuber endoparasites, the life cycle being similar to that of *G. rostochiensis*, with some important differences. The white, swollen females do not develop into persistent, brown cysts, but deposit all their eggs in an external egg mass. Hatching occurs when the physical conditions are suitable, and is not dependent on a hatching factor from host roots. Males may be present but are not necessary for reproduction.

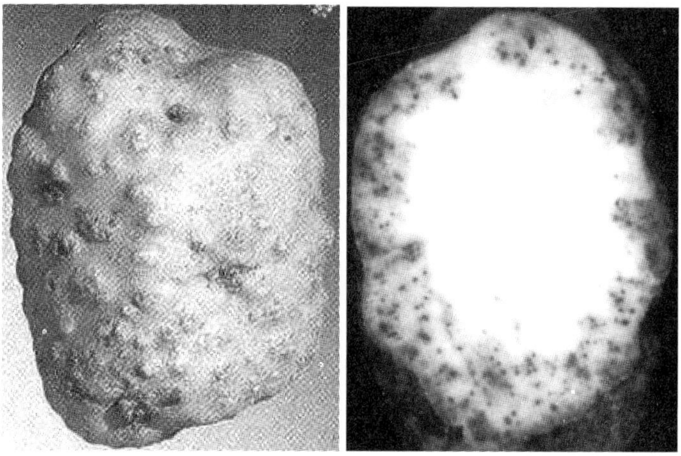

Plate 31.22. Blisters on potato tuber caused by *Meloidogyne incognita.*

31.7.2.3. Host range: Species of *Meloidogyne* usually have wide host range amongst crops and weeds in many botanical families.

31.7.2.4. Interactions with other pathogens: *Meloidogyne* species are associated with the fungi *Rhizoctonia solani* and *Verticillium* sp. and the bacterium *Pseudomonas solanacearum* in disease complexes.

31.7.2.5. Ecology: Root-knot damage to potatoes is usually associated with light soil. Aggravated attack by *Meloidogyne* on potatoes under irrigation is usually ascribed to spread of the nematodes in irrigation water. Soil temperature requirements of these nematodes vary with species. *M. hapla* requires cool climate, whereas *M. incognita* and *M. javanica* require higher temperatures for their development and reproduction.

31.7.2.6. Management

(a) Cultural methods

(i) Hygiene: Tubers from *Meloidogyne* infested crops should not be used as seed, as an even outwardly clean seed may be infested.

(ii) Crop rotation: Crop rotation with maize, wheat and barley effectively check the buildup of *Meloidogyne* population.

(iii) Trap crops: Tagetes patula planted in alternate rows with potato gives good control of root-knot nematodes.

(iv) Organic amendments: Application of organic amendments such as FYM, compost, green manures and crop residues suppress the nematode population.

This may promote fungal and other natural enemies of nematodes or release toxic compounds.

(b) Nematicides

Application of aldicarb at 5 kg a.i. per ha in two doses first at the time of planting and another at 50 days interval gave best results with no infestation either on roots or tubers.

(i) Seed treatment: *Meloidogyne* in tubers can be controlled by hot water treatment at 46-47.5°C for 2 hr. and by dry heat at 45°C for 48 hrs.

31.8. TUBEROSE, *POLIANTHES TUBEROSA*

31.8.1. Foliar nematode, *Aphelenchoides besseyi*

The infestation of foliar nematode, *Aphelenchoides besseyi* has emerged as a serious potential threat of great concern for profitable cultivation of tuberose in West Bengal, India. This nematode was reported for the first time from the Hawaii Islands (Holtzmann, 1968) on the leaves of tuberose. Later, the occurrence of 'floral malady' caused by *A. besseyi* in tuberose was found in the Ranaghat areas of the Nadia district of West Bengal, India (Chakraborti and Ghosh, 1993). The nematode has recently got spread to Odisha presumably through bulbs. *A. besseyi* is now-a-days a major limiting factor for cultivation of tuberose in Ranaghat, Haringhata and Panskura areas of West Bengal. The 'Calcutta single' cultivar of tuberose is more vulnerable to *A. besseyi* than the 'Calcutta double' cultivar. Research results confirmed that *A. besseyi* is the primary causal agent for malformed flowers and population of *A. besseyi* causing white tip disease in rice is the same population infecting tuberose causing floral disease (Khan, 2001). The nematode species is widespread in rice in West Bengal (Das and Khan, 2007). The foliar nematode problem is restricted to tuberose in WB and Odisha. The presence of nematode species in the cut flower stalk is one of the main constraints to export the flower to other countries.

31.8.1.1. Symptoms: Ectoparasitic feeding of *A. besseyi* on the scape (floral stalk) leads to the formation of prickle-like rough, irregular structure. Scape becomes curled and twisted. Scapes of the infected plants are many times smaller than healthy and normal scapes. Endoparasitic feeding of *A. besseyi* was observed in leaves, bracts and petals where brownish to black necrotic spots of varying shapes and sizes were developed (Plate 31.23). The nematode forms 'nematode wool' on dark brown spots. The ovary contains large number of nematodes. This nematode is generally more serious during rainy season from June to September and cent per cent loss of the second year crop of the 'Calcutta single' cultivar of tuberose was encountered. However, in 'Calcutta double' cultivar 30% to 40% flower stalk renders unsalable and individual flower stalk occur to the extent of 59% (Pathak and Khan, 2009).

31.8.1.2. Management: Out of the six tuberose cultivars viz. Calcutta Single, Calcutta Double, Suvasini, Swarna Rekha, Rajat Rekha and Shringar, only Suvasini was immune to attack by *A. besseyi*.

Fig. 31.23 Effects of foliar nematode infection on tuberose. (**a**) Healthy control stem with abundant white florets. (**b–d**) Infected plants. Note the browned stem, leaves and florets, as well as the distorted skin of the leaves on the stem

The treatments viz., pre-soaking of bulbs in plain water for overnight + hot-water treatment at 50ºC for 30 min. + three sprayings with monocrotophos 36 SL at 500 ppm at 15 days interval (repeated the same spraying in second year from March onwards) had low nematode infestation (26%) and PDI value (25%), reduction of nematode population (53%) and higher spike yield (mean 286/plot - 1.5 x 1.8 m^2) as compared to untreated plots (mean 207/plot) (Pathak and Khan, 2009).

With the adoption of nematode management practices, spike yield could be saved up to 38% and that could be to the extent of 59% when loose flower yield/plot was taken into consideration.

The following INM measures are recommended for the management of foliar nematode on tuberose:

- Pre-soaking of tuberose bulbs in plain water or in neem seed kernel extract (NSKE 4%) for overnight followed by dipping in monocrotophos 36SL at 500-700 ppm for 4-6 hours.
- After sprouting, three to four sprayings with monocrotophos 36SL at 500 ppm at 15 to 20 days interval is needed.
- In the second and third year crop, same spraying for 3-4 times with monocrotophos 36SL at 500 ppm at 15 to 20 days interval should start from April-May onward to reduce the nematode infestation.
- Field sanitation is most essential for reducing infestation. Infested plant parts should be burnt or buried into the soil.

- Grow nematode tolerant tuberose cultivars, Prajwal and Shringar (Khan and Ghosh, 2007)
- Growing tuberose away from rice field is encouraged to avoid contamination of tuberose field.

31.9. POLYHOUSE GROWN VEGETABLE AND FLOWER CROPS

In 1980s, the demand for high-quality export oriented horticultural products and need for the availability of horticultural crop produce throughout the year, especially during off-season, compelled growers to cultivate selected vegetable and flower crops under protected cultivation. Protected cultivation of high value vegetable and flower crops under controlled environmental conditions has gained popularity among farmers due to high productivity, better quality of produce and manifold increase in yield per unit area. Farming inside modern greenhouses is promoted in order to improve earning potential of farmers especially during harsh winters.

31.9.1. Vegetable crops

Tomato, capsicum, cucumbers, French bean, crucifers, gherkins, muskmelon, watermelon, etc. are some of the vegetable crops grown under polyhouse conditions.

31.9.2. Flower crops

Flower grown under polyhouse conditions include carnation, rose, gerbera, orchids and anthuriums.

31.9.3. Nematode problems

In most cases, problems arise from contaminated soil or soil mixture used as a component of the growing medium, mono-cropping and infested planting materials. Commonly followed nematode management practices such as summer plowing, fallowing and crop rotation which are meant for nematode suppression in field crops are not practicable in polyhouses. Therefore, nematode management here must be considered primarily as exclusion or avoidance. Once nematodes are introduced it is difficult to manage them.

The nematode incidence under protected cultivation became severe and led to complete crop losses because of congenial conditions of higher temperature, humidity and use of high agronomic inputs like fertilizers and plant growth promoters in polyhouses. The proliferation rates of nematodes in polyhouse cultivation reached up to 10–30-fold more than in the open field cultivation. Polyhouses were observed to be severely infested with major plant parasitic nematode genera viz., *Meloidogyne*, *Pratylenchus*, *Rotylenchulus, Helicotylenchus*, *Mesocriconema* and *Tylenchorhynchus*. The most economically important root-knot nematode causes characteristic galls on the roots resulting in damages and dislocation of xylem vessels which impair the movement of water and minerals. The characteristic symptoms are stunted growth and formation of small and chlorotic leaves. They feed on the roots devitalizing the plants ultimately affecting both qualitative and quantitative production. Nematode problems on all these crops under protected conditions have assumed alarming proportions leading to huge losses (up to 80%) in selected crops. The nematode infestations exacerbate severity of soil-borne fungal and bacterial diseases leading to complete crop failure. In view of damage potential of these tiny hidden enemies of crop plants, an urgent need is felt for developing nematode management strategies.

31.9.4. Management

The nematodes may be managed by the following integrated methods:

- Growing cowpea (trap crop for root-knot nematode) closely before the main vegetable/flower crop and removing the plants out of polyhouse at 45 days after sowing.
- Raising marigold (antagonistic to nematodes) and incorporating *in situ* after plucking the flowers.
- If the beds are highly infested with the nematodes, bring down the population of nematodes by applying neem/pongamia cake at 500 g + Carbofuran at 50 g/sq. m.
- Treat the seeds with biopesticide-*Pseudomonas fluorescens* at 10 g/kg seed.
- Treatment of seed (20 g biopesticide per kg seeds) and substrates with FYM, vermicompost, neem/pongamia cakes enriched with biopesticides (2 kg each of *Paecilomyces lilacinus*, *Trichoderma harzianum* and *P. fluorescens* per 2000 kg of FYM/500 kg of vermicompost/200 kg of neem/pongamia cakes).
- Raise the seedling in the coco-peat or any other substrate by treating with *P. fluorescens* at 1 kg + *T. harzianum* at 1kg + neem or pongamia cake at 50 kg/ton.
- Treat the bed with neem or pongamia cake at 50g, *T. harzianum* at 25 g or *Pochonia chlamydosporia* + *P. lilacinus* each at 25 g/sq. m at an interval of 2 months.
- Apply 2 tons of farmyard manure enriched with bio-pesticide - *T. harzianum* per acre before planting, along with 100-200 kg of neem or pongamia cake.
- Apply Carbofuran 3 G at 1kg a.i. /ha at planting and repeat after 45 days.

Chapter 32

PLANTATION AND SPICE CROPS

32.1. COFFEE, *Coffea* spp.

The lesion nematode, *Pratylenchus coffeae* and the root-knot nematode are the most important pest of coffee.

32.1.1. Lesion nematode, *Pratylenchus coffeae*

Zimmerman (1898) described a lesion nematode found infecting the roots of coffee plants in Java. This nematode has also been reported from coffee growing tracts of South India. *P. coffeae* destroyed over 95 per cent of arabica coffee plantations of Java. Over 1,000 hectares have been infected with this nematode in coffee growing tracts of South India. The annual loss to Indian coffee due to this nematode is estimated at $3 million.

32.1.1.1. Symptoms: In nurseries, *P. coffeae* causes death of coffee plants. The infested coffee plants show yellowing of leaves, loss of primary roots and stunting of the shoot followed by gradual wilting and death (Fig. 32.1). The infected plants can be easily pulled off by hand because of scanty root growth. Lesions on the roots were observed on the plants infected with *P. coffeae.*

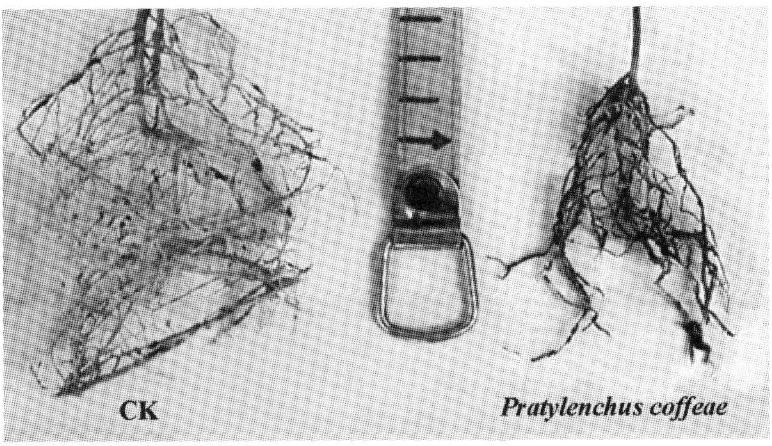

Fig. 32.1. *Right* - Coffee seedling infested by *Pratylenchus coffeae, Left* - Control.

32.1.1.2. Life cycle: *P. coffeae* is a migratory endoparasite of the root cortex where it feeds and multiplies. The favored place of entrance is slightly behind of the elongating zone in the piliferous region. A colony or `nest' is established showing brownish lesions on the root. When decay sets in due to secondary invasion by bacteria and fungi, the nematodes quit and migrate through the soil to attack fresh roots.

Zimmerman (1898) has discussed in detail the developmental stages of *P. coffeae* infesting coffee in Java. The first molt takes place within the egg and three molts occur outside. *P. coffeae* requires about 45 to 48 days for completion of the entire life cycle. Of this time, 15 to 17 days are necessary for eggs to hatch, 15 to 16 days are spent as larvae, and it takes 15 days from maturity to egg production.

32.1.1.3. Host range: *P. coffeae* is known to attack coffee, banana, citrus, grapevine, apples, strawberry, abaca, yam, cocoa, rubber, potato, sweet potato, peas, beans, cabbage, cauliflower, tomato, chrysanthemum, caladium, dahlia, marigold, cotton, lucerne, etc.

32.1.1.4. Management

a) Resistant rootstocks: *Coffea robusta* is relatively more resistant to *P. coffeae* than *C. arabica.* Grafting of Arabica scion on Robusta rootstock at 40 day-old 'topee stage' plants by using wedge-cleft method can be adopted (Kumar, 1988) (Fig. 32.2).

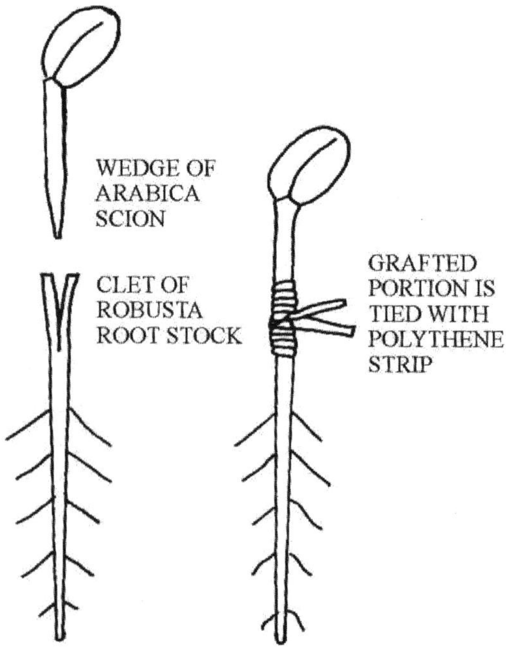

Fig. 32.2. Grafting of arabica coffee scion on robusta rootstock

b) Chemical methods: In field trials with young trees of Arabica coffee, phenamifos at 0.5 to 1.0 g/plant gave good control of *P. coffeae* and remained effective for 90 days after application. Repeated applications were necessary to maintain continuous control (Kumar, 1982).

c) Cultural methods

i) Organic amendments: Coffee pulp at 18 kg per sq. m. improved plant growth and reduced the nematode population.

ii) Fallowing: Lardello (1968) stated that the replanting of old nematode - infested plantations be preceded by a fallow period of at least two years after removing the old coffee roots.

32.1.2. Root-knot nematodes, *Meloidogyne exigua, M. coffeicola*

Srinivsan and D'Souza (1965) reported *M. exigua* on coffee from South India. *M. exigua* is a pest of *Coffea arabica* but not of *C. canephora* (robusta coffee) which is very resistant (Schieber and Sosa, 1960). A 5% loss in yield due to *M. coffeicola* damage has been reported in Hawaiian coffee plantations. Losses due to *Meloidogyne* spp. in Brazil are estimated at 15% of that country's production.

32.1.2.1. Symptoms: Affected trees look unthrifty and have chlorotic, wilting and dying leaves. *M. exigua* produces small elongated galls that are located primarily at the root tips (Fig. 32.3). In addition, roots may show cracks, necrosis and sloughing of cortical tissue. The effectiveness of root system is greatly diminished, with rootlets and root hairs practically eliminated. Trees may die, particularly during periods of drought and cold temperatures. Affected plants lack vigor and are unable to withstand adverse conditions.

Fig. 32.3. Symptoms caused by root-knot nematode on *Coffea arabica*

32.1.2.2. Life cycle: *M. exigua* takes 32-42 days at 25-30°C to complete the life cycle. The egg masses of *M. exigua* are mostly located under the epidermis of coffee roots.

32.1.2.3. Hosts: *M. exigua* is known to attack coffee, tea and bell pepper.

32.1.2.4. Survival and spread: *M. exigua* does not survive in the soil, in the absence of the host, for more than six months after eradication of infested plants. The nematode spreads through the infested seedlings used for transplanting in the main field.

32.1.2.5. Interaction with other pathogens: The disease complex involving *Rhizoctonia solani* and *M. exigua* caused more root necrosis and defoliation when both pathogens were inoculated. Extensive fungal colonization within the coffee root systems was observed.

32.1.2.6. Management

(a) Cultural methods: At the crop renovation time, the infested coffee area may be turned to pasture. Most graminicolous plants are non-hosts to *M. exigua* and *M. coffeicola*. The decision of growing forage crops and turning back again the area to coffee will certainly reduce drastically the population of *Meloidogyne* species in 6-12 months.

Rotation of coffee with cotton, soybean and maize in *M. exigua* infested areas for one year significantly reduced the nematode population, after which the farmer can return to coffee cultivation. Coffee can also be rotated with groundnut, castor (both immune), *Styzolobium deeringiatum* and *Crotalaria spectabilis* (both resistant).

(b) Chemical methods: Application of aldicarb, carbofuran and phenamiphos each at 1.6 to 6.0g a.i./plant, in one or two applications during the year gives effective control of root-knot nematodes. The first application should be in the beginning of the rainy season followed by the second 3 months later.

(c) Host resistance: *Coffea canephora* (line 2258 from Costa Rica) showed resistance to *M. exigua* and *M. incognita*. Planting of *Coffea arabica* var. Mundo Novo or Catuai Vermelho grafted on line 2258 in field showed good growth and production in Brazil.

32.2. TEA, *Camellia sinensis*

Root-knot nematodes are the most frequently encountered in the tea growing regions of India and most of the species of *Meloidogyne* except *M. brevicauda* are associated only with nursery plants.

32.2.1. Root-knot nematodes, *Meloidogyne* spp.

The first report of root-knot nematode in young tea was from South India, where large numbers of the seedlings were found infected (Barber, 1901). *Meloidogyne javanica, M. incognita* and *M. hapla* have been found associated with young tea, while *M. brevicauda* cause severe galling of the roots of mature tea bushes in India.

32.2.1.1. *Symptoms*

(a) On young tea: Young nursery plants, both seedling and vegetatively propagated clonal plants, are severely damaged by root-knot nematodes. Seedling plants in which both the tap root and lateral roots are severely attacked suffer greater damage than do clonal tea plants of equivalent age, probably because seedling tea plants possess less than half the root bulk of clonal plants of similar age. A marked increase in resistance is observed between 8 and 15 months.

(b) On mature tea: Root-knot nematodes (*M. brevicauda*) cause severe galling of the roots of mature tea bushes with resultant decline. Tea plants of all ages are susceptible to infection by this nematode which has been recorded in only Ceylon and South India.

The infected bushes grow more slowly and the leaves tend to become smaller, dull and yellow in appearance. The root show the characteristic swelling and pitting that is typical of *M. brevicauda* (Plate 32.4).

Plate 32.4. Mature tea bush roots infected with *Meloidogyne brevicauda.*

The effect of infection on the growth of tea bushes is most evident during the period of recovery from pruning. Healthy bushes are usually harvested about 16 weeks after pruning; but the harvesting of severely infected bushes is necessarily delayed up to a further 6 months. The production of new branches by infected bushes is very restricted, and as a consequence the cropping capacity of the bush is reduced. The most severely affected bushes fail to recover from pruning.

32.2.1.2. Management

(a) Management in nurseries

(i) Physical methods: Soil solarization in nurseries is adequate to manage root-knot and other nematodes as the plants develop complete resistance by about 9-15 months (Gnanapragasam *et al.*, 1989).

(ii) Cultural methods: Every effort should be made to raise the nurseries in nematode-free soil. Incorporation of organic matter into nematode infested soil is known to suppress the incidence of root-knot nematodes on tea.

The cultivation of marigold as a pre-plant crop considerably reduces the population of *Meloidogyne* spp. and *Pratylenchus loosi*. The reduction in the nematode population is significantly greater than when the land is left fallow. The interplanting of marigold in tea has resulted in a 7% increase in yield compared to 10% increase that followed soil fumigation with DBCP.

The rehabilitation crops such as Mana grass (*Cymbopogon confertiflorus*) or Guatemala grass (*Tripsacum laxum*) (which help to improve the soil) are resistant to *Meloidogyne* spp. The soil population of parasitic nematodes decline rapidly when these grasses are grown for one or more years before replanting tea.

(iii) Chemical methods: In field trials, aldicarb at 0.5 and 1.0g and carbofuran at 1g/tea seedling were found effective for the control of *M. incognita* and also caused significant increase in plant height (Basu and Gope, 1985).

(iv) Integrated methods: Combination of neem cake at 5 g/seedling + carbofuran 3G at 1.5 g/seedling improved plant growth characters and reduced number of galls, egg masses and eggs/egg mass (Kalita and Bora, 2006).

(b) Management under field conditions

(i) Cultural methods: The first step in the management of nematodes in replanted areas is by ensuring a very thorough uprooting of all old tea roots (up to pencil thickness). In Sri Lanka, such uprooted old tea lands are routinely reconditioned by planting deep-rooted grass species including Guatemala grass and Mana grass which are non-hosts to parasitic nematodes and bring down the residual nematode population (Kerr and Vythilingam, 1966).

Shade trees and green manure crops help to provide shade to the tea plants, aid in N fixation and provide regular supplies of mulch in the form of leaf litter and loppings which help in reducing the nematode population during their decomposition. The cultivation of shade trees and green manure crops susceptible to tea nematodes should be discouraged in tea plantations.

Plants such as marigolds (*Tagetes erecta* and *T. patula*), *Vetiveria zizanoides* and *Eragrostis curvula* have been found to help reduce nematode populations.

Large inputs of organic matter including cow dung and well decomposed plant residues have been reported to suppress nematode populations. In India, neem cake has been recommended to manage root-knot nematodes in the tea field (Muraleedharan, 1991). In Sri Lanka, castor oil cake has been proved to be the most effective one for suppressing nematodes, followed by mahua, neem and karanj cakes (Gnanapragasam, 1991).

(ii) Chemical methods: The recommended nematicides for the management of root-knot nematodes in mature tea include phenamiphos (Nemacur 5% G) and carbofuran (Furadan or Curaterr 3% G) at 10 g/plant applied after pruning (Gnanapragasam, 1987).

32.3. COCONUT, Cocoas nucifera

32.3.1. Burrowing nematode, *Radopholus similis*

The burrowing nematode was reported from coconut palms in Kerala, India by Weischer (1967). *R. similis* is the most important nematode pest of coconut and is responsible for considerable amount of root rotting (Koshy *et al.*, 1978). It causes 30% yield loss in coconut (Koshy and Geetha, 1992). The threshold inoculum density required to cause significant reduction in various growth parameters of coconut is 100 nematodes/625 ml sandy loam soil over a period of 5 years under field conditions (Koshy, 1986).

32.3.1.1. Symptoms: The burrowing nematode infested coconut palms exhibit general decline symptoms like yellowing, button shedding, reduction in number and size of leaves and leaflets, delay in flowering and reduced yield which are non-specific. Symptoms on the root are more specific. *R. similis* on infestation produces isolated elongate orange colored lesions on tender and semi-hard roots (Fig. 32.5). Consequent to nematode parasitization and multiplication, these lesions enlarge and coalesce to cause extensive rotting of roots. Tender roots on heavy infestation become spongy in texture. On semi-hard orange colored roots, surface cracks are commonly seen. Nematode populations as high as 4,000 were recovered from one gram (one inch length) of main roots. The drastic reduction in the number and mass of tertiary feeder roots on parasitization by the nematode limits plant growth (Koshy and Sosamma, 1987).

Fig. 32.5. Burrowing nematode symptoms on coconut plant and roots

32.3.1.2. Integrated management

(a) Botanicals and chemicals: Thirty per cent increase in yield and 5 to 10% decrease in disease indices of palms affected with root wilt disease was recorded by the application of *Hydnocarpus* oil cake and phorate granules at 10g a.i./palm in June-July and in October- November.

(b) Cultural, botanicals and chemicals: Application of farmyard manure or oil cakes, cultivation of *Crotolaria juncea* in the basin and interspaces (and used as green manure), and application of phorate 10G at 100 g/palm/twice a year (during May-June and September-October) is effective against burrowing nematode.

The following INM practices have been recommended for the management of burrowing nematodes in coconut:

- Application of cow dung, FYM, oil cakes and green manure to the basins. *Crotalaria juncea* may be cultivated in the basins and interspaces and used as a green manure.
- Application of phorate at 10g a.i./palm twice yearly.
- Avoiding banana as a shade crop in coconut nurseries.
- Use of nematode-free planting material of coconut and other intercrops.
- Use of less susceptible/tolerant cvs. such as Kenthali and Klappawangi or their hybrids in infested areas.

32.4. BLACK PEPPER, *Piper nigrum*

32.4.1. Burrowing nematode, *Radopholus similis*

The burrowing nematode, *R. similis* is found associated with "Slow-wilt" disease of black pepper in Kerala and Karnataka, India (Mohandas and Ramana, 1987; Ramana *et al*., 1987). Mohandas and Ramana (1991) reported that *R. similis* was responsible for 59% loss in yield of black pepper.

32.4.1.1. Symptoms: *R. similis* on black pepper is associated with pepper yellows (slow-wilt) disease, which appears as pale yellow or whitish-yellow drooping leaves on the vines. The number of such leaves increases gradually until large numbers of leaves, or even the entire foliage, become yellow (Fig. 32.6). Yellowing is followed by shedding of leaves, cessation of growth and dieback symptoms. The symptoms are well pronounced when soil moisture is depleted. Within 3-5 years of initiation of yellowing, all the leaves are shed and death of the vine takes place; hence the name slow-wilt disease.

The tender, thin, white feeding roots show typical orange- to purple-colored lesions (Fig. 32.6). Lesions are not clearly seen on older roots, being brown in color. The root system exhibits extensive rotting and this result in a lack of fine feeder roots from the main roots. Extensive necrosis of larger lateral roots develops subsequently.

Fig. 32.6. Yellows disease of black pepper plants caused by *R. similis* destroying roots.

32.4.1.2. Life cycle: The nematode completes its life cycle within 25-30 days at temperature range of 21-31°C. The black pepper isolate of *R. similis* can be cultured axenically on carrot discs (Koshy, 1986).

32.4.1.3. Host range: Crops like coconut and areca nut which are commonly used as live standards are good hosts of *R. similis*. Intercrops like banana, ginger, turmeric and cardamom are also susceptible to *R. similis* infestation.

32.4.1.4. Interaction with other pathogens: Yellows disease in Indonesia is caused by a nematode-fungus complex involving *R. similis* and *Fusarium* spp. and possibly other fungi. *R. similis* predisposed black pepper seedlings to attack by a weakly pathogenic isolate of *Fusarium solani* causing severe root damage.

32.4.1.5. Management

(a) Cultural methods: Use of healthy and nematode-free planting material can ensure better establishment, survival and higher yield. Soil solarization of nursery sites or potting mixture and random rotation of nursery sites are reported to be effective in black pepper. Mulching (with Guatemala grass) has an ameliorating effect on the symptoms of slow wilt/pepper yellows (*R. similis*). Planting nematode-free rooted cuttings, uprooting of affected vines and replanting after a period of 9-12 months, use of non-living standards/supports, avoiding *R. similis* susceptible intercrops like banana, ginger, turmeric and exclusion of susceptible trees as standards for trailing the vines reduces the nematode population. Application of organic amendments like neem cake (200 g), green foliage (3-5 kg) or FYM (1 kg per vine) and earthing up in September-October are recommended. Growing marigold as a trap crop and uprooting the trap crop at flowering stage and burning reduce the incidence of nematodes. *R. similis* was not encountered when black pepper was intercropped with coffee (Venkitesan, 1976).

Application of fertilizers at 400 kg N, 180 kg P, 480 kg K, 425 kg Ca and 112 kg Mg in combination with mulch gave effective control of yellows disease.

(b) Chemical methods: Aldicarb/carbofuran/phorate each at 3g a.i./vine applied in May-June and again in September-October resulted in reduction of the nematode population and remission of foliar yellowing. Among the above three nematicides, phorate proved to be the best (Ramana, 1986; Mohandas and Ramana, 1987). Aldicarb sulfone at 8 kg a.i. /ha gave best control of *R. similis* followed by fensulfothion (Venkitesan, 1976).

(c) Integrated management

(i) Cultural, botanicals and chemicals: Integrated methods of the burrowing nematode management that can be suggested include:

- Planting of nematode-free rooted cuttings.
- Uprooting of affected vines and replanting after a period of 9-12 months.
- Use of non-living supports or standards.
- Exclusion of *R. similis*-susceptible trees as standards for trailing black pepper vines.
- Exclusion of susceptible intercrops such as banana, ginger and turmeric.
- Application of organic amendments, such as neem oil cake, green foliage, or farm-yard manure.

- Earthing-up after application of nematicides, NPK fertilizers and organic amendments.

32.4.2. Root-knot nematodes, *Meloidogyne* spp.

Butler (1906) reported root-knot nematodes from black pepper in Wynaad, Kerala, India. *M. incognita* and *M. javanica* have been reported from India. *M. incognita* was responsible for 46% loss in yield of black pepper (Mohandas and Ramana, 1991).

32.4.2.1. Symptoms: Prominent symptoms of root-knot infestation on black pepper are unthrifty growth and yellowing of leaves. Inter-veinal yellowing of the foliage is also noticeable. The leaves exhibit dense yellowish discoloration of the inter-veinal areas making the leaf veins prominent with deep green color. Heavy galling of the root system is also present (Fig. 32.7).

Fig. 32.7. Black pepper vine and roots infected with root-knot nematodes

32.4.2.2. Host range: Among commercially used standards, *Oroxylon indicum, Erythrina lithosperma, Ceiba pentandra, Bombax malabaricum* are highly susceptible to root-knot nematodes (Koshy *et al*., 1977).

32.4.2.3. Life cycle: The life cycle under optimum conditions takes 3-4 weeks.

32.4.2.4. Interaction with other pathogens: *M. incognita* and *Fusarium solani* were found associated with black pepper vines. Both the organisms together were found to do more harm than either of them alone. Increased susceptibility of *M. incognita* and *M. javanica* infested pepper to *Phytophthora* infection has been reported.

32.4.2.5. Management

(a) Cultural methods: Live standards for black pepper such as *Artocarpus heterophyllus, A. hirsutus, Ailanthes malabarica, Mesopsis emini, Peltophorum* *pterocarpum, Swietenia macrophylla, Tamarindus indica, Garuga pinnata* and *Macaranga peltata* are resistant to *M. incognita*.

Growing of non-host cover crops like Siratro (*Macroptilium atropurpureus*) in the interspaces and mulching with Guatemala grass (*Imperata cylindrica*) are recommended to reduce the root-knot nematode (*M. incognita*) population on black pepper in the Amazonian region (Ichinohe, 1980).

Application of neem cake reduces nematode populations followed by increase in yield.

(b) Chemical methods: Fumigation of nursery potting mixture with methyl bromide is effective in checking the infestation.

Green berry yields can be doubled by four applications of carbofuran incorporated into mound soil at 114 g/vine per application in black pepper fields infested with *M. incognita* and *M. javanica*. Application of aldicarb at 1.25g a.i./vine or carbofuran at 10g a.i./vine twice a year, including at planting around cuttings, can reduce populations of *M. incognita* on black pepper. Phenamiphos at 1g a.i./vine followed by carbofuran and ethoprofos was effective in controlling nematodes.

(c) Biological methods

(i) Antagonistic bacteria: Root-knot nematodes in pepper can be managed by the application of *Bacillus macerans* at 1.2 x 10^8 cells/vine just before the monsoon period.

(ii) Antagonistic fungi: Application of *Paecilomyces lilacinus* and *Pochonia chlamydosporia* has potential to control nematodes which reduced the rate of infection to 15 and 12%, respectively. Soil application of *P. chlamydosporia* reduced nematode population and increased the yield (5.14 kg/vine) with cost: benefit ratio of 1: 7.12.

(iii) Arbuscular mycorrhizal fungi: AMF such as *Glomus fasciculatum, G. etunicatum* and *Acaulospora laevis* reduced root-knot nematodes and enhanced the growth of black pepper. Pre-inoculation of *A. laevis* in black pepper cv. Panniyur-1 reduced the root-knot index (0.75) followed by *G. mosseae/ Gigaspora margarita* (1.25) compared to control (3.5) (Anandraj *et al*., 1991).

(d) Host resistance: The root-knot nematode infestation was less in cvs. Veliakaniakadan, Karimunda, Kalluvalli, Balankotta, Narayakodi and Padapan.

Black pepper cv. 'Pournami' released by Indian Institute of Spices Research, Calicut, India, showed resistant reaction to *M. incognita* with a potential yield of 17,280 kg green berries/ha and has high oleoresin content (13.8%).

(e) Integrated management

(i) Two bioagents: Black pepper vines combinedly inoculated with *P. lilacinus* and *Pastueria penetrans* had put out 23-112 per cent increase in plant growth over control and were very effective in the management of root-knot nematodes (Sosamma and Koshy, 1995).

Under field conditions, plants treated with the consortial formulation of *Pseudomonas fluorescens* Pf 123 and *Bacillus subtilis* Bs 214 significantly enhanced the yield parameters and reduced nematode infestation both in soil and roots (Devapriyanga *et al.*, 2012).

(ii) Botanicals and chemicals: Integration of neem cake application at 1 kg/vine along with phorate/carbofuran at 3g a.i./vine during May-June and again during September-October gave effective control of nematodes infesting black pepper.

(iii) Cultural and chemicals: Application of aldicarb at 1g a.i./vine twice a year (May/June and October/November) integrated with fertilizers (N-100g, P- 40g, K-140g/vine) in two equal splits, earthing up to 50 cm radius at the base of the vines and mulching the base of the vines with leaves reduces foliar yellowing by 83% and *M. incognita* population by 33-38% (Venkitesan and Jacob, 1985).

(iv) Physical, bioagents, botanicals and AMF: Integrated management of foot rot (*P. capsici*) and nematodes (*M. incognita* and *R. similis*) on black pepper was achieved by (i) mixing AMF and *Trichoderma harzianum* in solarized nursery mixture to raise healthy and robust seedlings, (ii) application of *T. harzianum* and FYM in planting pit, (iii) field application of neem cake at 1 kg/vine mixed with 50 g of *T. harzianum* during August.

Chapter 33

COMMERCIAL CROPS

33.1. TOBACCO, *Nicotiana tabacum*

The root-knot nematodes are the most important nematode pests which cause serious losses to tobacco in India.

33.1.1. Root-knot nematodes, *Meloidogyne incognita; M. javanica*

The root-knot nematodes are economically the most important nematode pests of tobacco. The species most commonly recorded on tobacco are *Meloidogyne incognita* and *M. javanica*, the former occurring as the predominant species in temperate and the latter in subtropical and tropical zones. These species quite often occur together in the same field.

33.1.1.1. Symptoms: The above-ground symptoms are those typical of drought and nutrient deficiency. The plants are dwarfed or stunted in growth, wilt prematurely and the leaves are yellow and thin, scorching easily and eventually turning brown. *Meloidogyne* infestations occur typically as patches in tobacco plantations and can be recognized by the yellowing and wilting of the plants in these areas. The diagnostic symptoms of a *Meloidogyne* attacks on tobacco are the galls found on roots (Plate 33.1).

Plate 33.1. Tobacco plants and roots infected with *Meloidogyne javanica* showing wilting symptoms.

33.1.1.2. Life cycle: Species of *Meloidogyne* are sedentary root endoparasites. The life cycle on tobacco takes from 21 days in summer up to 56 days or more during cool weather.

33.1.1.3. Interaction with other pathogens: Sasser *et al.* (1955) conducting experiments with two black shank (*Phytophthora parasitica* var. *nicotianae*) resistant varieties (Dixie Bright 101 and 102) showed that in the presence of *M. incognita* and *M. javanica*, a high mortality due to black shank occurred. Infestation by the nematode is in fact obligatory for the development of the disease on black shank resistant varieties.

The root-knot nematode species *M. incognita* and *M javanica* predispose tobacco root tissue to attack by *Fusarium oxysporum* var. *nicotianae* wilt (Melendez and Powell, 1967).

Neither *Pythium ultimum* nor *Trichoderma harzianum* cause root decay of tobacco in the absence of nematodes, but when infection with *Meloidogyne* occurs four weeks before inoculation, tobacco suffers severe root decay from a combination of these soil pathogens.

Leaves of root-knot infested plants are more susceptible to attack by aerial pathogens such as *Alternaria tenuis* (brown spot), *Colletotrichum destructivum* (anthracnose), *Cercospora* leaf spot and *Erysiphe cichoracearum* (powdery mildew). Furthermore, a combination of root-knot nematode and Fusarium has been shown to increase considerably the occurrence of brown spot.

In experiments in which the bacterial wilt organism (*Ralstonia solanacearum*) and root-knot nematode (*M. incognita*) were added to tobacco plants, the infection with nematodes led to injury of the roots and consequent earlier and more severe wilt infection.

33.1.1.4. Ecology, spread and survival: *Meloidogyne* damage to tobacco is usually associated with light soils or peats.

The root-knot nematodes are spread through infected seedlings from one field to the other. The nematode can also spread through tillage implements, labors' feet, farm animals, and irrigation water.

The root-knot nematode can survive on volunteer weeds and grasses which are often good hosts. In Rhodesia, *M. javanica* was found viable in tobacco soils after more than four years of clean fallow.

33.1.1.5. Management

(a) Physical methods: Burning of 24 kg of rice husk per sq. m. on tobacco seed beds effectively frees soil of nematodes.

Pulling out of infested plants in sandy soils immediately after harvest so as to expose the roots to sun's heat significantly reduced the *M. incognita* population.

(b) Cultural methods: Crop rotation with cotton, maize, groundnut, marigold and sun hemp is useful in reducing the root-knot nematode population.

(c) Chemical methods: Spot application of aldicarb, oxamyl or fensulfothion, all at 2 kg a.i. per ha gives effective control of root-knot nematodes.

Oxamyl can also be applied as foliar spray which is translocated downwards from the leaves to the roots where it prevents invasion by *Meloidogyne* spp.

(d) Resistant varieties: The tobacco variety 'NC 95' is resistant to *M. incognita.* It also carries resistance to black shank, *Fusarium* wilt and to a certain extent bacterial wilt and brown spot. The mechanism of resistance of 'NC 95' to *M. incognita* apparently depends on a hypersensitive host root reaction. New strains of *M. incognita* were found which attack and breeds on 'NC 95' (Parvatha Reddy *et al.,* 1972a). Unfortunately 'NC 95' is not resistant to *M. javanica.* Other tobacco varieties resistant to *M. incognita* include 'Florida 22' which is, however, susceptible to black shank and bacterial wilt, and the Japanese variety 'RK 70' on which *M incognita* breeds very slowly. There is no commercial variety of tobacco which is resistant to *M. javanica.*

Tobacco cultivars Coker 86, 254, 258 and 347, Ga 1469, NC 79, 88, 98 and 2512, Speight G 23, G 28, G-33 and G-41, Virginia 080 and 770 were also found resistant to *M. incognita.*

33.2. COTTON, *Gossypium* spp.

Root-knot and reniform nematodes are the major cotton pests in loamy and clay soils of India.

33.2.1. Root-knot nematodes, *Meloidogyne incognita*

Root-knot nematodes cause varying degrees of damage to cotton. They are damaging to cotton as a single pest problem and as part of the Fusarium wilt race 1 and race 4 disease complex.

33.2.1.1. Symptoms: The most common symptoms in cotton would include stunting, yellowing, and wilting of plants (Fig. 33.2). The plants may be very irregular in plant height across a field or within certain areas. Premature death may occur late in the season resulting in cotton maturing at different times within a field. The root-knot nematode produces galls that are clearly visible on the roots making this nematode easily identifiable. Galling can be visible as early as 30-40 days after planting but may be detected anytime during the growing season (Fig. 33.2). Yield is impacted by nematodes and can range from very slight to 40-50% losses or more in parts of a field.

33.2.1.2. Management: Good field sanitation helps prevent infestations from spreading, and weed control is important in eliminating weed hosts supporting root-knot nematodes. If injury is severe enough and if the affected area is extensive, soil fumigation may be needed, or it may be necessary to plant a resistant cotton variety or a resistant variety of another crop such as tomato or alfalfa.

Fig. 33.2. Cotton plants and roots infected with root-knot nematodes

(a) Cultural methods: Clean fallowing (i.e., weed-free) during years when land is to be left unplanted is effective in controlling root-knot nematodes on cotton, but clean fallow can reduce beneficial mycorrhizal fungi to crop-damaging levels if the land is left plant-free for more than one year.

Crop rotation is effective in managing cotton root-knot nematode populations. Certain grass crops or pastures, grain sorghum, and peanut can effectively reduce cotton root-knot after a 2 to 3 year period. For successful control of nematodes in a rotation program, susceptible weeds must be controlled completely in the rotation crop. Rotation crops that help to reduce cotton root-knot nematodes include alfalfa, winter small grains, peanut, root-knot resistant cultivars of processing tomatoes, resistant cowpea (cvs. California Blackeye CB 46, CB 27, CB 50, and CB 5), sorghum and Sudan grass.

(b) Chemical methods: When nematodes are at levels requiring treatment, aldicarb at planting, can increase cotton yields.

(c) Host resistance: Field studies conducted indicate that varieties such as Fibermax 2011GT, Deltapine 174RF, Deltapine 1454RN B2RF, Phytogen 367WRF, and Stoneville 4946GLB2 possess root-knot resistance. Furthermore, Phytogen 417WRF greatly reduces nematode reproduction and may be an option for fields that are severely infested.

A resistant variety of Acala cotton, NemX HY, developed in the San Joaquin Valley of California, suppresses initial infection by southern root-knot nematode and thus reduces yield loss of cotton plants grown in infested soil. Because reproduction of the southern root-knot nematode on roots of NemX HY is limited, the population of this nematode in the soil in fall is much lower than the level in soil following susceptible cotton varieties.

The use of NemX HY is especially effective for managing root-knot nematode in rotations with other nematode susceptible crops. In addition to resisting the southern root-knot species, *Meloidogyne incognita,* it also acts as a non-host for the other common species of root- knot nematodes, including *M. arenaria, M. hapla,* and *M. javanica.* Therefore, the resistant cotton can effectively prevent the buildup of all the root-knot nematode species likely to be encountered; this will relieve the nematode infection and damage potential on other crops grown in rotation.

33.2.2. Reniform nematode, *Rotylenchulus reniformis*

The reniform nematode was first identified as a parasite of cotton by Steiner (Smith, 1940). Seshadri and Sivakumar (1963) have reported this nematode on cotton for the first time from India. In Louisiana, USA, *R. reniformis* was found associated with estimated reductions in cotton yield as high as 40-60 per cent in several thousand hectares. In addition to reducing cotton yield, this nematode also caused a delay in maturity, a reduction in size of boll and in some years a reduction in lint percentage. In India, the reniform nematode causes 10-15 per cent loss in cotton yield (Palanisamy and Balasubramanian, 1981).

33.2.2.1. Symptoms: The reniform nematode (Fig. 33.3) causes severe pruning of seeding roots and consequently stunted growth. Stunting, reduced yields and delayed maturity are common symptoms of damage caused by reniform nematodes. Under favorable conditions, damaged cotton may exhibit only slight stunting with fewer bolls. If heavily stressed, however, plants can be badly stunted and show symptoms of severe potassium deficiency.

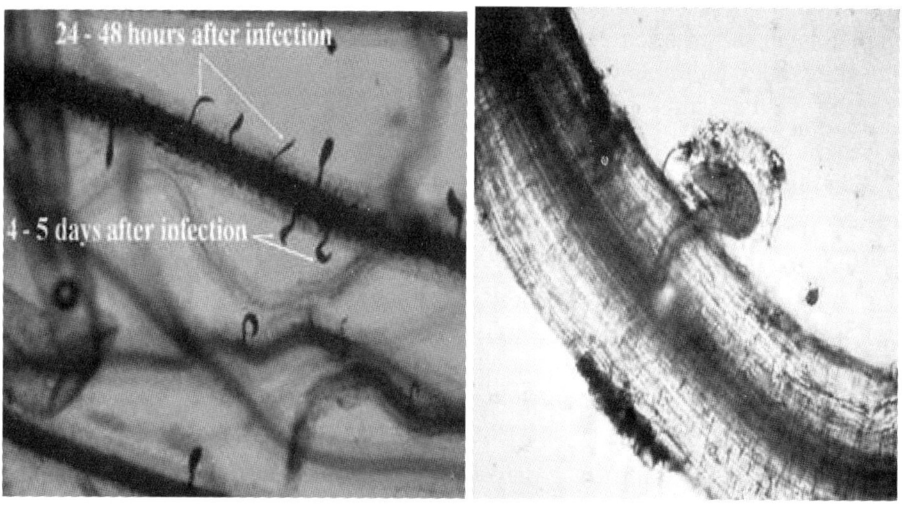

Fig. 33.3. Cotton root infected with reniform nematodes

33.2.2.2. Life cycle: The adult female of *R reniformis* is an obligate, sedentary, semi-endoparasite of roots while the male is non-parasitic. The species is bisexual and reproduces by amphimixis. The complete life cycle takes about 17 to 23 days on cotton. After infecting the roots the young female feeds, grows, becomes obese and starts laying eggs in 7 to 9 days; an average of 66 eggs per egg mass is recorded on cotton.

33.2.2.3. Host range: Besides infecting cotton, the reniform nematode also infects bananas, citrus, papaya, pineapple, mango, jack fruit, coffee, tea, clove, tobacco, potato, tomato, okra, cowpea, bean, cabbage, cauliflower, carrot, pea, lettuce, radish, cucumber, brinjal, groundnut, sugarcane, rice and maize.

33.2.2.4. Interaction with other pathogens: Neal (1954) found that the presence of the reniform nematode is an important factor in the incidence of *Fusarium* wilt of cotton in the region of Baton Rouge, Louisiana. In a glasshouse experiment, 81.4 per cent of the plants of a susceptible variety (Half and Half) developed wilt when both the nematode and the fungus were present in the soil, as against 10 per cent when only the fungus was present. However, the presence of the nematode did not appreciably increase the incidence of wilt in a wilt resistant variety (Delfos 425-920). *R. reniformis* is also an important factor in the incidence of Verticillium wilt of cotton (Birchfield, 1962).

33.2.2.5. Survival: *R. reniformis* survived in air dried soil (3.3 per cent moisture) stored for 7 months at 20-25°C.

33.2.2.6. Physiological races: At least two races, A and B, have been detected in India, the former multiplying on cowpea, castor and cotton, while the latter reproducing only on cowpea (Dasgupta and Seshadri, 1971).

33.2.2.7. Management

(a) Chemical control: Aldicarb lowered the reproduction rate of the nematode throughout the growing season and increased the cotton yield by almost 40-60 per cent (Oteifa, 1970). Spot application of aldicarb and carbofuran each at 1.0 kg a.i. per ha 15 days after sowing reduced nematode population and increased the yield by about 11 to 32 per cent.

(b) Resistant varieties: Resistant varieties of cotton such as Auburn 56, Plains, Delfos 425-920, Delfos 2169, Deltapine 6, Louisiana 33, Roxie, Stoneville 2B strains and Coker 100 wilt, have been shown to withstand the combined attacks of *Fusarium* and the reniform nematode.

(c) Crop rotation: Several non-hosts such as mustard, oats, onion, chili, rice, sorghum, sugarcane, turnip, peanuts, wheat, corn, rice, peas, cabbage, and cauliflower have been reported for *R. reniformis.* Growing of non-host crops for two years between susceptible crops may give fair to good control of the reniform nematode infecting cotton. A one- or two-year rotation with grain sorghum, corn, peanuts and other non-host crops is as effective as using nematicides.

33.3. CASTOR, *Ricinus communis*

The reniform nematode, *Rotylenchulus reniformis* is a major pest of castor in India.

33.3.1. Reniform nematode, *Rotylenchulus reniformis*

The reniform nematode was first reported on castor by Seshadri and Sivakumar (1963) from Tamil Nadu, India. The nematode infested plants producing seeds of inferior quality containing lesser amounts of oil. The quality of castor oil is also affected by nematode infestation.

33.3.1.1. Symptoms: The reniform nematode causes growth reduction, shedding of leaves, early flowering, malformation and discoloration of seeds. Seshadri and Sivakumar (1963) reported dieback symptoms and stunting in heavily infested castor fields.

33.3.1.2. Life cycle: The life cycle of *R. reniformis* on castor is studied in detail by Sivakumar and Seshadri (1971). The life cycle takes about 24-29 days from egg to egg. The nematode secretes a gelatinous matrix through the vulval opening in which about 83 eggs are laid.

33.3.1.3. Ecology: Soil moisture from 25 to 30 per cent and 30°C soil temperature were optimum for its reproduction on castor (Khan and Khan, 1973). Sivakumar and Seshadri (1972) found that while there is good multiplication of the nematode in sandy loam, brown loam and black clay loam, the symptom expression was more severe in black clay loam. Mukhopadhyaya and Haque (1980) reported that laterite (sandy clay loam) and alluvial (clay loam) soils supported maximum number of reniform nematodes while sandy soil supported the least. Vigorous plant growth in these soil types provided the nematodes with larger feeding areas which might account for higher rate of nematode multiplication.

33.3.1.4. Host range: Ayala and Ramirez (1964) compiled the literature on host range of *R reniformis* and reported 201 species of host plants of which 74 were new records. The reniform nematode is widespread throughout India.

33.3.1.5. Survival: Sivakumar and Seshadri (1976) reported that the reniform nematode (juvenile female) lives for about six months in desiccated soil and seven months in moist soil. There is an indication that the fourth stage larvae are better adapted for survival under adverse conditions, being protected by the previous larval cuticles which are not shed.

33.3.1.6. Management

(a) Crop rotation: Khan and Khan (1973) reported onion, garlic, turnip, bell pepper and carrot as non-hosts for *R. reniformis* in India. Growing of these non-host crops for two years between castors gives fair to good control of the reniform nematode.

(b) Resistant varieties: Nandwana *et al.* (1981) tested 70 varieties of castor against the reniform nematode and found only one variety. GH-3 to be highly resistant and 21 cultivars were tolerant.

■■■■

Chapter 34

FIELD CROPS

34.1. WHEAT, *Triticum* spp.

Two species of nematodes are of major importance on wheat at present - the cereal cyst nematode causing the `molya' disease, and the seed gall nematode causing the 'ear cockle' and 'tundu' diseases.

34.1.1. Cereal cyst nematode, *Heterodera avenae*

The cereal cyst nematode was first observed as a parasite of cereals in Germany by Kuhn (1874). Prasad *et al.* (1959) reported this nematode on wheat for the first time from India. The nematode can cause 6.3 to 95.3 per cent loss in wheat in India (Seshadri and Sethi, 1978), 60 per cent loss in Netherlands and 30 to 40 per cent loss in England (Gair, 1965). The cereal cyst nematode causes 'molya' disease of wheat and barley.

34.1.1.1. Symptoms: The affected fields usually give a patchy appearance. With continuous cropping of wheat or barley, such patches gradually spread and may cover the whole field in 3 to 4 years. Damage is most severe during dry seasons. Besides stunted growth and general chlorosis, the infected plants have stiffer, thinner and narrower leaf blades than the healthy ones. Tillering is greatly reduced, with culms getting thinner and weaker. The diseased plants flower prematurely, and the ears, if formed have very few grains.

The infected roots are short, with multiple branches, giving a bunchy appearance and often bearing small gall like formations. Under Rajasthan conditions, large numbers of cysts show on the roots in February (Plate 34.1). The fibrous root system is almost negligible. Such plants can be easily pulled out of the ground.

34.1.1.2. Life cycle: The nematode completes its life cycle within 9 to 14 weeks after larval invasion.

34.1.1.3. Host range: The host range of *H. avenae* is restricted to wheat, barley, oats, rye, spelt, maize and other graminaceous plants. The only non-graminaceous host recorded is *Senebiera pinnatifida,* belonging to Cruciferae.

34.1.1.4. Ecology: The critical threshold population was determined to be 2 larvae per 1 g of light soil. Gair (1965) showed that an initial inoculum level of

40 cysts per 100 g soil (32 eggs per g of soil) caused 30 to 40 per cent reduction in wheat yield in England.

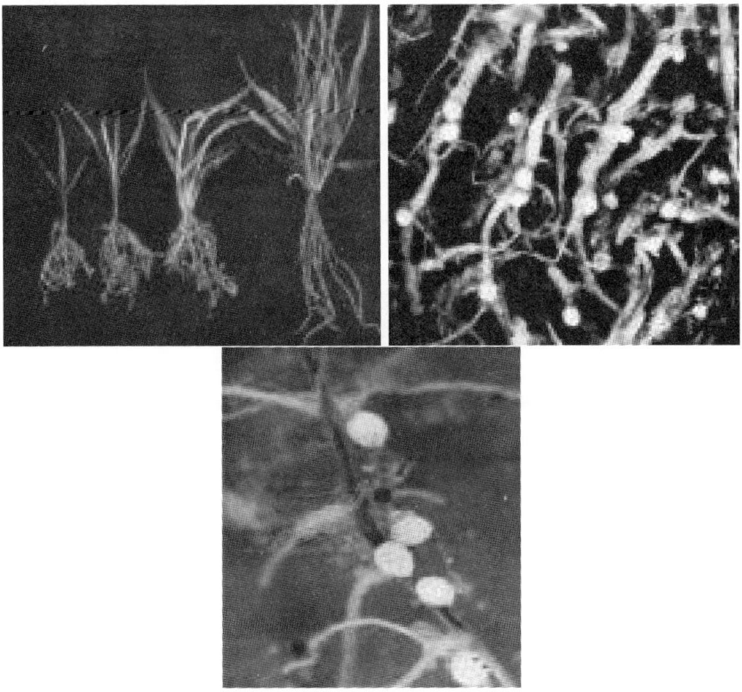

Plate 34.1. Wheat plants and roots infected with *Heterodera avenae*

34.1.1.5. Spread and survival: The cereal cyst nematode can spread through cultural operations, implements, labors' feet, and irrigation water.

The cysts can remain viable in the soil without a host for several years.

34.1.1.6. Physiological races: Studies conducted at Jaipur (Mathur *et al.,* 1974) indicated the presence of at least 5 biotypes in Rajasthan, characterized by the following host preferences:

- Biotype 1: Cysts formed on *Lolium perenne*, but not on barley 191, Herta, Drost, Sun II oat and *Dactylis glomerata*.
- Biotype 2: Cysts on barley 191 and Herta but not on Drost, *L. perenne, D. glomoerata* and Sun II oat.
- Biotype 3: Cysts on Herta but not on barley 191, Drost, *D. glomerata, L. perenne,* and Sun II oat.
- Biotype 4: Cysts on Herta and *D. glomerata* but not on barley 191. Drost. *L. perenne* and Sun.II oat.
- Biotype 5: Cysts on Drost and Herta but not on barley 191, *L. perenne, D. glomerata* and Sun II oat.

34.1.1.7. Management

(a) Crop rotation: As the cereal cyst nematode is highly host specific, there is much scope for adopting crop rotation as a method of control. Raising non-host crops like carrot, fenugreek, onion, mustard and gram have shown that the nematode population gets reduced from 47 to 55 per cent at the end of one year and up to about 75 per cent by the end of second year. Wheat yield increased by 83 per cent after one year and by 135 per cent after two years of crop rotation.

(b) Chemical methods: Aldicarb and Carbofuran each at 2 kg a.i. per ha gave 199.8 and 150.6 per cent increase in wheat yield. Carbofuran at 1.5 kg/ha in the soil also controls the disease to greater extent.

(c) Resistant varieties: Wheat varieties like Kalyan Sona and PV 18, and barley variety JB 226 were perhaps tolerant and could be raised on infested soil since they gave comparatively higher yields in spite of the high initial inoculum level of 200 cysts/kg soil (Mukhopadhyaya *et al.,* 1972).

(d) Integrated management: Integration of fenamiphos (4.0 kg a.i. /ha) and urea (600 kg/ha in three split doses) resulted in the best control. The combined application of urea and fenamiphos resulted in the most significant effect in decreasing the number of cysts/root system and increasing the growth of nematode-infected wheat plants (Ahmad *et al.,* 2014).

34.1.2. Seed gall nematode, *Anguina tritici*

The seed gall nematode was the first plant parasitic nematode reported by Needham as early as in 1743. This nematode is still one of the most important plant parasites of wheat in India. Besides being responsible for the ear-cockle disease, the nematode is often associated with a bacterium, *Rathayibacter tritici* in causing a yellow ear rot commonly referred to as the `tundu' disease (Gupta and Swarup, 1972). The sowing seed contamination with galls at 2.5, 6.5 and 8.5 per cent of total weight resulted in yield decreases of 30, 54 and 69 per cent, respectively.

34.1.2.1. Symptoms: Infested seedlings show slight enlargement of the basal part of the stem. Frequently, the leaves emerging from these seedlings are twisted and crinkled, often folded, with their tips held near the growing point. Severely infested seedlings are stunted and may even die prematurely. The infested plants generally show profuse tillering and may produce ears almost 30 to 40 days in advance of the healthy plants. The affected ears are usually shorter. Glumes are spread farther apart by the galls, which replace the kernels. Seed galls are smaller, proportionately shorter and of a darker color compared to healthy seeds (Plate 34.2).

Fig. 34.2. Symptoms of *Anguina tritici* infestation on wheat seedling (seedling leaves twisted and crinkled), ear heads (*Left* - healthy, *Right*-infected) and seeds (*Left*-infected, *Right* - healthy)

The `tundu' disease is characterized by the production of a light-yellow slime or gum on the leaf surfaces of young plants, and also on the abortive ears

(Fig. 34.3). This yellow slime, which can be seen trickling down the tissues in humid weather, becomes hard, brittle and brown on drying. Culms which are infested either die at the young stages or grow until heading. In the latter case either the emerging spike is narrower and shorter, with the grains partially or completely replaced by the bacterial mass, or it fails to emerge out of the boot-leaf. The stalk is always distorted when the ear shows the bacterial symptoms.

Fig. 34.3. Symptoms of 'tundu' disease of wheat (twisting and distortion of ear head and rotting of spikelets with profuse oozing of yellow liquid from the infected tissue)

34.1.2.2. Life cycle: The seed galls falls to the ground during harvest and breakdown in the soil to release second stage larvae. The number of larvae in a gall depends upon its size and may vary from 3,000 to 12,000, with an average of 6,000. These invade seedlings of the host plant, making their way up the shoot to live ectoparasitically round the growing point and its leaf sheath. When the embryonic flower tissues form, the nematodes invade them and change from ectoparasitic to an endoparasitic existence. They develop into adults in the developing seed, which becomes a gall. Mating occurs and the females lay thousands of eggs. The adults die and the eggs hatch. The larvae molt once in the gall, so that when they leave it to infect new plants they are in the second stage. The second stage larvae in seed galls can live for many years.

34.1.2.3. Host range: Wheat, rye, emmer (*Triticum dicoccum*) and spelt (*T. spelta*) are susceptible to injury by the seed gall nematode. For all practical purposes oats and barely are immune.

34.1.2.4. Spread and survival: The most important method of spread and the only one whereby this nematode is likely to be carried for any considerable distance is by galls in seed grains.

The second stage larvae can live for many years in dried seed galls. Needham (1743) claimed to have revived the larvae obtained from 25 years old galls. Fielding (1951) reported having revived larvae from galls that had been in storage for 28 years.

34.1.2.5. Management

(a) Regulatory methods

(i) Dry Cleaning of seeds or brine flotation: The source of inoculum for both ear cockle and `tundu' diseases being the seed material contaminated with nematode galls, the use of clean seed free of nematode galls is the surest way to control the diseases. This can be achieved by adopting dry-cleaning methods like sieving and winnowing, or by seed floating in plain water, or, preferably, in 20 per cent brine solution.

(ii) Seed certification: Seed supplying agencies, both in the private and public sectors, have an important role in preventing the recurrence of the disease by ensuring the distribution of seed certified to be free from nematode contamination.

(b) Physical methods

(i) Hot water treatment: The seed is pre-soaked for 4 to 6 hours in cold water and then treated for 10 minutes in hot water at 54°C.

(c) Cultural methods

(i) Crop rotation: Growing of non-host crops for one or two years is sufficient practically to eliminate nematodes remaining in the field after harvest.

(ii) Rogueing: In the field, infected plants can be easily detected in early stages and should be removed and destroyed.

(d) Host resistance: Use of resistant varieties like 'Kenred-hard' winter wheat is useful to control the seed gall nematode.

34.2. RICE, *Oryza sativa*

Investigations on the nematode diseases of rice have been concentrated on the five genera, *Aphelenchoides, Ditylenchus, Hirschmanniella, Meloidogyne* and *Heterodera*.

34.2.1. Root-knot nematode, *Meloidogyne graminicola*

The root-knot nematode, *Meloidogyne graminicola* prevalent in eastern, north-eastern and southern states of India has emerged as a serious threat for the successful rice production in nurseries, uplands, deep water and irrigated fields due to rice cropping intensification and increasing scarcity of water. In India, the losses due to *M. graminicola* have been estimated to be 16-32% and yield loss due to poorly filled kernels to be 17-30%.

34.2.1.1. Symptoms: Infested rice fields show patches of stunted and chlorotic rice plants with less vigor, yellowing and curling of leaves (Plate 34.4). The chlorophyll content of leaves is reduced (Swain and Prasad, 1988). The galls incited by *Meloidogyne* spp. on rice are terminal, spiral or horse shoe shaped (Plate 34.4). The crop has poor stand, less tillering and few number of effective tillers. The spikes become smaller in size and less in numbers with poorly filled grains, thus yield is reduced.

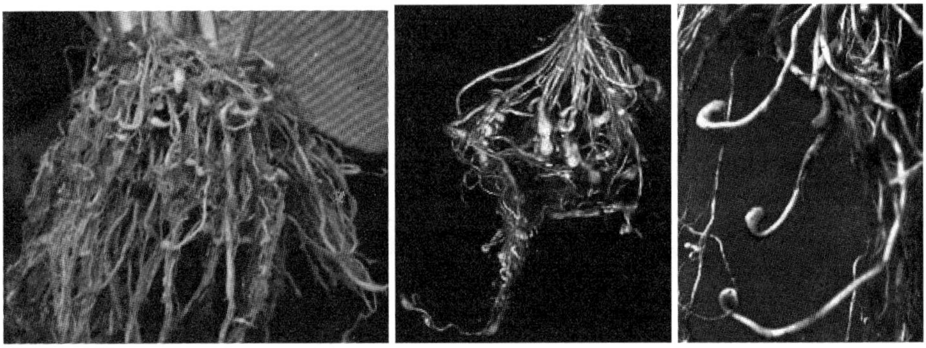

Plate 34.4. Rice roots showing terminal and horse-shoe galls caused by *Meloidogyne graminicola*.

34.2.1.2. Management

(a) Physical methods: The soil solarization in nursery bed may effectively reduce nematode populations. Prior to sowing, the nursery bed is covered with a plastic polyethylene tarp for 4-6 weeks. Soil solarization raises temperature by 4°C as compared to non-solarized soil to reach 40-50°C at 10 cm depth. The nematodes become undetectable in the upper 15 cm soil layer. It can give a reduction of *Meloidogyne* spp. population in rice beds by over 80% and improve seedling growth (Ganguly *et al.*, 1996).

(b) Cultural methods: Selection of root-knot nematode free area for nursery is important. The seedlings which are to be transplanted must be free from nematode galls.

(i) Crop rotation: Certain crops are resistant or poor hosts of *M. graminicola* and could be used in rotation to reduce nematode populations. Crop rotation with non-host crops like jute, mustard, chickpea, castor, cowpea, sweet potatoes, soybeans, sunflower, sesame, onion, cauliflower, turnip, beans, okra and resistant varieties (TKM-6, Patna 6, Dumai, Ch 47, Hamsa) reduces *M. graminicola* infestations (Rao *et al.*, 1984; Rao, 1985; Rao *et al.*, 1986). Long rotations, more than 12 months, will be needed to reduce *M. graminicola* soil populations to low levels. Introducing a fallow into the rotation will also give control of the nematodes but, to be effective, it needs to be a bare fallow free of weed hosts.

Rice following two crops of cowpea had increased yield by 34% when treated with carbofuran and by 26% in untreated plants. A single crop of cowpea increased the yield of the following rice crop by 85%. Cowpea cv. Iron Clay was shown to be an effective rotation crop to control some root-knot species because of its high degree of resistance. One or two consecutive crops of cowpea or seasons of fallow before a rice crop lowered nematode populations and improved rice yield by 30-80%. It is recommended that, to ensure higher rice yields, *M. graminicola* populations should be maintained at low density by non-host crop rotations or fallows, ideally for two seasons before planting rice (Soriano and Reversat, 2003).

An alternate rotation of rice with a non-host crop or a season of fallow seems to be the only solution at the moment for managing *M. graminicola* populations. Rice-mustard-rice crop sequence, followed by rice-maize-rice and rice-fallow-rice, were reported to be effective in reducing nematode development (Kalita and Phukan, 1990). *Tagetes* spp. and *Crotalaria* spp. (*C. incana* and *C. mucronata*) showed reduction in the population and development of *M. graminicola*. Planting *Tagetes* or *Crotalaria* species in nematode infested soil is therefore feasible and can be used in the overall management of *M. graminicola*.

(ii) Fallowing: One season of fallow (120 days) increased rice yield by 41% and two consecutive fallows (240 days) by only 31%. Two seasons of fallow, with carbofuran treatment before rice, proved to be the most efficient strategy to increase rice yields in the third season (Soriano and Reversat, 2003).

(iii) Flooding: *M. graminicola* will survive normal flooding but damage to the crop can be avoided by raising rice seedlings in flooded soils thus preventing root invasion by the nematodes (Bridge and Page, 1982). Continuous flooding is highly effective in controlling *M. graminicola* in Vietnam (Kinh *et al.*, 1982). The yield losses may be minimized when the rice crop is flooded early and kept flooded until a late stage of development (Garg *et al.*, 1995; Soriano *et al.*, 2000).

The root-knot problem in rice can be managed by growing healthy and disinfested nursery and following summer plowing and puddling of main fields before transplanting. Good puddling before transplanting helps in water retention in soil which reduces aeration and also reduces nematode movement and invasion of fresh roots by infective juveniles of *M. graminicola* in rice soil (Gaur, 2003; Pankaj *et al.*, 2006).

(iv) Organic amendments: Soil amendment with decaffeinated tea waste or water hyacinth compost (300 or 600g/4.5 kg soil) reduced root-knot nematode infestation and increased plant growth (Roy, 1976).

Marigold treatment (grown and incorporated into soil before planting rice) also suppressed nematode root galling, and increased rice grain yield by 46%.

Poultry manure can be utilized at least occasionally as a nematicide cum fertilizer to reduce nematode infestation level in rice plants and to improve soil fertility (Amarasinghe, 2012).

Seed treatment with neem-based pesticides can reduce populations and damage caused by *M. graminicola* (Das and Deka, 2002).

(c) Chemical methods: Several studies have supported the effectiveness of carbofuran to control *M. graminicola* in rice. Application of carbofuran increased rice yield by 10% for the first rice crop. During the second season, yield increase was observed when rice was planted after cowpea (85%); after cowpea with carbofuran application (75%); after fallow (41%); and after fallow with carbofuran (48%). In the third season, when all plots were planted to rice, 31 and 49% yield increases were obtained, without and with carbofuran, respectively, from plots that had been left fallow for two consecutive seasons (Soriano and Reversat, 2003).

With the non-availability of other nematicides, the possible treatments in the present scenario could be:

- Seed soaking in 500 ppm carbosulfan for 12hr; root-dip in 200 ppm of carbofuran for 12 hr.
- Soil application of carbofuran at 1 kg a.i. /ha in two equal split doses 15 and 45 days after sowing or transplanting.

(d) Biological methods: The biological control agents, *Trichoderma harzianum* and *T. virens* at 4 and 8 g/kg soil respectively, were significantly effective in improving plant growth and reducing number of galls and larval penetration per root system (Pathak and Kumar, 1996). *T. virens* was more suppressive to nematode population in host tissue than *T. harzianum* (Pathak and Kumar, 2003).

The extent of galling (12.80), gall index (1), yield (3260 kg/ha) and final soil population (209.28) were significantly best in *Trichoderma viride*. *T. viride* can be used as an alternative in biomanagement of root-knot nematode in aerobic rice (Shanmuga Priya, 2015).

In rice root tissues treated with *P. fluorescens* and challenge inoculated with *M. graminicola*, the nematode population was significantly lower. There was 40.7% reduction in the nematode population 7 days after inoculation in the *P. fluorescens* treated plants. Similarly there was 68.4% and 75.1% reduction in the nematode population 14 and 21 days after inoculation, respectively in the treated plants (Anita and Samiyappan, 2012). The present study implies that earlier and higher accumulation of phenols and defense enzymes viz., peroxidase, polyphenol oxidase, phenylalanine ammonia lyase and chitinase in rice root tissue treated with *P. fluorescens* in response to invasion by *M. graminicola* collectively contribute to induced systemic resistance and decrease in nematode infection (Anita and Samiyappan, 2012).

Seedlings raised from seeds treated with, *Pseudomonas fluorescens, Trichoderma viride, P. fluorescens + T viride* and *Bacillus subtilis* were superior at early stages. At the time of harvest *B. subtilis* showed superior effect in enhancing the shoot and root characters. *B. subtilis* was found to be equally superior as compared to carbofuran in controlling the nematode population in root and soil (Renitha Govind and Ushakumari, 2010).

(e) Host resistance: Soriano *et al.* (1998) observed that one accession of *Oryza longistaminata* represented by two individuals (WL02-2 and WL02-15) and three accessions of *O. glaberrima* (TOG7235, TOG5674 and TOG5675) were resistant to *M. graminicola*. There are number of cultivars reported resistant to *M. graminicola* (Prasad *et al*., 1986; Rao *et al*., 1986).

(f) Integrated management: Following methods could be used in combination along with the judicious use of chemicals against root-knot nematode in rice-wheat cropping system.

(i) Nursery-bed treatment

- Selection of root-knot nematode free area for nursery.
- Soil solarization of area for nursery beds, for 3-4 weeks in April-May (summer months).
- Application of carbofuran at 3.3 g/m^2 of nursery bed, before sowing.
- Soil solarization of nursery beds for 3–4 weeks in summer and nursery-bed treatment with carbofuran or phorate at 2 kg a.i. /ha was found very effective in reducing root-knot nematode infestation in rice nursery (Gaur, 2003).

(ii) Main field

- Proper puddling of main field before transplanting.
- Summer plowing of main field twice at 14 days interval.
- Avoid direct sowing of rice in the main field.
- Crop rotation with mustard or other non-host crops.
- Late sowing of wheat up to late November till mid-December using suitable variety.
- Green manuring with *Crotolaria* spp. or *Sesbania* spp. if time permits.

34.2.2. White tip nematode, *Aphelenchoides besseyi*

Dastur (1936) published an account of a widespread rice disease, known as 'white tip', in the Central Provinces of India, where it had reached epidemic proportions in 1934. The yield loss caused by *Aphelenchoides besseyi* amounts to as much as 17.4 to 54.1 per cent (Atkins and Todd, 1959).

34.2.2.1. Symptoms: Infected seeds emerge late in seed beds and produce small seedlings. The upper 2 to 5 cm leaf tips turn white or pale yellow in the tillering stage, then brown, necrotic and frayed (Plate 34.5). Flag leaves are characteristically shortened and twisted at their apical portions. General stunting of the plant accompanies leaf injury. Panicles are shorter and spikelets reduced in number. Terminal tissues of the panicle are reduced and produce small deformed kernels. Maturity of panicles is delayed, and secondary panicles may arise from lower nodes if the panicle is sterile.

Plate 34.5. Rice leaves infected with *Aphelenchoides besseyi* showing 'white tip' symptoms

34.2.2.2. ***Life cycle:*** The life cycle takes 8 days at 23° and 10 days at 21°C.

34.2.2.3. ***Host range:*** Besides rice, white tip nematode attacks *Setaria italica, S. viridis, Panicum sanguinale* and *Cyperus iria.* Other host plants are strawberry, tuberose, hydrangea, chrysanthemum, *Hibiscus bracheridgii, Vanda orchid, Boehmeria nivea, Ficus elastica, Sporobolus poirettii* and *Pennisetum.*

34.2.2.4. ***Spread and survival:*** The nematode is spread to long distances largely by planting infected seeds. The nematodes also spread through irrigation water for short distances.

Second stage larvae of the nematode will be in quiescent stage outside or inside the husk and can survive for 3 years in this stage but die in 4 months on grain left in the field and they do not survive in soil. Rice seedlings growing from shed grain may, however, allow the nematodes to survive from one season to the next. *A. besseyi* is carried beneath the hull of the rice kernel in a quiescent, immature, usually pre-adult stage. In this state the nematodes may lie dormant as long as 23 months.

34.2.2.5. Management

(a) Regulatory methods

(i) Seed certification

The use of certified seed can eliminate the white tip nematode.

(b) Physical methods

(i) Hot water treatment of seeds: Hot water treatment of seed at 55° to 61°C for 15 minutes gives good control of the nematode.

Seed fumigation with methyl bromide at 1.25 lb. /1000 cu. ft. eliminates the nematode. However, germination was reduced if the moisture content of seed was more than 14 per cent.

Chemical seed treatments with ethylthiocyanoacetate, phosphamidon or thiabendazole were successful in experiments at Madagascar.

(c) Cultural method: Cralley (1957) found that rice seeded in water was much less severely infested than that drilled and flooded. Probably the quiescent nematodes revive when immersed, begin moving about, leave the seed and are lost in the water before the seeds sprout.

(d) Chemical method: Chemical treatment of soil with fensulfothion, diazinon or disulfoton has given increased yields.

(e) Resistant varieties: Atkins and Todd (1959) reported that varieties like Blue Bonnet 50, Fortune and Texas Patna were very resistant to the white tip nematode. Goto and Fukatsu (1956) found that Toson 38 showed no symptoms of white tip during 3 years of field testing.

34.2.3. Rice stem nematode, *Ditylenchus angustus*

The rice stem nematode, causal agent of `ufra' disease, was first reported by Butler (1913) from rice producing region north of the Bay of Bengal and east of the Ganges River (Bangla Desh). Reduction in yield of as much as 50 per cent has been reported from India and 20 to 90 per cent in Thailand.

Fig. 34.6. Symptoms of rice stem nematode

34.2.3.1. Symptoms: The earliest sign in the field is chlorosis or streaks on the upper leaves of rice plants that are more than two months old. There are two distinct types of symptoms, one is the 'swollen ufra', where the panicle remains enclosed within the leaf sheath and there is a strong tendency towards branching of the stem in the infected portions, and the other type, called the 'ripe ufra', where the panicle emerges and produces normal grain only near the tip (Fig.

34.6). The peduncle turns dark brown, and flowers on the lower parts remain unfertilized. Infected plants are stunted and the leaves often wilt.

34.2.3.2. Life cycle: Butler (1913) and Vuong (1969) studied the biology of *D. angustus*. Cottony masses of preadult nematodes (Fourth stage larvae) are formed as the plants mature and dry. During humid periods, they become active, climb up on the stems, and invade the growing point. They feed ectoparasitically on the epidermal cells of the young seed head, the peduncle, the part of the stems just above the upper nodes, and the young leaves rolled about the bud above the growing point. Developing heads are found to contain enormous numbers of nematodes in all stages from eggs to adults. At no time are they found living endoparasitically in plant tissues.

34.2.3.3. Host range: *D. angustus* is usually confined to *Oryza* spp. *Leersia hexandra* has been recorded as a host in Madagascar.

34.2.3.4. Spread and survival: At the end of the growing season, cottony masses of nematode in a stage of anabiosis are formed as the plants mature and dry. Some of these nematodes remain on the stubble when the crop is harvested. When the rainy season starts and the humidity is high they revive and infect the new crop. They can also be spread in irrigation water. The nematode can survive desiccation for more than 15 months.

34.2.3.5. Management

(a) Cultural methods: Rotation of rice with jute (Dastur, 1936), improving drainage as the disease increase in water logged areas, destruction of stubbles, elimination of volunteer plants, using nematode-free seeds are some of the methods for cultural control.

(b) Chemical methods: Spraying Diazinon 100 ppm on the soil in the rice crop completely controlled the nematodes within 72 hr.

(c) Resistant varieties: In Thailand, the 'Khao Tah Oo' variety resistant to the leaf blast disease was most resistant to the rice stem nematode.

34.3. PIGEON PEA, *Cajanus cajan*

The pigeon pea cyst nematode, *Heterodera cajani* is a major pest of pigeon pea in India.

34.3.1. Cyst nematode, *Heterodera cajani*

The pigeon pea cyst nematode was first reported by Koshy (1967).

34.3.1.1. Symptoms: Unthrifty growth of pigeon pea was associated with *H. cajani* infestation. The main symptoms include stunting of plants, abnormal color of leaves, reduced size of pods, reduced root system and cysts on root surface (Fig. 34.7). The infestation initiated the production of lateral roots slightly away from the site of infection.

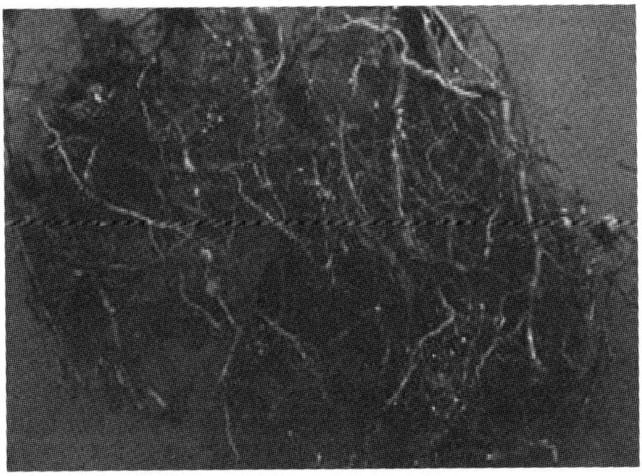

Fig. 34.7. Pigeon pea roots infected with cyst nematode

34.3.1.2. Host range: Pigeon pea, hyacinth bean (*Dolichos lab-lab),* moth bean (*Phaseolus aconitifolius*), green gram, *Phaseolus lathyroides,* black gram, cowpea and sesame were the most favored hosts. Bhatti and Gupta (1973) recorded guar (*Cyamopsis tetragonoloba*) as an additional host of *H. cajani*.

34.3.1.3. Survival: The cysts stored in air dried soil under laboratory conditions remained viable for at least two years (Koshy and Swarup, 1971a).

34.3.1.4. Management: Pigeon pea varieties P -1397, P-1071, P-1-1072, P-4555 and P-9-143-3 were found less susceptible to *H. cajani*.

Green gram varieties PIMS-4, ML-80, and MH-1 were found least susceptible as only few cysts could be recovered from roots.

Cowpea cultivars Culture-2, V-16 and PBT-1 were found least susceptible.

34.4. GROUNDNUT, *Arachis hypogea*

34.4.1. Root-knot nematodes, *Meloidogyne arenaria, M. javanica*

The peanut root-knot nematode, *Meloidogyne arenaria* is the most common and severe species infecting peanut (Dickson, 1998). In addition, *M. javanica* also has been reported to infect peanut from several parts of the world including India (Patel *et al.*, 1988; Sakhuja and Sethi, 1985). *Meloidogyne arenaria* and *M. javanica* are the major constraints to groundnut damage in Gujarat, India (Patel *et al.*, 1996). *Meloidogyne arenaria* and *M. javanica* are responsible for 13-50% and 10-23% of yield losses, respectively.

34.4.1.1. Symptoms: Areas of root-knot nematode infected peanuts are usually round to oblong in shape. The plants are stunted, less green, and will wilt more readily in the heat of the day than plants in less infested areas of the field. Rows of infected plants may never, or not so quickly, meet as those of healthy plants (Plate 34.8).

Plate 34.8. *Left* - Peanut fields exhibiting symptoms of extensive damage due to a severe infestation of the root-knot nematode. *Right and Bottom* - Galling of peanut pods and pegs due to infection by *Meloidogyne arenaria.*

Symptoms of peanut root-knot nematodes include galls (knots) on both roots and pods (Plate 11.8). The presence of galls on the pods appears as single or multiple wart-like growths that may or may not be discolored. As root-knot nematode infection progresses, secondary root and pod rotting causes further damage and eventual death of the plants.

Root-knot nematode affected fields show stunting, reduced number of branches and internodes, burning of leaf margins, reduced leaf size, and yellowing of foliage. The damage thresholds of *Meloidogyne* spp. are 0.1-1.0 juveniles cm^{-3} soil at the time of sowing.

34.4.1.2. Management

(a) Cultural Methods

(i) Crop rotation: Rotation of a peanut crop with grasses such as Bahia grass, Bermuda grass, millet, and sorghum are among the most effective crops in reducing soil populations of peanut root-knot nematodes and should be grown for at least one year, preferably two years, before planting fields to peanut. Though some root-knot nematode reproduction could be expected on field corn, this crop is considerably less susceptible than peanut and is generally effective in the reduction of root-knot nematode soil populations. It is a suitable crop to grow in a year following a root-knot nematode-infected peanut crop. Cotton is not a host for peanut root-knot nematodes and is a very good rotational crop with peanut. Favorably, peanut is not a host for the southern root-knot nematode that infects cotton. Thus a two-year monoculture of peanut and cotton is helpful in maintaining the root-knot nematode pests of both crops at manageable population levels.

A two-year Bahia grass rotation is sufficient to manage plant-parasitic nematodes in a future peanut crop provided weeds in the Bahia grass planting are controlled early and regularly in the first year Bahia grass and this continued throughout the cycle of the rotation.

Two year rotation with cotton, Bahia grass, or velvet bean (Rodríguez-Kábana *et al.*, 1991a, b) was also found effective.

(ii) Cover crops: The use of winter cover crops is helpful to provide competition against volunteer peanuts and spring weeds and also the cover crop planting process helps to destroy those plants growing in the fall. Cover crops, however, must be a poor or non-host of the problem nematode. Winter cereals are most suitable for managing root-knot nematode in this regard.

(iii) Organic amendments: Organic cakes of castor, mustard and neem (1 ton ha^{-1} and above) significantly reduce the root-knot disease and increase pod and haulm yields. Neem cake is more effective than other cakes. Soil application of neem cake at 1 ton/ha and above gave effective control of root-knot nematodes.

(b) Chemical methods: Management of root-knot nematodes on peanut has traditionally relied primarily on treatment of infested fields with nematicides such as granular aldicarb (Temik). A split application of Temik 15G with a band or in-furrow application at planting followed by a band application at peg initiation is acceptable usage.

Seed treatment with 3-6% aldicarb sulfone or 6% carbofuran reduces *M. arenaria* damage (Patel *et al.*, 1986). Groundnut seeds soaked for 12 hr. in phenamiphos, monocrotophos, phosalane or oxamyl (each at 125, 250 and 500 ppm) inhibits the penetration of *M. arenaria* larvae in roots and increases root biomass. Phenamiphos at 250 and 500 ppm is highly effective.

Application of carbofuran at 2 kg ha^{-1} in *M. arenaria* or *M. javanica* infested soils reduces the root-knot disease by 16-96% and increases the pod yields by 15-33% (Patel, 1983).

(c) Biological methods: The biocontrol agent, NemOut, is a formulation of spores of the fungus, *Paecilomyces lilacinus*, and this product has shown a moderate level of nematode suppression in University of Florida trials. It should only be used in conjunction with other recommended nematode management practices including good crop rotation and weed control. For best activity of the product, NemOut should be used both as an in-furrow and as an at-pegging treatment.

(d) Host resistance: The source of resistance against root-knot nematodes on groundnut in Gujarat has been identified (Table 34.1) which can be used in breeding programs.

Table 34.1. Groundnut lines identified as resistant to *Meloidogyne arenaria* or *M. javanica* pathotype 2 in Gujarat, India

Root-knot nematode species	Highly resistant	Resistant
Meloidogyne arenaria	None	A3, Abuarbaa, Ah 25, Ah 3328, Ah 4515, Ah 6902, Ah 6719, Ah 7188, Ah 7299, Ak 10-2, 55-437, C 83, C 149, EC 24118, Kigung, Khargaon 3, NC Ac 50.
M. javanica	ICG 5341, ICG 6330	Apexy, B 1, C 162, C166, Dangi, EC 85994, ICGs 411, 852, 859, 1268, 2248, 2496, 3053, 3104, 6323, 6826, 10047, JH 223, KG 61-22, PI 268594, PI 270787, S 7-2-1, S 7-24-3, U 4-7-3, No 923, No 523.

Groundnut cv. Tifguard is available in Florida, USA that contains resistance to the peanut root-knot nematode and also has good resistance to tomato spotted wilt and other diseases prevalent to Florida growing conditions. The variety has a good yield potential and governed by single dominant gene which provides excellent resistance to the peanut root-knot nematode.

The cultivar COAN with resistance to *M. arenaria* and *M. javanica* was released by the Texas Agricultural Experiment Station (Simpson and Starr, 2001). In 2002, the cultivar NemaTam, which has greater yield potential than COAN with the same level of resistance to *M. arenaria* and *M. javanica*, was released by the Texas Agricultural Experiment Station. The availability of these nematode-resistant cultivars gives growers an additional option for management of *M. arenaria* and *M. javanica*, and will reduce the growers' reliance on nematicides.

Neither COAN nor NemaTAM have the yield potential of the best susceptible cultivars in the absence of root-knot disease. Therefore, these resistant cultivars are recommended only for fields known to be infested with *M arenaria* or *M. javanica*.

An additional benefit of the resistance of COAN and NemaTAM is that rcproduction of thc root-knot ncmatodc is inhibitcd rclativc to rcproduction on a susceptible cultivar (Koenning and Barker, 1992). Therefore, these resistant cultivars not only protect the peanut crop from yield loss due to nematode parasitism, but they also suppress the nematode's population density at crop harvest. Any crop planted after the resistant peanut will be subjected to less disease pressure than it would experience following a susceptible peanut crop.

(e) Integrated management: Use of organic amendments, neem or castor cakes at 1 ton/ha seven days prior to sowing has been found to reduce root-knot nematode damage in groundnut. Their combinations with seed treatment with carbosulfan (25 DS) at 3% a.i. (w/w) further improved efficacy in reducing the nematode population and enhancing the yield significantly.

The following practices have been recommended for the integrated management of root-knot nematodes in groundnut:

- Farmers of South Saurashtra region of Gujarat have been advised to sow groundnut with castor as an intercrop (row ratio 2 : 1) along with soil application of carbofuran at 1.0 kg a.i. /ha to reduce the population of root-knot nematodes infecting groundnut.
- Late maturing varieties have more potential for damage than short-season Spanish market types of varieties

34.4.2. Stunt nematode, *Tylenchorhynchus brevilineatus*

A severe nematode disease of groundnut, popularly called 'Kalahasti malady', caused by *Tylenchorhynchus brevilineatus*, has been prevalent since 1976 in certain parts of Andhra Pradesh, India.

34.4.2.1. Distribution: Andhra Pradesh, India.

34.4.2.2. Symptoms: Infected plants appear in patches in the field, and are stunted with greener than normal foliage. Small brownish-yellow lesions appear on the pegs, and on young developing pods. The margins of the lesions are slightly elevated because of the proliferation of host cells around the lesions. Peg length is reduced and in advanced stages of the disease the entire pod surface becomes blackened (Fig. 34.9). Discoloration can also be observed on roots, but this is less conspicuous than pod discoloration. Kernels from diseased pods are apparently healthy, although commonly smaller than normal.

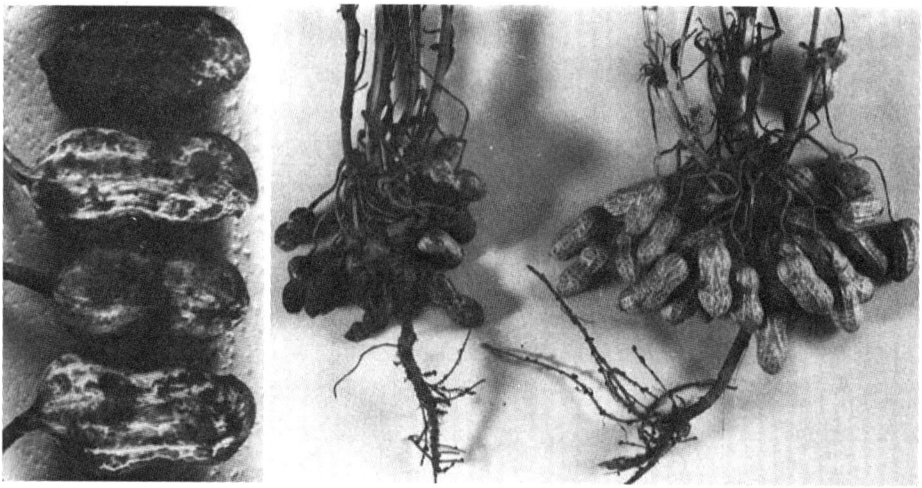

Fig. 34.9. *Left* – Discolored groundnut pods. *Right* – Groundnut plants infected with *Tylenchorhynchus brevilineatus* showing small severely discolored pods (left) as compared to healthy plants (right)

34.4.2.3. Host range: *Tylenchorhynchus brevilineatus* has been reported on several crops viz., *Cicer arietinum, Citrus sinensis, Cuminum cyminum, Dahlia* spp., and *Sorghum* spp.

34.4.2.4. Management: Carbofuran at 4 kg a.i./ha decreased both soil populations of *T. brevilineatus* and of the percentage of disease pods, increased plant height and yield of pods, pod and kernel weights.

One of the resistant genotypes is a high-yielding breeding line (TCG 1518) and this is being released for use in disease-affected areas of Andhra Pradesh.

REFERENCES

Abad, P. and Williamson, V.M. 2010. Plant nematode interaction: A sophisticated dialogue. *Advances in Botanical Research.* 53: 147-912.

Abu-Gharbieh, W.I. and Perry, V.G. 1970. Host differences among Florida populations of *Belonolaimus longicaudatus* Rau. *Journal of Nematology* 2(3): 209-216.

Adams, R.E. and Eichenmuller, J.J. 1963. A bacterial infection of *Xiphinema americanum* (Abstr.). *Phytopathology* 53: 745.

Agrios, G.N. 1987. *Plant Pathology, 2nd edn.* Academic, New York.

Ahmad S.A. and Ahmed A.M.D. 2014. Effect of urea and certain NPK fertilizers on the cereal cyst nematode (*Heterodera avenae*) on wheat. *Saudi Journal of Biological Sciences* 21: 191–196.

Alam, M.M., Khan, A.M. and Saxena, S.K. 1974. Reaction of some cultivated varieties of egg plant, pepper and okra to the root-knot nematode, *Meloidogyne incognita. Indian Journal of Nematology* 4: 64-68.

Al-Banna, L., Williamson, V. and Gardner, S.L. 1997. Phylogenetic analysis of nematodes of the genus *Pratylenchus* using nuclear 26S rDNA. *Mol. Phylogenet. Evol.* 7: 94-102.

Alexander, M.P., Paravatha Reddy, P. Srinivasan, K. and Singh, D.B. 1979. Differential staining of endoparasitic nematodes. *Curr Sci* 49: 71.

Amarasinghe, L.D. 2012. An integrated approach to the management of rice root knot nematode, *Meloidogyne graminicola* in Sri Lanka. *Journal of Science of the University of Kelaniya Sri Lanka* 6: 55–63.

Anandraj, M., Ramana, K.V. and Sarma, Y.R. 1991. Interaction between vesicular arbuscular mycorrhizal fungi and *Meloidogyne incognita* in black pepper. In D.J. Bagyaraj and A. Manjunath (eds.) *Proc. of IInd Natl. Conf. on Mycorrhizal Symbiosis and Plant Growth.* Univ. of Agri. Sci., Bangalore, pp. 110-112.

Andersen, S. 1956. Collection of cysts of *Heterodera major* and estimation of the cyst content. *Nematolgica* 1: 303-306.

Anita, B. and Samiyappan, R. 2012. Induction of systemic resistance in rice by *Pseudomonas fluorescens* against rice root knot nematode *Meloidogyne graminicola. J Biopest* 5 (Suppl.): 53-59.

Anon, 1971. *Estimated Crop Losses Due to Plant Parasitic Nematodes in the United States*. Special Publ. No. 1, Supple J. Nematol, 7 p.

Atkins, J.G. and Todd, E.H. 1959. White tip disease of rice, III Yield tests and varietal resistance. *Phytopathology* 49: 189-191.

Atkinson, G.F. 1892. Some diseases of cotton. *Albama Polytech Inst Agr Exp Sta Bull No* 41: 61-65.

Atwal, A.S. and Mangar, A. 1969. Repellant action of root exudates of *Sesamum orientale* against the root-knot nematode. *Meloidogyne incognita* (Heteroderdae: Nematoda). *Indian J Ent* 31: 286.

Avelar-Mejia, J.J., Teliz-Ortiz, D. and Zavaleta-Mejia, E. 2001. Patógenos asociados con el "declinamiento del guayabo". *Revista Mexicana de Fitopatologia* 19: 223-229.

Ayala, A. and Ramirez, C.T. 1964. Host range, distribution and bibliography of the reniform nematode, *Rotylenchulus reniformis* with special reference to Puerto Rico. *J. Agri Uni. Rice.* 48: 142-161.

Ayyar, P.N.K. 1926. A preliminary note on the root nematode, *Heterodera radicicola* Mitter and its economic importance in South India. *Science and Culture* 22: 391-393.

Baghel, P.P.S. and Bhatti, D.S. 1983a. Evaluation of pesticides for the control of phytonematodes on citrus. *Third Nematology Symposium*, Himachal Pradesh Agricultural University, Solan, pp. 38-39.

Baghel, P.P.S., Bhatti, D.S. and Chauhan, K.S. 1980. *Haryana Journal of Horticultural Science* 9: 136-137.

Baines, R.C., Cameron, J.W. and Soost, R.K. 1974. Four biotypes of *Tylenchulus semipenetrans* in California identified, and their importance in the development of resistant citrus rootstocks. *J. Nematol.* 6: 63-66.

Barber, C.A. 1901. *Dept Land Records and Agriculture*. Madras Agricultural Branch 2, Bull No 45, pp. 227-234.

Barbercheck, M.E. and von Broembsen, S.L. 1986. Effects of soil solarization on plant-parasitic nematodes and *Phytophthora cinnamomi* in South Africa. *Plant Disease* 70: 945-950.

Barker, K.R., Townshend, J.L., Michell, R.E. *et al.* (1978). Determining nematode population responses to control agents. In: E.I. Zehr, G.W. Bird, K.D. Fisher, K.D. Hickey, F.H. Lewis, R.F. Line, S.F. Rickards (eds.) *Methods for Evaluating Plant Fungicides, Nematicides and Bactericides*. American Phytopathological Society, St Paul, Mn., pp. 114-125.

Barnes, R.D. 1968. *Invertebrate Zoology. Second edn.* Saunders, Philadelphia, Pennsylvania, 743 pp.

Barrons, K.C. 1940. Root-knot resistance in beans. *J. Hered* 3: 35-38.

Basu, S.D. and Gope, B. 1985. Control of root-knot nematode *Meloidogyne incognita* (Kofoid and White) Chitwood in tea seedlings with some systemic nematicides. *Two Leaves and a Bud* 32: 31-35.

Bazan de Segura, C. and Aguilar, F.P. 1955. Nematodes and root rot diseases of Peruvian cotton. *Pl Dis Reptr* 39: 12.

Bending, G.D. and Lincoln, S.D. 1999. Characterization of volatile sulphur-containing compounds produced during decomposition of *Brassica juncea* tissues in soil. *Soil Biol Biochem* 31: 95-103.

Berkeley, M.J. 1855. Vibrio forming excrescences on the root of cucumber plants. *Gardener's chronicle,* April, p. 220.

Bessey, E.A. 1911. *Root Knot and its Control.* USDA Bull 217, 89 pp.

Bhatti, D.S. and Gupta, D.C. 1973. Guar an additional host of *Heterodera cajani. Indian J Nematol* 3: 160.

Bhatti, D.S. and Jain, R.K. 1977. Estimation of loss in okra, tomato and brinjal yield due to *Meloidogyne incognita. India J Nematol.* 7: 37-41.

Birchfield, W. 1962. Host parasitic relations of *Rotylenchulus reniformis* on *Gossypium hirsutum. Phytopathology* 52: 862-865.

Bird, G.W. 1981. Integrated nematode management for plant protection. In B.M. Zuckerman and R.A. Rohde (eds.) *Plant Parasitic Nematodes.* Vol III. Acad Press, New York, pp 355-375.

Blake, C.D. 1966. The histological changes in banana roots caused by *Radopholus similis* and *Helicotylenchus multicinctus. Nematologica* 12: 129-137.

Bogantes-Arias, A. and Mora-Newcomer, E. 2010. Evaluation of four rootstocks for graft in guava (*Psidium guajava* L.). *Agronomia Mesoamericana* 21: 103-111.

Boosalis, M.G. and Mankau, R. 1965. Parasitism and predation of soil microorganisms. In K.F. Baker and W.C. Snyder (eds.) *Ecology of Soil-borne Plant Pathogens.* Univ Calif Press, Berkeley, pp 374-391.

Borradiale, L.A., Eastham, E.S., Potts, F.A. and Saunders, J.T. 1961. *The Invertebrata: A Manual for the Use of Students.* Fourth edn., Cambridge Univ Press, London and New York, 1374 pp.

Boydston, R.A., Mojtahedi, H., Crosslin, J.M., Brown, C.R. and Anderson, T. 2008. Effect of hairy nightshade (*Solanum sarrachoides*) presence on potato nematodes, diseases, and insect pests. *Weed Sci.* 56:151-154.

Bridge, J. and Page, S.L.J. 1982. The rice root-knot nematode, *Meloidogyne graminicola,* on deep water rice (*Oryza sativa* sub sp. *indica*). *Revue de Nematologie* 5: 225-232.

Brunt, A.A., Crabtree, K., Dallwitz, M.J., Gibbs, A.J., Watson, L. and Zurcher, E.A. (eds.). 1996 (onward). *Plant Viruses Online*: Descriptions and Lists from the VIDE Database. http://biology.anu.edu.au/ Groups/MES/vide/

Butler, E.J. 1906. The wilt disease of pigeon-pea and pepper. *Agriculture Journal of India* 1: 25-36.

Butler, E.J. 1913. *An Eelworm Disease of Rice.* Bull 34, Agr Res Inst., Pusa, India.

Cadman, C.H. 1963. Biology of soil borne viruses. *Ann Rev Phytopath* 1: 143-172.

Cai, D.G., Kleine, M., Kifle, S., Harloff, H.J. *et al.* 1997. Positional cloning of a gene for nematode resistance in sugar beet. *Science* 275: 832–834.

Cannon, O.S. 1941. Potato nematode disease. *Nossu County Farm and Home Bureau News* 27 (8): 3.

Carneiro, R.M.D.G., Cirotto, P.A., Silva, D.B. and Carneiro, R.G. 2007. Resistance to *Meloidogyne mayaguensis* in *Psidium* spp. accessions and their grafting compatibility with *P. guajava* cv. 'Paluma'. *Fitopatol. Bras.* 32: 281-284.

Carter, W. 1943. A promising new soil amendment and disinfectant. *Science (New York)* 97: 383-384.

Casassa, A.M., Matheus, J., Crozzoli, R., Bravo, V. and Gonzalez, C. 1997. Response of some selections of guava to *Meloidogyne incognita*, in Mara County, Zulia State, Venezuela. *Fitopatologia Venezolana* 10: 5-8.

Castagnone-Sereno, P., Piotte, C., Uijthof, J., Abad, P., Wajnberg, E., Vanlerberghe-Masutti, F., Bongiovanni, M. and Dalmasso, A. 1993. Phylogenetic relationships between amphimictic and parthenogenetic nematodes of the genus *Meloidogyne* as inferred from repetitive DNA analysis. *Heredity* 70: 195-204.

Castillo, P., Vovlas, N., Subbotin, S. and Troccoli, A. 2003. A new root-knot nematode: *Meloidogyne baetica* n. sp. (Nematoda: Heteroderidae) parasitizing wild olive in Southern Spain. *Phytopathology* 93: 1093-1102.

Cetintas, R., Kaur, R., Brito, J.A., Mendes, M.L., Nyczepir, A.P. and Dickson, D.W. 2007. Pathogenicity and reproductive potential of *Meloidogyne mayaguensis* and *M. floridensis* compared with three common *Meloidogyne* spp. *Nematropica* 37: 21–31.

Chakraborti, H.S. and Ghosh, S.C. 1993. Studies on the floral malady of *Polianthes tuberosa* L. *Journal of Mycopathology Research* 33: 25-28.

Charles, J.S.K., Venkitesan, T.S. and Thomas, Y. 1983. Field screening of banana against the burrowing nematode, *Radopholus similis* (Cobb) Thorne. *Third Nematology Symposium*, Himachal Pradesh Agricultural University, Solan, p. 28.

Chhabra, H.K., Bindra, O.S. and Singh, I. 1977. Further studies on the control of the citrus nematode, *Tylenchulus semipenetrans* Cobb, 1913 infecting *Citrus maxima* Merrill (Abstr.). *Proceedings of the International Symposium on Citriculture,* Bangalore, P. 38.

Chitwood, B.G. and Chitwood, M.B. 1974. *Introduction to Nematology*. Revised edn. Univ. Press, Baltimore, Maryland, 334 pp.

Chitwood, B.G. and Toung, M.C. 1960. Host parasite interactions of the Asiatic pyroid citrus nematode. *Pl Dis Reptr* 44: 848-854.

Christie, J.R. 1945. Some preliminary tests to determine the efficacy of certain substances when used as soil fumigants to control the root knot nematode, *Heterodera marioni* (Cornu) Goodey. *Proc. Helminth Soc. Wash* 12 (2): 14-19.

Christie, J.R. and Perry, V.G. 1951. A root disease of plants caused by a nematode of the genus *Trichodorus*. *Science* 113: 491-493.

Cobb, N.A. 1893. *Nematodes, mostly Australian and Fijian.* Linn Soc New South Wales, Macleay Memorial Vol, 59 pp.

Cobb, N.A. 1914 a. Nematodes and their relationships. USDA Year Book, p. 457-490.

Cobb, N.A. 1914b. The citrus root nematode. *J. Agric Res* 2: 217-230.

Cohn, E. 1965. The development of the citrus nematode on some of its hosts. *Nematologica* 11: 593-600.

Cook, R.J. and Baker, K.F. 1983. *The Nature and Practice of Biological Control of Plant Pathogens*. American Phytopathological Society, St. Paul, Minn.

Cooper, B.A. 1955. A mechanical device for concentrating *Heterodera* cysts (Nematoda). In D.K.M. Kevan (ed.) *Soil Zoology*. Butterworths, London, pp 385-389.

Cralley, E.M. 1957. The effect of seeding methods on the severity of white tip of rice (Abstr.), *Phytopathology* 47: 7.

Dallimore, C.E. 1966. Culture of *Ditylenchus destructor* on aseptic potato plugs (Abstr.). *Phytopthology* 56: 874-875.

Dareker, K.S. and Patil, B.D. 1982. Reaction of some grapevine varieties to root-knot and citrus nematode. *Indian Journal of Nematology* 12: 390-392.

Das, P. and Deka, B.C. 2002. Efficacy of neem-based pesticides against *Meloidogyne graminicola* on rice as seed treatment. *Indian Journal of Nematology* 32: 203-205.

Das T.K. and Khan, M.R. 2007. Occurrence and distribution of white tip nematode, *Aphelenchoides besseyi* in West Bengal, India. *Indian Journal of Nematology* 27: 94–97.

Dasgupta, D.R. and Seshadri, A.R. 1971. Races of the reniform nematode. *Rotylenchulus reniformis* Linford and Oliveira, 1940. *Indian J Nematol* 1: 21-24.

Dastur, J.P. 1936. A nematode disease of rice in the Central Provinces. *Proc Indian Acad Sci* 4: 108-121.

Delaquis, P.J. and Mazza, G. 1995. Antimicrobial properties of isothiocyanates in food preservation. *Food Technol.* November, pp. 73-84.

Devapriyanga, R., Jonathan, E.I., Sankari Meena K. and Kavitha, P.G. 2012. Bioefficacy of *Pseudomonas* and *Bacillus* isolates against root-knot nematode, *Meloidogyne incognita* in black pepper cv. Panniyur 1. *Indian Journal of Nematology* 42(1): 57-65.

Dickson, D.W. 1998. Peanut. In K.R. Barker, G.A. Peterson, and G.L. Windham (eds.) *Plant and Nematode Interaction.* American Society of Agronomy, Madison, WI, pp. 523–566.

Dickson, D.W., Sasser, J.N. and Huisingh, D. 1970. Comparative disc electrophoretic protein analysis of selected *Meloidogyne, Ditylenchus, Heterodera*, and *Aphelenchus* spp. *Journal of Nematology* 2: 286–293.

Dimock, A.W. and Ford, C.H. 1950. Control of foliar nematode of chrysanthemum with parathion sprays (Abstr.) *Phytopathology* 40: 7.

Dropkin, V.H. 1969. The necrotic reaction of tomatoes and other hosts resistant to *Meloidogyne:* Reversal by temperature. *Phytopathology* 59: 1632-1637.

Ducharme, E.P. and Birchfield, W. 1956. Physiologic races of the burrowing nematode. *Phytopathology* 46: 615-616.

Duddington, C.L., Everard, C.O.R. and Duthoit, C.M.G. 1961. Effect of green manuring and a predaceous fungus on cereal root eelworm in oats. *Pl Path* 10: 108-109.

Duddington, C.L., Jones, F.G.W and Moriarty, F. 1956. The effect of predaceous fungus and organic matter upon the soil population of beet eelworm, *Heterodera schachtii* Schm. *Nematologica* 1: 344-348.

Eapen, S.J. 1995. *Investigations on Plant Parasitic Nematodes Associated with Cardamom.* Final Report of the Project Nema I (813), Indian Inst. of Spices Research, Calicut, 39 pp.

Elbadri, G., De Ley, P., Waeyenberge, L., Verstraete, A., Moens, M. and Vanfleteren, J. 2002. Intraspecific variation in *Radopholus similis* isolates assessed with restriction fragment length polymorphism and DNA sequencing of the internal transcribed spacer region of the ribosomal RNA cistron. *Parasitol. Int.* 32: 199-205.

Ellenby, C. 1951. Mustard oils and control of the potato root eelworm, *Heterodera rostochiensis* Wollenweber: Further field and laboratory experiments. *Ann Appl Biol.* 38: 859-875.

Epps, J.M. and Golden, A.M. 1967. Biological differences in population of the soybean cyst nematode, *Heterodera glycines* (Abstr.) *Nematologica* 13: 141.

Esbenshade, P. and Triantaphyllou, A. 1990. Isozyme phenotypes for the identification of *Meloidogyne* species. *J. Nematol.* 22: 10-15.

Esser, R.P and Sobers, E.K. 1964. Natural enemies of nematodes. *Proc Soil Crop Sci Soc Fla* 24: 326-353.

Evans, A.A.F. 1971. Taxonomic value of gel electrophoresis of proteins from mycophagous and plant-parasitic nematodes. *International Journal of Biochemistry* 2: 72–79.

Evans, T.A., Miller, L.C., Vasilas, B.L., Taylor, R.W. and Mulrooney, R.P. 2007. Management of *Xiphinema americanum* and soybean severe stunt in soybean using crop rotation. *Plant Dis*. 91: 216-219.

Fargette, M., Phillips, M., Blok, V., Waugh, R. and Trudgill, D. 1996. An RFLP study of relationships between species, populations, and resistance breaking lines of tropical species of *Meloidogyne*. *Fundam. Appl. Nematol.*19: 193-200.

Fassuliotis, G. 1970. Resistance of *Cucumis* spp to the root-knot nematode, *Meloidogyne incognita*. *J. Nematol* 2: 174-178.

Fassuliotis, G. and Sparrow, A.H. 1955. Preliminary report of X-ray studies on the golden nematode. *Pl Dis Reptr*. 39: 572.

Feder, W.A. 1960. Osmotic destruction of plant parasitic and saprophytic nematodes by the addition of sugars to soil. *Pl. Dis Reptr*. 44: 883-885.

Feder, W.A., Eichdron, J.L. and Hutchins, P.C. 1962. Sugar induced osmotic dehydration of nematodes enhanced by the addition of detergents (Abstr.) *Phytopathology* 52: 9.

Ferris, V.R. 1971. Taxonomy of the Dorylaimids. In B.M. Zuckerman, W.F. Mai and R.A. Rohde (eds.) *Plant Parasitic Nematodes* Vol. 1. Acad Press, New York and London, pp. 164-189.

Fielding, M.J. 1951. Observation on the length of dormancy in certain plant infecting nematodes. *Proc Helminth Soc. Wash* 18: 110-112.

Fullaondo, A., Barrena, E., Viribay, M., Barrena, I., Salazar, A. and Ritter, E. 1999. Identification of potato cyst nematode species *Globodera rostochiensis* and *G. pallida* by PCR using species specific primer combinations. *Nematology* 1: 157-163.

Gair, R. 1965. Cereal root eelworm. In J.F. Southey (ed.) *Plant Nematology*. Tech Bull No. 7, Minist, Agric Fish Fd., London, pp. 199-211.

Gamliel, A. and Stapleton, J.J. 1993. Characterization of antifungal volatile compounds evolved from solarized soil amended with cabbage residues. *Phytopathology* 83: 899- 905.

Ganguly, A.K., Pankaj, S. and Sirohi, A. 1996. Effect of soil solarization of rice nursery beds to suppress plant parasitic nematodes. *International Rice Research Notes* 21: 80-81.

Ganguly, S., Singh, M. and Ganguly, A.K. 2010. Record of potato cyst nematode, *Globodera rostochiensis* and *G. pallida* in Shimla, Himachal Pradesh, India. *Indian J Nematol* 40: 96-102.

Garg, R.N., Gaur, H.S. and Singh, G.C. 1995. Effect of tillage practices and water regimes on the soil physical properties and plant parasitic nematodes in rice. *Annals of Plant Protection Science* 3: 121-126.

Gaugler, R. 2002. *Entomopathogenic Nematology*. CABI Publications, UK, 388 pp.

Gaur, H.S. 2003. *Root-Knot Disease of Rice and Wheat: Problem and Management*. Tech. Bull. (TB – ICN: 1/2003), IARI, New Delhi, 23 pp.

Gnanapragasam, N.C. 1987. Integrated strategies to manage nematodes in tea. *Proceedings of the International Conference on Tea Quality and Human Health*, Hangzhou, China, pp. 117-183.

Gnanapragasam, N.C. 1991. Influence of soil amendments in reducing pathogenicity to tea by the root lesion nematode, *Pratylenchus loosi. Proceedings of the International Symposium on Tea Science,* Shizuoka, Japan, pp. 684-688.

Gnanapragasam, N.C., Dharmasena, W.A.M. and Liyanage, D.D. 1989. Development of age resistance in young tea to different species of *Meloidogyne. International Nematology Network Newsletter* 6: 4.

Godfrey, G.H. Oliveria, J.M. and Hoshine, H.M. 1934. Increased efficiency of chloropirin for nematode control. *Phytopathology* 24: 1332-1336.

Golden, A.H. 1971. Classification of the genera and higher categories of the order Tylenchida (Nematoda). In B.M. Zuckerman, W.F. Mai and R.A. Rohde (eds.) *Plant Parasitic Nematodes* Vol 1. Acad Press, New York and London, p. 191-232.

Golden, A.M., Epps, J.M., Riggs, R.D., Duclos, L.A., Fox, J.A. and Bernard, R.L. 1970. Terminology and identity of infraspecific forms of the soybean cyst nematode (*Heterodera glycines*). *Plant Dis. Rep*. 54: 544-546.

Gomes. V.M., Souza, R.M., Corrêa. F.M. and Dolinski, C. 2010. Management of *Meloidogyne mayaguensis* in commercial guava orchards with chemical fertilization and organic amendments. *Nematologia Brasileira* 34: 23-30.

Goode, M.J. and McGuire, J.M. 1967. Relationship of root-knot nematodes to pathogenic variability in *Fusarium oxysporum* f. sp. *lycopersici*. *Phytopathology* 57: 812.

Goodey, T. 1963. *Soil and Fresh Water Nematodes*. Revised by J.B. Goodey, Methuen and Co Ltd., London, 544 pp.

Goto, K. and Fukatsu, R. 1956. *Studies on the White Tip of Rice Plant*. Inst. Agri Sci (Japan) Ser C, No. 6.

Griffiths, B., Donn, S., Neilson, R. and Dniell, T. 2006. Molecular sequencing and morphological analysis of a nematode community. *Appl. Soil Ecol*. 32: 325-337.

Griffiths, B.S., Robertson, W.M. and Trudgill, D.L. 1982. Nuclear changes induced by the nematodes *Xiphinema diversicaudatum* and *Longidorus elongatus* in root-tips of perennial ryegrass, *Lolium perenne. Histochem. J.* 14: 719–730.

Gupta, P. and Swarup G. 1972. Ear-cockle and ear rot diseases of wheat. II Nematode - bacterial association. *Nematologica* 18: 320-324.

Harrison, B.D. 1964. The transmission of plant viruses in soil. In M.K. Corbett and H.D. Sisler (eds.) *Plant Virology*. Univ Fla Press, Gainesville, pp 118-147.

Hartman, K.M. and Sasser, J.N. 1985. Identification of *Meloidogyne* species on the basis of differential host test and perineal pattern morphology. In: K.R. Barker, C.C. Carter and J.N. Sasser (eds.) *An Advanced Treatise on Meloidogyne. Volume II: Methodology*. A cooperative publication of the Department of Plant Pathology and the United States Agency for International Development, North Carolina State University Graphics, Raleigh, North Carolina, pp. 69–77.

Hechler, H.C. 1963. Description, development biology, and feeding habits of *Seinura tenuicaudata* (de Man) JB Goodey, 1960 (Nematoda: Aphelenchoididae), a nematode predator. *Proc Helminth Soc Wash* 30: 182-195.

Hewitt, W.B., Raski, D.J. and Goheen, A.C. 1958. Nematode vector of soil-borne fanleaf virus of grapevines. *Phytopathology* 48: 586-595.

Hirschmann, H. 1962. The life cycle of *Ditylenchus triformis* (Nematoda: Tylenchida) with emphasis on post embryonic development. *Proc. Helminth Soc Wash* 29: 30-43.

Hollaway, G., Ophel-Keller, K., Taylor, S., Burns, R. and McKay, A. 2004. Effect of soil water content, sampling method and sample storage on the quantification of root lesion nematode (*Pratylenchus* sp.) by different methods. *Australas. Plant Pathol.* 32: 73-79.

Holtzmann, O.V. 1965. Effect of soil temperature on resistance of tomato to root-knot nematode (*Meloidogyne incognita). Phytopathology* 55: 990-992.

Holtzmann, O.V. 1968. A foliar disease of tuberose caused by *Aphelenchoides besseyi. Plant Disease Reporter* 52: 56.

Huettel, R.N., Dickson, D.W. and Kaplan, D.T. 1984. *Radopholus citrophilus* n.sp. (Nematoda), a sibling species of *Radopholus similis. Proceedings of the Helminthological Society of Washington* 51: 32-35.

Hussey, R.S., and Roncadori, R.W. 1978. Interaction of *Pratylenchus brachyurus* and *Gigaspora margarita* on cotton. *Journal of Nematology* 10: 16- 20.

Hyman, L.H. 1951. *The Invertebrates: Acanthocephoda, Aschelminthes, and Entro-procta. The Pseulococlomate bilateria.* Vol III, McGraw Hill, New York, 472 pp.

Ichniohe, M. 1980. Studies on the root-knot nematode of black pepper plantation in Amazon. *Annual Report of the Society of Plant Protection, North Japan* 31: 1-8.

Ikram, A. 1990. Beneficial soil microbes and crop productivity. *The Planter* 66: 640-648.

Jain, R.K., Bhatti, D.S., Bhutani, R.D. and Kalloo, G. 1983. Screening of germplasm of some vegetable crops for resistance to root-knot nematode *Meloidogyne incognita. Indian Journal of Nematology* 13: 212-215.

Jatala, P. 1986. Biological control of nematodes. *Annual Review of Phytopathology* 24: 453-489.

Jenkins, W.R. and Coursen, B.W. 1957. The effect of root-knot nematodes, *Meloidogyne incognita acrita* and *M. hapla,* on *Fusarium* wilt of tomato. *Pl Dis Reptr* 41: 182-186.

Johnson, L.F., Chambers, A.Y. and Reed, H.E. 1967. Reduction of root-knot of tomatoes with crop residue amendments in field experiments. *Pl Dis Reptr* 51: 219-222.

Jones, F.G.W. 1961. The potato root eelworm *Heterodera rostochiensis* Woll in India. *Curr. Sci* 30: 187.

Jones, F.G.W. 1969. Integrated control of the potato cyst nematode. *Proc. Fifth Br Insectic Fungi Conf.* 3: 646-356.

Kalita, M. and Phukan, P.N. 1990. Reactions of some rice cultivars to *Meloidogyne graminicola. Indian Journal of Nematology* 20: 215-216.

Kalita, S. and Bora, B.C. 2006. Effect of nematicides and organic amendments in the management of *Meloidogyne incognita* in nursery tea. *Indian Journal of Nematology* 36: 148-149.

Kampfe, L. and Dietze, H. 1972. Histochemical demonstration of enzymes in normal nematodes and those influenced by nematicides. *Proc XIth Intn Symp Nematol,* Reading, UK, pp 34-36.

Kamra, A. and Gaur, H.S. 1998. Control of nematodes, fungi and weeds in nursery beds by soil solarization. *Indian Journal of Nematology* 28: 46-52.

Kanwar, R.S. and Bhatti, D.S. 1990. Yield performance of resistant and susceptible tomato cultivars in monoculture and related reproduction of *Meloidogyne javanica. International Nematology Network Newsletter* 7(3): 16-18.

Kerr, A. and Vythilingam, M.K. 1966. Replanting eelworm infested areas. *Tea Quarterly* 37: 67-72.

Khan, A., Shaukat, S.S. and Siddiqui, I.A. 2005. A survey of nematodes of pomegranate orchards in Balochistan province, Pakistan. *Nematol Medit* 33: 25-28.

Khan, A.M., Saxena, S.K. and Khan, M.W. 1971a. Interaction of *Rhizoctonia solani* Kuhn and *Tylenchorhychus brassicae* Siddiqi, 1961 in pre-emergence damping off of Cauliflower seedlings. *Indian J Nematol.* 1: 85-86.

Khan, A.M., Saxena, S.K. and Siddiqi, Z.A. 1971b. Reaction of some cultivated and wild species of *Lycopersicon* and cucurbits to *Meloidogyne incognita* (Kofoid and White) Chitwood. *Indian Phytopathology* 24: 197-199.

Khan, F.A. and Khan, A.M. 1973. Studies on the reniform nematode, *Rotylenchulus reniformis,* I. Host range and population changes. *Indian J Nematol* 3: 24-30.

Khan, M.R. 2001. White-tip nematode, *Aphelenchoides besseyi* in rice-tuberose cropping system. In S.K. Mukhopadhyay, D.C. Ghosh and G.C. De (eds.) *Proceedings of National Seminar on Frontiers of Crop Management.* Visva-Bharati, Sriniketan, India.

Kinh, D.N., Huong, N.M. and Ut, N.U. 1982. Root-knot disease of rice in the Mekong Delta, Vietnam. *Int. Rice Res. Newsl.* 7(4): 15.

Kirkpatrick, J.D. Mai, W.F., Fisher, E.G. and Parker, K.G. 1959a. Population levels of *Pratylenchus penetrans* and *Xiphinema americanum* in relation to potassium fertilization of Montmorency sour cherries on Mazzard rootstock (Abstr.). *Phytopathology* 49: 543.

Kirkpatrick, J.D., Mai, W.F., Fisher, E.G. and Parker, K.G. 1959b. Relation to nematode population to nutrition of sour cherries (Abstr.). *Phytopathology* 49: 543.

Koenning, S.R. and Barker, K.R. 1992. Relative damage functions and reproductive potentials of *Meloidogyne arenaria* and *M. hapla* on peanut. *J. Nematol.* 24: 187-192.

Koshy, P.K. 1967. A new species of *Heterodera* from India. *Indian Phytopath* 20: 272-274.

Koshy, P.K. 1986. The burrowing nematode, *Radopholus similis* (Cobb, 1893) Thorne, 1949. In G. Swarup and D.R. Dasgupta (eds.) *Plant Parasitic Nematodes of India-Problems and Progress*. Indian Agri. Res. Inst., New Delhi, India, pp. 223-248.

Koshy, P.K. and Geetha, S.M. 1992. Nematode pests of palms and cocoa. In D.S. Bhatti and R.K. Walia, (eds.) *Nematode Pests of Crops*. CBS Publishers & Distributors, Delhi, pp. 214-227.

Koshy, P.K. and Sosamma, V.K. 1978. Studies on the population fluctuations of *Radopholus similis* in coconut and areca nut roots. *Indian Phytopathology* 31: 180-185.

Koshy, P.K. and Sosamma, V.K. 1987. Pathogenicity of *Radopholus similis* on coconut (*Cocos nucifera* L.) seedlings under greenhouse and field conditions. *Indian Journal of Nematology* 17: 108-118.

Koshy, P.K., Sosamma, V.K. and Sundararaju, P. 1977. Screening of plants used as pepper standards against root-knot nematode. *Indian Phytopathology* 30: 128-129.

Koshy, P.K. and Swarup, G. 1971. On the number of generations of *Heterodera cajani,* the pigeon-pea cyst nematode in a year. *Indian J Nematol* 1: 88-90.

Krishnappa, K., Setty, K.G.H. and Krishna Prasad, K.S. 1981. Crop loss assessment in brinjal due to root-knot nematode *Meloidogyne incognita* (Abstr.) *Nematol Soc India Symp.,* Coimbatore, p. 1.

Krishna Prasad, K.S. 1986 Potato nematodes. In G. Swarup and D.R. Dasgupta (eds.) *Plant Parasitic Nematodes in India*. Indian Agri. Res. Inst., New Delhi.

Kuhn, J. 1857. Uber des Vorkommen von Anguillule an erkrankten Bluhten koppen von *Dipsacus fullonum L. Zeitscher Wiss Zool* 9: 129-137.

Kuhn, J. 1874. Uber des Vorkommen von Ruben - Nematoden an den wurzeln der Halmfruchte. *Landw Jbr* 3: 47-50.

Kuhn, J. 1881. Die Ergebnisse der versuche zue Ermittelung der Ursache der Ruben mudigkeit und zur Erforschung der Natur der Nematoden. *Ber Physiol Lab Univ Halle* 3: 1-53.

Kumar, A.C. 1982. Evaluation of Nemacur 5G against coffee root-lesion nematode, *Pratylenchus coffeae*. *Journal of Coffee Research* 12: 1-7.

Kumar, A.C. 1988. Nematode problem of coffee and its management. *Indian Coffee* 52: 12-19.

Lakshmanan, P.L. and Sivakumar C.V. 1981. Control of *Rotylenchulus reniformis* in Bhendi by integrated method (Abstr.) *Nematol Soc India Symp.*, Coimbatore, p 54.

Linford, M.B. and Oliveira J.M. 1937. The feeding of hollow spear nematodes on other nematode. *Science* 35: 295-297.

Linford, M.B. and Oliveira, J.M. 1938. Potential agents of biological control of plant parasitic nematodes. *Phytopathology* 28: 14.

Linford, M.B., Yap, F. and Oliveira, J.M. 1938. Reduction of soil populations of the root-knot nematode during decomposition of organic matter. *Soil Sci* 45: 127-141.

Loof, P.A.A. 1975. Taxonomy of Trichododridae. In F. Lamberti, C.E. Taylor and J.W. Seinhorst (eds.) *Nematode Vectors of Plant Viruses*. Plenum Press, London and New York, p. 103-127.

Loos, C.A. 1961. Eradication of the burrowing nematode, *Radopholus similis,* from bananas. *Pl Dis Reptr.* 45: 457-461.

Loos, C.A. and Loos, S.B. 1960. Preparing nematode free banana “seed”. *Phytopathology* 50: 383-386.

Lordello, L.G.E. 1968. *Nematodes das plantas Cultivadas.* Livaria Nobel SA, Sao Paulo, Brazil.

Lowenberg, J.R., Sullivan, T. and Schuster, M.L. 1959. A virus disease of *Meloidogyne incognita*, the southern root-knot nematode. *Nature* 184: 1896.

MacFarlane, S.A., Zasada, I., Lemaire, O. and Demangeat G. 2016. Nematode-borne plant viruses. In J.K. Brown (ed.) *Vector-Mediated Transmission of Plant Pathogens*. The American Phytopathological Society, pp. 365-378.

Madani, M., Subbotin, S. and Moens, M. 2005. Quantitative detection of the potato cyst nematode, *Globodera pallida*, and the beet cyst nematode, *Heterodera schachtii*, using Real-Time PCR with SYBR green I dye. *Mol. Cell. Probes* 19: 81-86.

Mani, A. and Murty, V.D. 1989. *National Symposium on Insect Pests and Disease Management on Citrus*, Jawaharlal Nehru Krishi Viswa Vidyalaya, Chindwara.

Mani, A. and Reddy, G.S. 1986. Studies on the varietal reaction of some species of *Citrus* and *Poncirus* to *Tylenchulus semipenetrans* Cobb. *Indian Journal of Nematology* 16: 267-268.

Mankau, R. and Minteer, R.J 1962. Reduction of soil populations of the citrus nematode by the addition of organic materials. *Pl. Dis Reptr* 46: 375-378.

Mathur, B.N., Arya, H.C, Mathur, R.L. and Handa, D.K. 1974. The occurrence of biotypes of the cereal cyst nematode *(Heterodera avenae*) in the light soils of Rajasthan and Haryana, India, *Nematologica* 20: 19-25.

Matthews, J.D. 1919. Report on the work of the W.B. Randall, Research Assistant. *Ann Rept Nurs and Mark Gar Ind Dev Coc Exp and Res Sta, Chechunt, Herts* 5: 18-21.

Mazzola, M., Granatstein, D.M., Elfving, D.C. and Mullinis, K. 2001. The suppression of specific apple root pathogens by *Brassica napus* seed meal amendment regardless of glucosinolate content. *Phytopathology* 91: 673-679.

McLeod, R.W. and Steel, C.C. 1999. Effects of Brassica-leaf green manures and crops on activity and reproduction of *Meloidogyne javanica. Nematology* 1: 613-624.

Meijnke, C.A.R. and Oostenbrink, M. 1958. *Tagetes* ter bestriding van aaltjesaanat-astingen. *Meded Dir Tuinb* 21: 283-290.

Mekete, T., Grey, M. and Niblack, T. 2009. Distribution, morphological description, and molecular characterization of *Xiphinema and Longidorous* spp. associated with plants (*Miscanthus* spp. and *Panicum virgatum*) used for biofuels. *GCB Bioenergy* 1: 257-266.

Melendez, P.L. and Powell, N.T. 1967. Histological aspects of *Fusarium* wilt root-knot complex in flue-cured tobacco. *Phytopathology* 57: 286-292.

Milan, A.R. 2007. Breeding of *Psidium* species for root-knot nematode resistance in Malaysia. *Acta Hort.* 735: 61-69.

Milan, A.R. 2010. Collection and evaluation of guava (*Psidium guajava* L.) for nematode resistance in Malaysia. *Acta Hort.* 849: 357-362.

Miller, L.I. 1951. A report on the effect of ethylene dibromide soil treatment on root-knot control, nodulation, and yield of peanuts. *Virginia* J *Sci New Series* 2: 109-112.

Minton, N.A. and Minton, E.B. 1963. Infection relationship between *Meloidogyne incognita* acrita and *Fusarium oxysporum* f. *vasinfectum* in cotton (Abstr.) *Phytopathology* 53: 624.

Mohandas, C. 2001. Nematode diseases of tuber crops and their management. *Natl. Cong. on Centenary of Nematol. in India – Appraisal & Future Plans*, Indian Agri. Res. Inst., New Delhi, pp. 35-36.

Mohandas, C. and Ramana, K.V. 1987. Slow wilt disease of black pepper and its control. *Indian Cocoa, Areca nut and Spices Journal* 11: 10-11.

Mohandas, C. and Ramana, K.V. 1991. Pathogenicity of *Meloidogyne incognita* and *Radopholus similis* on black pepper (*Piper nigrum* L.). *Journal of Plantation Crops* 19: 41-43.

Morgan, D.O. 1925. Investigations on eelworm in potatoes in South Lincolnshire. *J. Helminth.* 3: 185-192.

Mountain, W.B. and Patrick, Z.A. 1959. The peach replant problem in Ontario. 7. The pathogenicity of *Pratylenchus penetrans* (Cobb, 1917) Filip and Stek 1941. *Can J Bot* 37: 459-470.

Mukhopadhyaya, M.C., Dalal, M.R., Saran S. and Kharub, S.S. 1972. Studies on the 'molya' disease of wheat and barley. *Indian J. Nematol.* 2: 11-20.

Mukhopadhyay, M.C. and Haque, M.S. 1980. Influence of soil types on *Rotylenchulus reniformis* on castor. *Indian J. Nematol.* 10: 89-91.

Muraleedharan, N. 1991. Non-arthropod pests – Nematodes and rodents. In Director (ed.) *Pest Management in Tea.* United Planter's Association of Southern India, Tea Res. Inst., pp. 77-81.

Murphy, P.W. and Doncaster C.C. 1957. A culture method for soil microfauna and its application to the study of nematode predators. *Nematologica* 2: 202-214.

Nagesh, M., Janakiram, T., Rao, T.M. and Parvatha Reddy, P. 1995a. Evaluation of China aster varieties/pure lines for resistance to *Meloidogyne incognita* (Kofoid & White, 1919) Chitwood, 1949. *Pest Management in Horticultural Ecosystems* 1: 101-104.

Nagesh, M., Meenakshi, S., Murthy, N. and Parvatha Reddy, P. 1995b. Screening of some tuberose cultivars/hybrids for resistance against *Meloidogyne incognita* race-1. *Indian Perfumer* 39: 138-140.

Nair, M.R.G.K., Das, N.M. and Menon, M.R. 1966. On the occurrence of the burrowing nematode, *Radopholus similis* (Cobb, 1893) Thorne, 1949 on banana in Kerala. *Indian J. Ent* 28: 553-554.

Nandwana, R.P., Pathak, A.K. and Yadav, B.S. 1981. Susceptibility of varieties of certain oil seed crops to the reniform nematode *Rotylenchulus reniformis. Nematol Soc India Symp*, Coimbatore, p. 49.

Nasira, K., Shaheen, N. and Shahina, F. 2011. Root-knot nematode *Meloidogyne incognita* Wartellei on pomegranate in Swat, KPK. *Pakistan J Nematol* 29(1): 117-118.

Navas, A., Castagnone-Sereno, P., Blazquez, J. and Esperrago, G. 2001. Genetic structure and diversity within local populations of *Meloidogyne* (Nematoda: *Meloidogynidae*). *Nematology* 3: 243-253.

Neal, D.C. 1954. The reniform nematode and its relationship to the incidence of Fusarium wilt in cotton at Baton Rouge, Louisiana. *Phytopathology* 44: 447-450.

Neal, J.C. 1889. *The Root-Knot Disease of the Peach, Orange and Other Plants in Florida Due to the Work of Anguillula.* Bull. U.S. Bur Ent No. 20, 31 pp.

Needham, T. 1743. A letter concerning chalky tubulous concretions called malm: with some microscopical observations on the fauna of the red lily and of worms discovered in smutty corn. *Philos Trans of Roy. Soc* 42: 173-174, 634-641.

Netscher, C. 1973. Étude sur la variabilité de la longueur des larves chez *Meloidogyne incognita* Chitwood, 1949, *et. Meloidogyne javanica* Chitwood, 1949. *Cah. ORSTOM, Sér. Biol.* 21: 91-95.

Newhall, A.G. 1958. The incidence of Panama disease of banana in the presence of the root-knot and burrowing nematodes (*Meloidogyne* and *Radopholus*). *Pl. Dis Reptr.* 42: 853-856.

Nour El-Deen, A.H., Al-Barty Amal, F., Darwesh Hadeer, Y. and Al-Ghamdi, A.S. 2016. Eco-friendly management of root-knot nematode, *Meloidogyne incognita* infecting pomegranate at Taif Governorate, KSA. *Research Journal of Pharmaceutical, Biological and Chemical Sciences* 7(1): 1070-1076.

O'Bannon, J.H. and Tomerlin, A.T. 1969. Population studies on two species of *Pratylenchus* on Citrus (Abstr.). *J. Nematol* 1: 299-300.

Oschae, J.J. and Brewton, W.S. 1954. Preliminary report on *Crotalaria versus* nematodes. *Proc. Fla Sta Hort Soc* 67: 218-219.

Oostenbrink, M., Kuiper, M.K. and s'Jacob. J.J. 1957. *Tagetes* als Feindpflanzen von *Pratylenchus* Asten. *Nematologica* 2: 424-437.

Oteifa, B.A. 1970. The reniform nematode problem in Egyptian cotton production (Abstr.). *J. Parasitol* 56: 255.

Palanisamy, S. and Balasubramanian, P. 1981. Assessment of avoidable yield loss in cotton variety `Suvin' (*Gossypium barbadense* L.) by fumigation with metham sodium (Abstr.). *Nematol Soc India Symp,* Coimbatore, p. 52.

Pankaj, S., Ahlawat, J.S. and Saha, M. 2006. Predominant nematode pests of rice nursery in North-Western India. In: *2nd International Rice Congress*, ICAR, New Delhi.

Paramonov, A.A. 1963. On factors determining the formation of the structures of nematodes. *Helminthologia* 4: 366-376.

Parvatha Reddy, P. 1981. Pathogenicity of *Meloidogyne incognita* to passion fruit. *Indian Journal of Nematology* 11: 80-81.

Parvatha Reddy, P. 1983. Reaction of selected citrus rootstocks to the citrus nematode. *Proceedings of International Citrus Symposium*, Bangalore, pp. 379-381.

Parvatha Reddy, P. and Agarwal, P.K. 1987. Resistance in citrus rootstocks to the citrus nematode, *Tylenchulus semipenetrans*. *Indian Journal of Horticulture* 44: 111-114.

Parvatha Reddy, P., Govindu, H.C. and Setty, K.G.H. 1975a. Studies on the effect of amino acids on the root-knot nematode *Meloidogyne incognita* infecting tomato. *Indian J. Nematol* 5: 36-41.

Parvatha Reddy, P., Govindu, H.C. and Setty, K.G.H. 1975b. Further studies on the action of DL methionine on *Meloidogyne incognita*. *Indian J. Nematol* 5: 200-206.

Parvatha Reddy, P., Khan, R.M. and Rao, M.S. 1991. Integrated management of the citrus nematode, *Tylenchulus semipenetrans* using oil cakes and *Paecilomyces lilacinus*. *Afro-Asian Journal of Nematology* 1: 221-222.

Parvatha Reddy, P. and Nagesh, M. 2002. A*vermectins – Isolation, Fermentation, Preliminary Charecterization and Screening for Nematicidal Activity.* Tech. Bull.17, Indian Inst. of Hortil. Research, Bangalore, 28 pp.

Parvatha Reddy, P., Setty, K.G.H. and Govindu, H.C. 1972. Susceptibility of flue-cured tobacco variety NC 95 to southern root-knot nematode. *Mysore J. Agric Sci* 6: 192.

Parvatha Reddy, P. and Singh, D.B. 1978. Evaluating the reaction of some species and varieties of *Citrus* and *Poncirus* to the citrus nematode. *Indian J. Nematol.* 8: 82-84.

Parvatha Reddy, P. and Singh, D.B. 1980. Further studies on the reaction of some species and varieties of citrus to the citrus nematode. *Indian J. Nematol* 10: 249-250.

Parvatha Reddy, P. and Singh, D.B. 1981. Assessment of avoidable yield loss in okra, brinjal, French bean and cowpea due to root knot nematodes (Abstr.). *IIIrd Intn Symp Pl Path,* New Delhi, pp. 93-94.

Parvatha Reddy, P. and Singh, D.B. 1984. Reaction of selected cultivars/rootstocks of grapevine to *Meloidogyne incognita. Indian Journal of Nematology* 14: 73-74.

Parvatha Reddy, P., Singh, D.B. and Nagalingam, B. 1981. Fighting citrus nematodes in Andhra Pradesh. *Indian Horticulture* 26(3): 24-25.

Patel, B.A., Patel, D.J., Sharma, S.B. and Patel, H.V. 1996. Nematode problems of groundnut and their management in Gujarat, India. *International Arachis Newsletter* 16: 38-39.

Patel, D., Patel, B.A., Patel, S.K., Patel, R.L. and Patel, R.G. 1999. Root-knot nematode, *Meloidogyne indica* on kagzi lime in North Gujarat. *Indian J Nematol.* 29(2): 185-205.

Patel, D.J., Patel, B.A., Chavda, J.C. and Patel H.V. 1988. Record of *Meloidogyne javanica* on groundnut in Gujarat, India. *International Arachis Newsletter* 3: 16–17.

Patel, G.J., Shah, H.M. and Patel, D.J. 1979. Reaction of some tomato cultivars to root-knot nematode. *Indian Journal of Nematology* 9: 60-61.

Patel, H.R. 1983. *Studies on Root-Knot Nematodes of Groundnut.* Ph. D. thesis, Gujarat Agri. Univ., Sardarkrushinagar, Gujarat, India, 125 pp.

Patel, H.R., Vaishnav, M.U. and Dhruj, I.U. 1986. Efficacy of aldicarb sulfone and carbofuran flowable seed treatment on plant growth and against *Meloidogyne arenaria* on groundnut. *Pesticides* 20: 29-31.

Pathak, B. and Khan, M.R. 2009. Yield loss potential of foliar nematode, *Aphelenchoides besseyi* in tuberose. *Indian J. Nematol.* 39: 38-40.

Pathak, K.N. and Kumar, B. 1996. Effect of culture filtrates of *Gliocladium virens* and *Trichoderma harzianum* on the penetration of rice roots by *Meloidogyne graminicola. Indian Journal of Nematology* 25: 223- 224.

Pathak, K.N. and Kumar, B. 2003. Effect of culture filtrates of *Gliocladium virens* and *Trichoderma harzianum* on the penetration of rice roots by *Meloidogyne graminicola. Indian Journal of Nematology* 33: 149- 151.

Pawar, S.A., Mhase, N.L., Kadam, D.B. and Chandele, A.G. 2013. Management of root-knot nematode, *Meloidogyne incognita*, race-II infesting pomegranate by using bioinoculants. *Indian Journal of Nematology* 43(1): 92-94.

Petersen, D., Zijlstra, C., Wishart, J., Blok, V. and Vrain, T. 1997. Specific probes efficiently distinguish root-knot nematode species using signature sequence in the ribosomal intergenetic spacer. *Fundam. Appl. Nematol.* 20: 619-626.

Pinkerton, J.N. and Martin, R.R. 2006. Management of tomato ringspot virus in red raspberry with crop rotation. *Int. J. Fruit Sci.* 5:55-67.

Ploeg, A.T. and Stapleton, J.J. 2001. Glasshouse studies on the effects of time, temperature and amendment of soil with broccoli plant residues on the infestation of melon plants by *Meloidogyne incognita* and *M. javanica*. *Nematology* 3: 855-861.

Poornima, K., Kumar, N. and Balamohan, T.N. 2004. Biological control of lesion nematode in banana. *National Symposium on Paradigms in Nematological Research for Biodynamic Farming*, Univ. of Agri. Sci., Bangalore, p. 99.

Potter, M.J., Davies, K. and Rathjen, A.J. 1998. Suppressive impact of glucosinolates in *Brassica* vegetative tissues on root lesion nematode *Pratylenchus neglectus*. *J. Chem. Ecol.* 24: 67-80.

Poucher, C., Ford, H.W., Suit, R.F. and Du Charme, E.P. 1967. *Burrowing Nematode in Citrus*. Fla Dept Agri Divn Plant Ind Bull. No. 7, 63 pp.

Powell, C.A., Forer, L.B. and Stouffer, R.F. 1982. Reservoirs of tomato ringspot virus in fruit orchards. *Plant Dis*. 66:583-584.

Powell, N.T. and Batten, C.K. 1967. The influence of *Meloidogyne incognita* on *Rhizoctonia* root rot in tobacco (Abstr.). *Phytopathology* 57: 826.

Powell, N.T. and Nusbaum C.J. 1958. The effect of root-knot nematode resistance in the incidence of black shank in tobacco. *Phytopathology* 48: 344.

Prasad, J.S., Panwar, M.S. and Rao, Y.S. 1986. Screening of some rice cultivars against the root-knot nematode, *Meloidogyne graminicola*. *Indian Journal of Nematology* 16: 112-113.

Prasad, N. and Mankau, R. 1969. Studies on a sporozoan endoparasite of nematodes (Abstr.). *J. Nematol* 1: 301-302.

Prasad, N, Mathur, R.L. and Sehgal, S.P. 1959. Molya disease of wheat and barley in Rajasthan. *Curr Sci* 28: 453.

Radewald, J.D., O'Bannon, J.H. and Tomerlin, A.T. 1971. Anatomical studies of *Citrus jambhiri* roots infected by *Pratylenchus coffeae*. *J. Nematol* 3: 409-416.

Raj, D. and Gill, J.S. 1983. Reaction of some potato varieties to root-knot nematode *Meloidogyne incognita. Third Nematology Symposium,* Himachal Pradesh Agricultural University, Solan, pp. 27-28.

Rajendran, G., Naganathan, T.G. and Sivagami, V. 1979. Studies on banana nematodes. *Indian Journal of Nematology* **9**: 54.

Ramana, K.V. and Mohandas, C. 1988. Occurrence of potato cyst nematode *Globodera pallida* in Kerala. *Indian J Nematol* 18: 141.

Ramana, K.V., Mohandas, C. and Balakrishnan, R. 1987. Role of plant parasitic nematodes in the slow wilt disease complex of black pepper (*Piper nigrum* L.) in Kerala. *Indian Journal of Nematology* 17: 225-230.

Rangaswamy, S.D., Parvatha Reddy, P. and Nanje Gowda, D.N. 1999. Management of root-knot nematode, *Meloidogyne incognita* in tomato by intercropping with marigold and mustard. *Pest Management in Horticultural Ecosystems* 5: 118-121.

Rao, V.R. and Tikoo, S.K. 1973. Resistance of tomato cultivars to the root-knot nematode, *Meloidogyne incognita* (Kofoid and White, 1919) Chitwood, 1949. *Current Science* 44: 282-283.

Rao, Y.S. 1985. Research on rice nematodes. In: S.Y. Padmanabhan (ed.) *Rice in India.* ICAR Monograph, pp. 591-615.

Rao, Y.S., Prasad, J.S. and Panwar, M.S. 1986. Nematode problems in rice: Crop losses, symptomatology and management. In G. Swarup, and D.R. Dasgupta, (eds.) *Plant Parasitic Nematodes of India – Problems and Progress*, pp. 279-299. Indian Agricultural Research Institute, New Delhi.

Rao, Y.S., Prasad, J.S., Yadav, C.P. and Padalia, C.R. 1984. Influence of rotation crops in rice soils on dynamics of plant parasitic nematode populations. *Biology Agriculture Horticulture* 2: 69-78.

Raski, D.J. and Allen, M.W. 1948. Spring dwarf nematode. *Calif Agri.* pp. 23-24.

Reddy, D.D.R. 1981. Analysis of crop losses in tomato due to *Meloidogyne incognita* (Abstr.). *Nematol Soc India Symp,* Coimbatore, p 60.

Renitha Govind and Ushakumari, R. 2010. Efficacy of different bioagents in controlling the root-knot nematode (*Meloidogyne graminicola* Golden and Birchfield) of rice under dry condition. *National Conf. on Innovations in Nematological Research for Agricultural Sustainability – Challenges and A Roadmap Ahead*, Tamil Nadu Agri. Univ., Coimbatore.

Reynolds, H.W. and Hanson, R.G. 1957. *Rhizoctonia* disease of cotton in the presence or absence of the cotton root-knot nematode in Arizona. *Phytopathology* 47: 256-261.

Rhoades, H.L. and Forbes, R.B. 1986. Effects of fallow, cover crops, organic mulches, and fenamiphos on nematode populations, soil nutrients, and cash crop growth. *Nematropica* 16: 141–151.

Risbeth, J. 1960. Factors affecting the incidence of banana with Panama disease. *Emp J. Expt. Agric* 28: 109-113.

Rodriguez-Kabana, R., Pinochet, J., Robertson, D.G., Weaver, C.F. and King, P.S. 1992. Horse bean (*Canavalia ensiformis*) and crotalaria (*Crotalaria spectabilis*) for the management of *Meloidogyne* spp. *Nematropica* 22: 29–35.

Rodríguez-Kábana, R., Robertson, D.G., Weaver, C.F. and Wells, L. 1991a. Rotations of bahiagrass and castorbean with peanut for the management of *Meloidogyne arenaria. Suppl. J. Nematol.* 23: 658-661.

Rodríguez-Kábana, R., Robertson, D.G., Wells, L., Weaver, C.F. and King, P.S. (1991b). Cotton as a rotation crop for management of *Meloidogyne arenaria* and *Sclerotium rolfsii* in peanut. *Suppl. J. Nematol.* 23: 652-657.

Rohde, R.A. 1990. Mechanisms of resistance to plant parasitic nematodes. In J.N. Sasser and W.R. Jenkins (eds.) *Nematology*. Univ N Carolina Press, Chappel Hill, pp 447-453.

Rohde, R.A. and Jenkins W.K. 1958. Basis for Resistance of *Asparagus officinalis* var. *altilis* L to the Stubby Root Nematode *Trichodorus christiei* Allen. 1957. Bull. Univ. Mad Agric Exp Sta A-97, 19 pp.

Rossi, M., Goggin, F.L., Milligan, S.B., Kaloshian, I., Ullman, D.E. and Williamson, V.M. 1998. The nematode resistance gene Mi of tomato confers resistance against the potato aphid. *Proc. Natl. Acad. Sci.* USA 95: 9750–9754.

Roy, A.K. 1976. Effect of decaffeinated tea waste and water hyacinth compost on the control of *Meloidogyne graminicola* on rice. *Indian Journal of Nematology* 6(1): 73-77.

Sakhuja, P.K. and Sethi, C.L. 1985. Frequency of occurrence of various plant-parasitic nematodes and root rot fungi on groundnut in Punjab. *Indian Journal of Nematology* 15: 191–194.

Sanwal, K.C. 1959. A simple method for rearing pure populations of the foliar nematode, *Aphelenchoide ritzemabosi*, in the laboratory. *Canadian Journal of Zoology* 37: 707–711.

Sarah, J.L. 1989. Nematological review - banana nematodes and their control in Africa. *Nematropica* 19: 199-216.

Sasser, J.N. and Carter, C.C. (eds.). 1985. *An Advanced Treatise on Meloidogyne. Volume I: Biology and Control*. A cooperative publication of the Department of Plant Pathology and the United States Agency for International Development, North Carolina State University Graphics, Raleigh, North Carolina.

Sasser, J.N., Lucas, G.B. and Powers, H.R. 1955. The relationship of root-knot nematodes to black shank resistance in tobacco. *Phytopathology* 45: 459-461.

Sasser, J.N. and Freckman, D.W. 1987. A world perspective on Nematology: The role of the Society. In J.A. Veech and D.W. Dickson (eds.) *Vistas on Nematology*. Society of Nematologists, Hyattsville, Maryland, pp. 7-14.

Sayre, R.M. 1969. A method of culturing a predacious tradigrade on the nematode *Pangrellus redivivu*. *Trans Am Microsc Soc* 88: 266-274.

Sayre, R.M. and Keeley, L.S. 1969. Factors influencing *Catenaria anguillulae* infections in a free living and plant parasitic nematode. *Nematologica* 15: 492-502.

Schacht, H. 1859. Ueber einige Feinde and Krankheithen der Zuckerribe. *Zeitschr Ver Ruben ercker Ind Zoolver* 9: 390.

Schaerffenberg, B. and Tendl, H. 1951. Untersuchungew Uber das Verhatten der Enchytraciden gegenbuber dem Zuckerruben-nematoden, *Heterodera schachtii* (Schm) *Z. Aagew Entomol* 32: 476-388.

Schieber, E. and Sosa, O.N. 1960. Nematodes on coffee in Guatemala. *Plant Disease Reporter* 44: 722-723.

Seinhorst, J.W. and Bels, P.J. 1951. *Ditylenchus destructor* Thorne 1945 in champignons. *Tijdschr. PlZiekt.* 57: 167-169.

Seshadri, A.R. and Sethi, C.I. 1978. Nematode Problems. In "*Wheat Research in India*". ICAR, New Delhi, pp 168-187.

Seshadri, A.R. and Sivakumar, C.V. 1963. A preliminary note on the reniform nematode (*Rotylenchulus reniformis* Linford and Oliveira, 1940) on a number of cultivated crops in South India. *Madras Agric J* 50: 134-137.

Setterquist, R., Smith, G., Jones, R. and Fox, G. 1996. Diagnostic probes targeting the major sperm protein gene that may be useful in the molecular identification of nematodes. *J. Nematol.* 28: 414-421.

Shanmuga Priya, M. 2015. Biomanagement of rice root-knot nematode, *Meloidogyne graminicola* Golden and Brichfield in aerobic rice. *International Journal in Management and Social Science* 3(4): 591-598.

Sharma, N.K. and Sethi, C.L. 1976. Reaction of certain cowpea varieties to *Meloidogyne incognita* and *Heterodera cajani*. *Indian Journal of Nematology* 6: 99-102.

Shi-Bin, W., Liang, L., Ju, W., Yong, J. and Si-Wei, J. 2008. PCR-DGGE analysis of nematode diversity in Cu-contaminated soil. *Pedosphere* 18: 621-627.

Siddiqi, M.R. 1961. Occurrence of the citrus nematode, *Tylenchulus semipenetrans* Cobb, 1913, and the reniform nematode, *Rotylenchulus reniformis* in India (Abstr.). *Proc. 48th Indian Sci Cong* Part III, p. 504.

Siddiqi, M.R. 1964. Studies on nematode root rot of citrus in Uttar Pradesh, India. *Proc. Zool Soc., Calcutta.* 17: 67-75.

Sikora, R.A. 1978. Einfluss der endotrophen Mykorrhiza (*Glomus mosseae*) auf das Wirt-Parasit-Verh~iltnis von *Meloidogyne incognita* in Tomaten. *Zeitschrift ffir Pflanzenkrankheiten und Pflanzenschutz* 85:197-202.

Sikora, R.A. 1992. Management of the antagonistic potential in agricultural ecosystems for the biological control of plant-parasitic nematodes. *Ann Rev Phytopathol* 30: 245–270.

Simpson, C.E. and Starr, J.L. 2001. Registration of 'COAN' peanut. *Crop Science* 41: 918.

Singh, D.B. and Parvatha Reddy, P. 1982. Cultivars of cowpea resistant to root-knot nematode, *Meloidogyne incognita* in India. *Tropical Pest Management* 28: 185-186.

Singh, D.B., Parvatha Reddy, P. and Rao, V.R. 1981. Reaction of some varieties and selections of French bean to *Meloidogyne incognita. Indian Journal of Nematology* 11: 81-83.

Singh, R.S. and Sitaramaiah, K. 1966. Incidence of root-knot of okra and tomato in oil cake amended soil. *Pl Dis Reptr.* 50: 668-672.

Sivakumar, C.V. and Seshadri, A.R. 1971. The life history of the reniform nematode, *Rotylenchulus reniformis* Linford and Oliveira, 1940. *Indian J. Nematol* 1: 7-20.

Sivakumar, C.V. and Seshadri, A.R. 1972. Effect of soil texture on the reniform nematode, *Rotylenchulus reniformis. Indian J. Nematol* 2: 83-86.

Sivakumar, C.V. and Seshadri, A.R. 1976. Longevity of the reniform nematode, *Rotylenchulus reniformis* in host free soil. *Indian J. Nematol* 6: 138-144.

Sivakumar, M. and Marimuthu, T. 1987. Preliminary studies on the effect of soil solarization on phytonematodes of betel vine. *Indian J. Nematol.* 17: 58-59.

Smith, A.L. 1940. Distribution and relation of meadow nematode, *Pratylenchus pratensis* to Fusarium wilt of cotton in Georgia. *Phytopathology* 30: 710.

Smith, G.S. 1987. Interactions of nematodes with mycorrhizal fungi. In J.A. Veech, and D.W. Dickson (eds.) *Vistas on Nematology*. Hyatlsville, Maryland, pp. 292-300.

Sol, H.H. 1963. Some data on the occurrence of rattle virus at various depths in the soil and on its transmission. *Neth J Pl Path* 69: 209-214.

Soriano, I.R., Espiritu, M.J., Schmit, V., Brar D. and Reversat G. 1998. Resistance to rice root-knot nematode *Meloidogyne graminicola* in *Oryza longistaminata* and *Oryza glaberrima. Philippine Journal of Crop Science* 23: 89.

Soriano, I.R.S., Prot, J.C. and Matias, D.M. 2000. Expression of tolerance for *Meloidogyne graminicola* in rice cultivars as affected by soil type and flooding. *J. Nematol.* 32: 309–317.

Soriano, I.R. and Reversat, G. 2003. Management of *Meloidogyne graminicola* and yield of upland rice in South-Luzon, Philippines. *Nematology* 5(6): 879-884.

Srinivasan, C.S. and D'souza, G.I. 1965. A note on the occurrence of Brazilian pyroid coffee nematode, *Meloidogyne exigua* Goeldi in South India. *Indian Coffee* 29(6): 13.

Stapleton, J.J., DeVay, J.E. and Elmore, C.L. 1998. Modes of action of solarization and biofumigation. In: *Soil Solarization and Integrated Management of Soil-borne Pests*. Plant Production and Protection Paper 147. FAO/UN 1998, Rome, pp.78-88.

Stirling, G., Griffin, D., Ophel-Keller, K. *et al.* 2004. Combining an initial risk assessment process with DNA assays to improve prediction of soil borne diseases caused by root-knot nematode (*Meloidogyne* sp.) and *Fusarium oxysporum* f. sp. *lycopersici* in the Queensland tomato industry. *Australas. Plant Pathol.* 33: 285-293.

Stirling, G.R. 1991. *Biological Control of Plant-Parasitic Nematodes: Progress, Problems, and Prospects*. CAB International, Wallingford, UK.

Stock, S., Campbell, J and, Nadler, S. 2001. Phylogeny of *Steinernema* Travassos, 1927 (Cephalobina: Steinernematidae) inferred from ribosomal DNA sequences and morphological characters. *J. Parasitol.* 87: 877-889.

Stokes, D.E. 1969. *Andropogon rhizomatus* parasitized by a strain of *Tylenchulus semipenetrans* not parasitic to four citrus rootstocks. *Pl. Dis Reptr*. 53: 882-885.

Subbotin, S., Halford, P. and Perry, R. 1999. Identification of populations of potato cyst nematodes from Russia using protein electrophoresis, rDNA-RFLP and RAPDS. *Russ. J. Nematol.* 7: 57-63.

Subbotin, S., Sturhan, D., Chizhov, V., Vovlas, N. and Baldwin, J. 2006. Phylogenetic analysis of Tylenchida Thorne, 1949 as inferred from D2 and D3 expansion fragments of the 28S rRNA gene sequences. *Nematology* 8: 455-474.

Subbotin, S., Sturhan, D., Vovlas, N., Castillo, P., Tanyi, T., Moens, M. and Baldwin, J. 2007. Application of the secondary structure model of rRNA for phylogeny: D2-D3 expansion segments of the LSU gene of plant parasitic nematodes from the family Hoplolaimidae Filipjev, 1934. *Mol. Phylogenet. Evol.* 43: 881-890.

Subbotin, S., Vovlas, N., Crozzoli, R., Sturhan, D., Lamberti, F., Moens, M. and Baldwin, J. 2005. Phylogeny of Criconematina Siddiqi, 1980 (Nematoda: Tylenchida) based on morphology and D2-D3 expansion segments of the 28S-rRNA gene sequences with application of a secondary structure model. *Nematology* 7: 927-944.

Suit, R.F. and Du Charme, E.P. 1953. The burrowing nematode and other parasitic nematodes in relation to spreading decline of citrus. *Pl. Dis Reptr* 37: 379-383.

Sundararaju, P., Cannayane, I. and Thangavelu, R. 2002. Biomanagement of *Pratylenchus coffeae* and *Meloidogyne incognita* on banana by using endophytic fungi. *Global Conference on Banana and Plantain*, Bangalore, p. 191.

Swain, B.N. and Prasad, J.S. 1988. Chlorophyll content in rice as influenced by the root-knot nematode, *Meloidogyne graminicola* infection. *Current Science* 57: 895-896.

Taylor, A.L. and Sasser, J.N. 1978. *Biology, Identification and Control of Root-knot Nematodes* (*Meloidogyne* species). N. Carolina Sta Univ Graphics, 111 pp.

Taylor, C.E. and Raski, D.J. 1964. On the transmission of grape fan leaf by *Xiphinema index*. *Nematologica* 10: 489-495.

Taylor, C.E. and Robertson, W.M. 1969. The location of raspberry ring spot and tomato blackring viruses in the nematode vector *Longidorus elongatus* (deMan). *Ann. Appl. Biol* 64: 233-237.

Taylor, D.P. and Wyllie, T.D. 1959. Inter-relationship of root-knot nematode and *Rhizoctonia solani* on soybean emergence (Abstr.). *Phytopathology* 49: 552.

Teliz, D., Grogan, R.G. and Lownsberry, B.F. 1966. Transmission of tomato ring spot, peach yellow bud mosaic, and grape yellow vein viruses by *Xiphinema americanum*. *Phytopathology* 56: 658-663.

Thirumala Rao, V. 1956. Stray notes on some crop pest outbreaks of South India. *Indian J. Ent* 18: 123-126.

Thomas, C. and Cottage, A. 2006. Genetic engineering for resistance. In: R.N. Perry and M. Moens (eds.) *Plant Nematology*. CABI, King's Lynn, UK, pp. 255–*272*.

Thorne, G. 1927. The life history habits and economic importance of some Mononchs. *J. Agric. Res* 34: 265-286.

Thorne, G. 1961. *Principles of Nematology*. McGraw Hill, New York and London, p. 553.

Tollsten, L. and Bergström, G. 1988. Head space volatiles of whole plants and macerated plant parts of Brassica and *Sinapis*. *Phytochem*. 27: 4013-4018.

Triffitt, M.J. 1930. On the bionomics of *Heterodera schachtii* on potatoes with special reference to the influence of mustard on the escape of the larvae from the cysts. *J. Helminth* 8: 19-48.

Trudgill, D.L. 1991. Resistance to and tolerance of plant parasitic nematodes in plants. *Annu. Rev. Phytopathol*. 29: 167–193.

Uehara, T., Mizukubo, T., Kushida, A. and Momota, Y. 1998. Identification of *Pratylenchus coffeae* and *P. loosi* using specific primers for PCR amplification of ribosomal DNA. *Nematologica* 44: 357-368.

Vadivelu, R.S., Rajendran, G., Naganathan, T.G. and Jayaraj, S. 1987. Studies on the nematode pests of banana and their management - A review. *Third Group Discussion on the Nematological Problems of Plantation Crops,* Sugarcane Breeding Institute, Coimbatore, pp. 32-33.

Van Der Vossen, E.A., Van Der Voort, J.N., Kanyuka, K. *et al.* 2000. Homologues of a single resistance-gene cluster in potato confer resistance to distinct pathogens: A virus and a nematode. *Plant J.* 23: 567–576.

Van Gundy, S.D. 1958. The life history of the citrus nematode *Tylenchulus semipenetrans*. *Nematologica* 3: 283-294.

Van Gundy, S.D. and Kirkpatrick J.D. 1965. Factors explaining citrus nematode resistance. *Calif Citrog*. 50: 235-241.

Van Hoof, H.A. 1964. Serial transmission of rattle virus by a single male of *Trichodorus pachydermis* Seinhorst. *Nematologica* 10: 141-144.

Vargas-Ayala, R., Rodriguez-Kabana, R., Morgan-Jones, G., McInroy, J.A. and Kloepper, J.W. 2000. Shifts in soil micro-flora induced by velvet bean (*Mucuna deeringiana*) in cropping systems to control root-knot nematodes. *Biol Control* 17: 11–22.

Venkitesan, T.S. 1976. Studies on the Burrowing Nematode *Radopholus similis* (Cobb, 1893). Thorne, 1949 on Pepper (*Piper nigrum* L.) and its Role in Slow Wilt Disease. Ph.D. thesis, Univ. Agri. Sci., Bangalore, 122 pp.

Venkitesan, T.S. and Jacob, A. 1985. Integrated control of root-knot nematode, infecting pepper vines in Kerala (Abstr.). *Indian Journal of Nematology* 15: 261-262.

Vilardebo, A. and Luc, M. 1961. Slow decline of citrus caused by the nematode, *Tylenchulus semipenetrans* (In French). *Fruits* 16: 445-454.

Vuong, H.H. 1969. The occurrence in Madagascar of the rice nematodes, *Aphelenchoides besseyi* and *Ditylenchus angustus*. In J.E. Peachey (ed.) *Nematodes of Tropical Crops*. Commonw Agr Bur, Farnham Royal, Bucks, England, pp. 274-288.

Waeyenberge, L., Ryss, A., Moens, M., Pinochet, J. and Vrain, T. 2000. Molecular characterization of *Pratylenchus* species using rDNA Restriction Fragment Length Polymorphism. *Nematology* 2: 135-142.

Weber, A.P., Zwillenberg, I.O. and Van der Laan, P.A. 1951. A predacious amoeboid organism destroying larvae of the potato root eelworm and other nematodes. *Nature* 169: 834-835.

Webster, J.M. 1969. The host-parasite relationships of plant parasitic nematodes. *Advan Parasitol* 7: 1.

Weischer, B. 1967. *Report to the Government of India on Plant Parasitic Nematodes*. PL 24/52, Report No. TA 2332, FAO, Rome, 15 pp.

Wergin, W.P. 1981. Scanning electron miscroscopic techniques and applications for use in nematology, In B.M. Zuckerman and R.A. Rhode (eds.) *Plant Parasitic Nematodes*, Vol. III. Academic Press, New York, pp. 175-204.

Whitehead, A.G. 1968. Taxonomy of *Meloidogyne* (Nematoda, Heteroderidae) with descriptions of four new species. *Trans Zool Soc Lond* 31: 263-401.

Williams, J.R. 1960. Studies on the nematode soil fauna of sugarcane fields in Mauritius. 5. Notes upon a parasite of root-knot nematodes. *Nematologica* 5: 37-42.

Williams, T.D. 1958. Potatoes resistant to root eelworm. *Proc Linn Soc Lond* 169: 93-104.

Williams, T.D. 1963. The development of females of the potato-root eelworm (*Heterodera rostochiensis* Woll.) on 'tuber-pieces' of susceptible and resistant potatoes. *Euphytica* 12: 277-284.

Wilski, A. and Giebel, J. 1966. β-glucosidase in *Heterodera rostochiensis* and its significance in resistance of potato to this nematode. *Nematologica* 12: 219.

Winslow, R.D. and Willis, R.J. 1972. Nematode diseases of potatoes. In J.M. Webster (ed.) *Economic Nematology*. Academic Press, New York, pp 17-48.

Wood, F.C. and Goodey, J.B. 1957. Effects of gamma ray irradiation on nematodes infesting cultivated mushrooms beds. *Nature* 180: 760-761.

Yamashita, T. and Akiya, T. 1968. Studies on amino acids generated from rape seed cake in the soil. *Soil Sci Pl Nutr*. 14: 225-230.

Yeates, G.W. 1969. Predation by *Mononchoides pothikus* in laboratory culture. Nematologica 15: 1-9.

Zijlstra, C., Donkers-Venne, D. and Fargette, M. 2000. Identification of *Meloidogyne incognita, M, javanica* and *M. arenaria* using sequence characterized amplified region (SCAR) based PCR assays. *Nematology* 2: 847-853.

Zimmeman, A. 1898. De nematodes der Koffieworiels. *Meded Pl Tuin Batalva* 27: 16-41.

Annexure I

QUOTABLE QUOTES

"If all the matter in the universe except the nematodes were swept away, our world would still be dimly recognizable…………we would find its mountains, hills, valleys, rivers, lakes and oceans represented by a film of nematodes".

- N.A. Cobb (1914)

"Each year these minute organisms (nematodes) exact an ever-increasing toll from almost every cultivated acre in the world: a bag of rice in Burma, a pound of tea in Ceylon, a ton of sugar beets in Germany, a bag of potatoes in England, a bale of cotton in Georgia, a bushel of corn in Iowa, a box of apple in New York, a sack of wheat in Kansas, or a crate of oranges in California…….which the farmers can ill afford to lose and which the undernourished millions on earth would welcome if made available to them ".

- G. Thorne (1961)

"Some say roundworms, others thread or eelworms
Sexes are separate, producing eggs and sperms
They reproduce sexually, have habits strange
A millimeter to a few meters, is their body range
Esophagus with nerve ring, lumen triradiate
Body cavity false, hence pseudocoelomate
Circulatory organs lacking, yet triploblastic
Respiration through cuticle, which is elastic
Amphids and phasmids are their eyes and ears
Plant feeders possess – either stylets or spears".

R.S. Kanwar

"Whatever approaches we take to achieve the goal of controlling diseases associated with plant nematodes, knowledge of the nematodes themselves as well as the plants which they parasitize, will increase the chances of success…… In

other words, there must always be room for the specialist who studies different aspects of the biology of plant nematodes".

H.R. Wallace

"The extraction of food from one organism, the host, by another organism, the parasite, inevitably entails some disturbance of the host's integrity".

- H.R. Wallace

"Plants are rarely, if indeed ever, subject to association with only one potential pathogen".

- N.T. Powell

"The most severe nematode problems occur where good host crops are grown too frequently for too long on the same land".

- B.G. Chitwood

"If at all there was one single factor that triggered nematological research in India, I would say it was this discovery (relating to the report of potato cyst nematode from Ooty, Tamil Nadu by Jones, 1961)".

- Abrar M. Khan (1974)

"A convergence of technical, environmental and social forces is moving agriculture towards more non-pesticide pest management alternatives like biological control, host plant resistance and cultural management".

- Michael Fitzner, National IPM Program Leader, USDA Extension Service

Annexure II

COMMON NAMES OF NEMATODES

American dagger nematode - *Xiphinema americanum*

Awl nematodes - *Dolichodorus* spp.

Banana root-lesion nematode - *Pratylenchus coffeae*

Bud and leaf nematodes - *Aphelenchoides* spp.

Bulb and stem nematodes - *Ditylenchus* spp.

Burrowing nematode - *Radopholus similis.*

Cereal cyst nematode - *Heterodera avenae.*

Chrysanthemum foliar nematode - *Aphelenchoides ritzemabosi.*

Citrus nematode - *Tylenchulus semipenetrans.*

Coconut red ring nematode - *Bursaphelenchus cocophilus.*

Cyst nematodes - *Heterodera* spp., *Globodera* spp.

Cystoid nematodes - *Meloidodera* spp.

Dagger nematodes - *Xiphinema* spp.

False root-knot nematode - *Nacobbus batatiformis.*

Golden nematode of potato - *Globodera rostochiensis.*

Javanese root-knot nematode - *Meloidogyne javanica*

Lance nematodes - *Hoplolaimus* spp.

Lesion or meadow nematodes - *Pratylenchus* spp.

Needle nematodes - *Longidorus* spp.

Pin nematodes - *Paratylenchus* spp.

Potato rot nematode - *Ditylenchus destructor.*

Reniform nematodes - *Rotylenchulus* spp.

Rice stem nematode - *Ditylenchus angustus.*

Rice root nematodes – *Hirschmanniella* spp.

Rice white tip nematode - *Aphelenchoides besseyi.*

Ring nematodes –*Criconemoides* spp.

Root-knot nematodes - *Meloidogyne* spp.

Seed gall or leaf gall nematodes - *Anguina* spp.

Sessile nematodes - *Cacopaurus* spp.

Sheath nematodes - *Hemicycliophora* spp.

Sheathoid nematodes - *Hemicriconemoides* spp.

Southern root-knot nematode - *Meloidogyne incognita*

Sphaeroid nematodes – *Sphaeronema* spp.

Spine nematodes - *Criconema* spp.

Spiral nematodes - *Helicotylenchus* spp.

Spring crimp or dwarf nematode - *Aphelenchoides fragariae.*

Stem and bulb nematode - *Ditylenchus dipsaci*

Sting nematodes - *Belonolaimus* spp.

Stubby root nematodes - *Trichodorus* spp.

Stunt or stylet nematodes - *Tylenchorhynchus* spp.

Yam nematode - *Scutellonema bradys*

Annexure III

NEMATOLOGICAL JOURNALS

Afro-Asian Journal of Nematology (1990-1996)
Fundamental and Applied Nematology (until 1991 Revue de Nématologie)
Helminthological Abstracts, Series B. Plant Nematology
Indian Journal of Nematology (1971)
Japanese Journal of Nematology (1972)
Journal of Nematology (1969)
Journal of Nematode Morphology and Systematics (1999)
Nemapix (1998)
Nematologica (1956-1998)
Nematologia Mediterranea (1973)
Nematology: International Journal of Fundamental and Applied Nematology (1999)
Nematropica (1971)
Pakistan Journal of Nematology (1983)
Revue de Nematologie (1978-1991)
Russian Journal of Nematology (1993)

■■■■

Annexure IV

PROFESSIONAL SOCIETIES FOR NEMATOLOGISTS

Afro/Asian Society of Nematologists
American Phytopathological Society
American Society of Parasitologists
Association Peruana de Fitopatologia
Association of Applied Biologists
Brazilian Society of Nematologists
British Society of Parasitologists
Chinese Society of Nematologists
European Society of Nematologists
Japanese Nematological Society
Mediterranean Nematological Society
Nematological Society of India
Nematological Society of Southern Africa
Organization of Tropical American Nematologists
Pakistan Society of Nematologists
Russian Society of Nematologists
Sociedad Brasileira de Nematologia
Societa Italina Di Nematologia
Society of Nematologists
The Helminthological Society of Washington

■■■■

Annexure V

GLOSSARY

a. A ratio calculated by dividing the length of the nematode by its maximum width.

Adanal. Pertaining to bursa which does not envelop the entire tail.

Alae. Wing-like expansions or projections formed by longitudinal thickening of the cuticle.

Allotype. A paratype of the opposite sex to the holotype.

Ambifenestrate. The occurrence of two openings in the vulval cone which are separated by the vulval bridge. Found in some species of *Heterodera.*

Amphidelphic. Having two ovaries, one proceeding anteriorly and one proceeding posteriorly.

Amphids. Chemo- sensory organs, occurring laterally in pairs, located in the lip region of the nematode.

Ampulla. Widening in canal, forming a reservoir.

Annulation. A transverse depression, appearing as a line, on the cuticle.

Annule. The interval between adjacent annulations on the cuticle.

Apophysis. A swelling or expansion as found in the organe-Z, in species of *Xiphinema.*

Arch. That portion of the perineal pattern in the genus *Meloidogyne* which is located dorsal to the lateral field or lines.

Arcuate. Curved like a bow.

Areolated. Divided into small spaces or areolations; usually pertains to cuticle.

Areolation. The condition whereby transverse body striae extend into the lateral cuticle.

b. A ratio calculated by dividing the total body length by the length of the esophagus measured from the anterior end of the body.

Basal bulb. Posterior portion of esophagus which may be a distinct bulb or a lobe overlapping the intestine.

Basement membrane. A layer which separates the innermost layer of the cuticle from the hypodermis.

Bilateral symmetry. Referring to organisms whose morphology is such that the body can be divided into mirror image halves by passing a dorso-ventral plane through the body axis.

Bioassay. Use of appropriate hosts to determine relative population levels. Bioassays are used primarily for endoparasites such as root-knot and cyst nematodes, but can be used for any nematode under suitable propagation conditions. They are usually done in a greenhouse.

Biotype (Race). A subgroup within a species usually characterized by the possession of a single or a few characters. Used to describe those individuals within a species that are capable of reproducing on a resistant variety.

Blastula. An early stage in embryogeny in which the embryo is a fluid-filled sphere bounded by a single layer of cells.

Bloat. The common name for the disease caused on onion by *Ditylenchus dipsaci.*

Buccal aperture. The oral aperture; the anterior opening to the digestive system.

Buccal cavity. The stoma; connects oral opening with anterior portion of esophagus.

Bullae. Knob-like structures located within the vulval cone of cysts of *Heterodera* near the under bridge or fenestra.

Bursae (Caudal alae). Lateral cuticular extensions present at the posterior end of males used to clasp the female during copulation.

c. A ratio calculated by dividing the total length by the length of the tail.

Capitulum. Medial ventral sclerotization of the spicular pouch.

Cardia (Cardiac valve or esophago-intestinal valve). Valvular apparatus connecting the esophagus and intestine.

Caudal. Pertaining to or located at or on the tail.

Caudal alae. Occur in the posterior region of males in a number of genera.

Caudal glands. Three to five glands located in the tail.

Cephalic. Pertaining to or located near the head.

Cephalic framework. A subcuticular framework that supports the lip region and to which are attached the stylet protractor muscles.

Cervical. Pertaining to or located near the neck region.

Cervical alae. Confined to the anterior region of the nematodes parasitic on animals.

Circumesophageal commissure (Nerve ring). The central portion of the nematode nervous system to which are connected anteriorly and posteriorly directed nerves and nerve chords. This structure encircles the esophagus.

Circumfenestrate. The condition in some species of *Heterodera* in which a vulval bridge across the vulval cone is not present producing only a single opening.

Cloaca. Common cavity in males, in which the digestive and reproductive systems terminate.

Coelomyarian. A type of muscle cell with its protoplasmic portion extending into the pseudocoel with the fibrillar part extending along the hypodermis.

Corpus. Anterior elongate cylindrical part of the esophagus. The basal portion of the corpus is swollen to form a bulb, in which case the slender anterior portion is termed a procorpus and the bulb termed the metacorpus.

Cortex. The outermost layer of the cuticle.

Cuticle. The non-cellular external layer of a nematode produced by the hypodermis. Cuticular material occurs as linings of natural openings in the body.

Cyst. The oxidized cuticle of dead adult females of cyst nematodes (*Heterodera, Globodera, Punctodera, etc.*) which confers protection to the eggs that may be contained within.

Deirids. Paired, pore-like organs located in the lateral fields; believed to be sensory in nature.

De Man's values. A series of measurements and ratios which express a nematode's length, its relative width, the relative lengths of the esophagus and tail and the position of the vulva.

Detection level. Minimum population density of a nematode species that can be identified routinely from a soil or tissue sample. This level varies with each sampling and extraction technique.

Didelphic. Possessing two ovaries.

Dieback. Progressive death of shoots, branches and roots generally starting at the tip.

Dioecious. Having the male reproductive organs in one individual, the female in another; sexually distinct.

Diorchic. Having two testes.

Disease. Harmful physiological processes due to continuous irritation by a primary causal agent, resulting in morbific cellular activity and expressed in characteristic pathological responses called symptoms.

Diverticulum. Pouch or lobe containing the esophageal glands.

Dorsal gland outlet. The point at which the dorsal gland empties into the lumen of the esophagus.

Ectoparasite. A parasite feeding on a host from the exterior.

Egg. The first stage of the life cycle containing a zygote or a larva.

Endoparasite. A parasite which enters host tissue and feeds from within.

Epiboly. The growth of one structure around another during embryo development.

Epithelium. A sheet like layer of cells lining the internal or external surface of an organ or tissue.

Esophagus (Pharynx). That portion of the digestive system between the stoma and the intestine.

Excretory pore. Exterior opening of the excretory system. It is generally located on the ventral side of the body near the basal region of the esophagus.

Extracorporeal. Occurring outside of the body.

Fallow. Maintenance of land with no plant growth. Weeds are controlled by cultivation or the application of herbicides and no crop planted.

Fenestra. A window-like opening in the vulval cone of *Heterodera* spp.

Free living nematode. Those nematodes which are not known to be parasite on animals or higher plants.

Fumigation. The application of a toxic gas (frequently applied as a liquid) to an area to control various pests.

Ganglion. A well-defined mass of nerve cells forming a nerve center.

Gastrula. An early embryo composed of a hollow two layered cup shaped structure derived from the blastula.

Genital papillae. Tactile or sensory organs located on the tail.

Giant cell (Syncytium). Host responses in which 40-60 adjacent cells coalesce to form multi nucleate mass of protoplasm.

Glottoid apparatus. Sclerotized structure at the base of the stoma.

Gonopore. The exterior opening of gonads - the vulva in females, the anus or cloacal opening in males.

Granek's ratio. The ratio of the distance of the anus from the edge of the vulval fenestra and the diameter of the fenestra; useful in differentiating some species of cyst nematodes.

Gravid. Containing an egg or eggs.

Gubernaculum. Spicule guide, sclerotized accessory piece.

Guiding ring. A cuticularized structure which surrounds and guides stylet.

Hatching factor. A material produced by the roots of certain hosts that increases the hatching percentage of eggs within cyst nematodes.

Head. That portion anterior to the base of the stoma or stylet.

Hemizonid. Lens like structure situated between the cuticle and hypodermis on the ventral side of the body just anterior to the excretory pore; generally believed to be associated with the nervous system.

Hermaphrodite. An individual bearing both functional male and female reproductive organs.

Hyperlasia. An abnormal increase in the number of cells of a tissue.

Hypertrophy. An abnormal increase in the size of cells of a tissue.

Hypodermis. Thin tissue layer beneath the cuticle which thickens to form the dorsal, ventral and lateral cords which extend the length of the body.

Incisures (Involutions). Longitudinal cuticular clefts which divide the lateral fields.

Infective larvae. That stage in the life cycle of a parasite capable of invading host tissue and establishing infection.

Intersex. An individual possessing both male and female characteristics.

Isthmus. Relatively narrow portion of the esophagus between the metacorpus and the basal bulb.

L. The overall length of a nematode in millimeters.

Labial. Pertaining to or located on the lips.

Labial disc. A raised area in the cuticle of the most anterior annule of some species of *Criconemoides*.

Labium. A lip.

Larva. Any life stage between the embryo and the adult; an immature nematode.

Lateral field. A pair of laterally located longitudinal cuticular markings located above the lateral hypodermal cords.

Lesion. A localized spot of diseased tissue.

Longitudinal alae. Usually 4 in number and extend the length of the body sublaterally.

Lumen. Triradiate canal of duct of the esophagus.

Matrix. The gelatinous substance into which eggs of *Meloidogyne* and some other nematodes are deposited to form an egg mass.

Mesenteron. The intestine.

Metacorpus (Median bulb). An enlarged portion of the esophagus located near its mid-point frequently containing a valve.

Micropyle. Minute opening in the membrane of an egg through which the spermatozoa enter.

Migratory ectoparasite. A nematode which feeds on the host from the outside while remaining capable of free movement.

Migratory endoparasite. A nematode which enters a host, feeds internally, moves freely through the tissues, and never becomes sedentary.

Molt. To shed or cast off the cuticle.

Monoculture. The planting of the same crop for several consecutive years.

Monodelphic. Having one ovary.

Monogenic. Producing of only one sex.

Monorchic. Having one testis.

Mucro. A small pointed projection as occurring on the tail tip of certain nematodes.

Neck. That portion of the body occupied by the esophagus.

Necrosis. The death of cells or tissues.

Nematicide. A chemical lethal to nematodes.

Nematodes. Triploblastic, bilaterally symmetrical, unsegmented, pseudocoelomate and vermiform animals.

NEPO. Nematode transmitted viruses with polyhedral (spherical) particles.

NETU. Nematode transmitted viruses with tubular (rod) particles.

Odontostylet (Onchiostylet). Buccal stylet; a large tooth which originates in the esophageal wall. Generally found among some members of the Adenophorea.

Oocyte. An immature female gamete.

Oogonium. An early stage in the development of an oocyte.

Opisthodelphic. Having a single ovary situated posterior to the vulva.

Orange-Z. A structure in certain *Xiphinema* species located between the oviduct and uterus.

Ovary. Female sexual gland in which the ova or eggs are formed.

Oviduct. Portion of the female reproductive system between the ovary and uterus.

Oviparous. Producing eggs that hatch after being laid.

Ovoviviparous. Producing eggs that hatch within the uterus.

Papillae. Tactile, sensory organs found on various body regions, e.g. labial or cephalic papillae.

Parasite. An organism deriving all or a portion of its nutrition from another organism, the host, while permanently or temporarily attached to or within the host.

Parthenogenesis. Reproduction from eggs without fertilization by sperm cells.

Pathogen. A disease causing organism.

Perineal pattern. Fingerprint like pattern formed by cuticular striae surrounding the vulva and anus of the mature *Meloidogyne* female.

Phasmid. Pore-like structures located in the lateral fields of the posterior region of a nematode belonging to the class Secernentea; function is believed to be sensory.

Platymyarian. A type of muscle cell in which the fibers are adjacent and perpendicular to the hypodermis.

Population density. Number of nematodes per unit volume of soil or weight of plant tissue.

Post uterine sac. The uterus extending posterior to the vulva which may serve to store the spermatozoa.

Preadult. The larval stage just before adulthood; the fourth stage larva.

Predator. An organism which lives by preying upon animals.

Prerectum. Portion of the digestive system between the intestine and the rectum present in *Xiphinema.*

Procorpus. The anterior subdivision of the corpus of the esophagus located between the base of the stoma and the metacorpus.

Prodelphic. Ovary located anterior to the vulva.

Pseudocoel (Body cavity). A false coelom; a body cavity not lined with an epithelium of mesodermal origin containing a liquid in which the various internal organs are suspended. The body fluid apparently functions as a respiratory and circulatory system.

Quiescent. A stage of relative biological inactivity induced by physical factors such as low temperature.

Rachis. The central or axial chord of the ovary of a nematode.

Rectal glands. Glands in the rectal region, which in the genus *Meloidogyne,* produce the gelatinous matrix into which eggs are deposited.

Renette. A ventral cell present in the excretory system of some nematode.

Retrorse. Backward-projecting.

Rhizosphere. The area in the soil immediately surrounding plant roots and influenced by them.

s. Stylet length/body diameter measured at base of stylet.

Sclerotized. A term which refers to a hardened condition, generally of cuticle.

Scutellum. A greatly enlarged phasmid found on some species of the subfamily Hoplolaiminae.

Sedentary. Staying in one place; stationary.

Sedentary endoparasite. Describing those nematodes which enter a host, migrate to a feeding site and then become non-motile.

Semifenestrae. A part of openings in the vulval cone on cysts of the genus *Heterodera* separated by the vulval bridge.

Seminal vesicle. That portion of the male reproductive tract functioning in the temporary storage of sperm.

Sessile. Permanently attached; not capable of moving about.

Setae. Elongated cuticular structures usually located around the oral opening, tactile sensory organs.

Sexual dimorphism. Species in which the two sexes differ morphologically.

Sheath. A covering of a structure. Refers to a retained or extra cuticle in some nematode species.

Species inquirendae. Species of doubtful status because of inadequate descriptions and lack of preserved material.

Spermatheca. An enlarged portion of the female gonad between the oviduct and uterus functioning in the storage of sperm.

Spermatocyte. A cell giving rise to sperm cells or spermatozoa.

Spicule. A pair of heavily cuticularized male copulatory organs.

Stomatostylet. A stylet evolved from a fusion of the walls of the stoma commonly found in nematodes belonging to the order Tylenchida.

Stomodeum. The anterior portion of the digestive system formed by the invagination of the body wall.

Stylet (Spear). A relatively long, slender, axially located, hollow feeding structure of sclerotized cuticle located in the buccal cavity.

Synergism. Co-parasitism of a host to produce symptoms or other effects of greater magnitude than the sum of the effects caused by both parasites acting alone.

Syngonic. Pertaining to hermophroditic reproduction in which both sperm and eggs are produced by the same gonad.

Tail. That portion of the body between the anus and the posterior terminus.

Telemon. Rigid sclerotized portion of the cloacal wall which apparently guides the spicules from the spicular pouch into the cloaca.

Telegonic. That type of gonad in which germ cells are produced only at the distal end.

Testis. That portion of the male reproductive system producing sperm.

Transovarial transmission. The ability of a virus vector to transmit the virus to its progeny by means of eggs.

Trap crop. A crop plant which is penetrated by a parasite but in which it cannot be or is not allowed reaching maturity.

Triploblastic. Possessing three germ layers: ectoderm, mesoderm and endoderm.

Triradiate. Having three radiating members.

Under bridge. A structure extending across the vulval cone of *Heterodera* cysts below and parallel to the vulval bridge.

Uterus. The portion of the female reproductive system between the oviduct and the vagina; a structurally modified portion of the oviduct in which some egg development may occur.

v. The location of the vulva expressed as a percentage of the total body length as measured from the anterior end.

Vagina. The cuticle lined canal leading from the uterus to the vulva.

Vas deferens. The portion of the male reproductive system leading from the vas efferentia to the seminal vesicle and the ejaculatory duct.

Vector. An agent of dissemination and/or inoculation e.g. a nematode that transmits a virus.

Vermiform. Relatively long and slender; worm shaped.

Virulence. State of being pathogenic.

Viviparous. Bearing living young.

Vulva. The female gonophore; external opening of the female reproductive system.

Vulval bridge. A narrow connection across the fenestra of vulval cones of some *Heterodera* cysts that forms two semifenestrae.

Vulval cone. The conspicuous elevation on the posterior of *Heterodera* cysts.

Wool. A mass of desiccated *Ditylenchus dipsaci* found in some plant tissues.

SUBJECT INDEX